SIGNALLING PATHWAYS IN ACUTE OXYGEN SENSING

The Novartis Foundation is an international scientific and educational charity (UK Registered Charity No. 313574). Known until September 1997 as the Ciba Foundation, it was established in 1947 by the CIBA company of Basle, which merged with Sandoz in 1996, to form Novartis. The Foundation operates independently in London under English trust law. It was formally opened on 22 June 1949.

The Foundation promotes the study and general knowledge of science and in particular encourages international co-operation in scientific research. To this end, it organizes internationally acclaimed meetings (typically eight symposia and allied open meetings and 15–20 discussion meetings each year) and publishes eight books per year featuring the presented papers and discussions from the symposia. Although primarily an operational rather than a grant-making foundation, it awards bursaries to young scientists to attend the symposia and afterwards work with one of the other participants.

The Foundation's headquarters at 41 Portland Place, London W1B 1BN, provide library facilities, open to graduates in science and allied disciplines. Media relations are fostered by regular press conferences and by articles prepared by the Foundation's Science Writer in Residence. The Foundation offers accommodation and meeting facilities to visiting scientists and their societies.

Information on all Foundation activities can be found at http://www.novartisfound.org.uk

Novartis Foundation Symposium 272

SIGNALLING PATHWAYS IN ACUTE OXYGEN SENSING

2006

John Wiley & Sons, Ltd

Published in 2006 by John Wiley & Sons Ltd,
The Atrium, Southern Gate,
Chichester PO19 8SQ, UK

National 01243 779777
International (+44) 1243 779777
e-mail (for orders and customer service enquires): cs-books@wiley.co.uk
Visit our Home Page on http://eu.wiley.com

Other Wiley Editorial Offices

John Wiley & Sons Inc., 111 River Street, Hoboken, NJ 07030, USA

Jossey-Bass, 989 Market Street, San Francisco, CA 94103-1741, USA

Wiley-VCH Verlag GmbH, Boschstr. 12, D-69469 Weinheim, Germany

John Wiley & Sons Australia Ltd, 33 Park Road, Milton, Queensland 4064, Australia

John Wiley & Sons (Asia) Pte Ltd, 2 Clementi Loop #02-01, Jin Xing Distripark, Singapore 129809

John Wiley & Sons Canada Ltd, 22 Worcester Road, Etobicoke, Ontario, Canada M9W 1L1

Wiley also publishes its books in a variety of electronic formats. Some content that appears in print may not be available in electronic books.

Novartis Foundation Symposium 272
xii + 288 pages, 56 figures, 2 tables

British Library Cataloguing in Publication Data

A catalogue record for this book is available from the British Library

ISBN-13 978-0-470-01457-8
ISBN-10 0-470-01457-1

Typeset in 10½ on 12½ pt Garamond by SNP Best-set Typesetter Ltd., Hong Kong
Printed and bound in Great Britain by T. J. International Ltd, Padstow, Cornwall.
This book is printed on acid-free paper responsibly manufactured from sustainable forestry, in which at least two trees are planted for each one used for paper production.

Contents

Symposium on Signalling pathways in acute oxygen sensing, held at the Novartis Foundation, London, 25–27 January 2005

Editors: Derek J. Chadwick (Organizer) and Jamie Goode

This symposium is based on a proposal made by Jeremy Ward

Participants

Philip I. Aaronson Department of Asthma, Allergy and Respiratory Science, 5th Floor, Thomas Guy House, Guy's Hospital Campus, London SE1 9RT, UK

Helmut Acker Cellvis, Harkordstr 92, Dortmund, D-44225, Germany

Stephen L. Archer Cardiology Division, University of Alberta, WMC 2C2.36, 8440 112th St, Edmonton, Alberta, T6G 2B7, Canada

Keith J. Buckler University Laboratory of Physiology, University of Oxford, Oxford OX1 3PT, UK

Navdeep S. Chandel Division of Pulmonary and Critical Care Medicine, Department of Medicine, Northwestern University Medical School, McGaw 2–2334, 240 East Huron Avenue, Chicago, IL 60611-2909, USA

Michael Duchen (*Chair*) Department of Physiology, University College London, Gower Street, London WC1E 6BT, UK

A. Mark Evans School of Biology, University of St Andrews, Bute Medical Buildings, St Andrews, Fife KY16 9TS, UK

Constancio Gonzalez Departamento de Bioquimica y Biologia Molecular & Fisiologia, Facultad de Medicina, Instituto de Biologia y Genetica Molecular, Universidad de Valladolid y CSIC, Valladolid 47005, Spain

Alison Gurney Department of Physiology and Pharmacology, Strathclyde Institute for Biomedical Sciences, University of Strathclyde, Glasgow G4 0NR, UK

Adrian L. Harris Cancer Research UK Oxford Cancer Centre, Medical Oncology Unit, Churchill Hospital, Headington, Oxford OX3 7LJ, UK

Paul J. Kemp School of Biosciences, University of Cardiff, P O Box 911, Museum Avenue, Cardiff CF1 3US, UK

Prem Kumar Department of Physiology, The Medical School, University of Birmingham, Birmingham B15 2TT, UK

Wolfgang Kummer Institute of Anatomy and Cell Biology, Justus-Liebig-University, Aulweg 123, 35385, Giessen, Germany

José López-Barneo Laboratorio de Investigaciones Biomedicas, Edificio de Laboratorios, 2 planta, Hospital Universitario Virgen del Rocio, Avda Manuel Siurot s/n, Sevilla 41013, Spain

Michael P. Murphy MRC Dunn Human Nutrition Unit, Wellcome Trust/MRC Building, Hills Road, Cambridge CB2 2XY, UK

Colin A. Nurse Department of Biology, McMaster University, 1280 Main Street West, Hamilton, Ontario, L8S 4KI, Canada

Chris Peers School of Medicine, University of Leeds, Leeds LS2 9JT, UK

Daniel Peet School of Molecular and Biomedical Science and ARC Special Research Centre for Molecular Genetics of Development, The University of Adelaide, Adelaide, SA 5005, Australia

Jacques Pouysségur Institute of Signaling, Developmental Biology and Cancer Research, CNRS UMR 6543, Centre Antoine Lacassagne, 33 Avenue Valombrose, Nice 06189, France

Nanduri Prabhakar Department of Physiology and Biophysics, Case Western Reserve University, 10900 Euclid Avenue, Cleveland, OH 44106, USA

Peter J. Ratcliffe Nuffield Department of Medicine, NDM offices, level 7, John Radcliffe Hospital, Oxford OX3 9DU, UK

Peter R. Rich Department of Biology, University College London, Gower Street, London WC1E 6BT, UK

Paul T. Schumacker Department of Pediatrics, Division of Neonatology, Feinberg School of Medicine, Northwestern University, Chicago, IL 60611, USA

Gregg L. Semenza Institute for Cell Engineering, Departments of Pediatrics, Medicine, Oncology, Radiation Oncology and Institute of Genetic Medicine,

The Johns Hopkins University School of Medicine, Broadway Research Building, Suite 671, 733 North Broadway, Baltimore, MD 21205, USA

Ineke P. Stolze The Henry Wellcome Bulding for Molecular Physiology, Roosevelt Drive, Oxford, OX3 7BN, UK

James T. Sylvester Division of Pulmonary and Critical Care Medicine, The John's Hopkins Asthma and Allergy Center, Baltimore, MD 21224, USA

Rodrigo Varas Orozco (*Novartis Foundation Bursar*) University Laboratory of Physiology, Parks Rd, Oxford OX1 3PT, UK

Jeremy P. T. Ward Department of Asthma, Allergy and Respiratory Science, c/o Cardiovascular Medicine & Biology, 2nd Floor, New Hunt's House, King's College London, Guy's Hospital Campus, London SE1 1UL, UK

Kenneth Weir Department of Medicine and Physiology, V A Medical Center, University of Minnesota, 1 Veteran's Drive, 111C, Minneapolis, MN 55417, USA

Norbert Weissmann Department of Internal Medicine, Justus-Liebig-University, Klinikstr 36, 35392 Giessen, Germany

John Westwick Novartis Institutes for Biomedical Research Wimblehurst Rd, Horsham, West Sussex RH12 5A6, UK

Chair's introduction

Michael Duchen

Department of Physiology, University College London, Gower Street, London WC1E 6BT, UK

This looks to be a very exciting meeting, and I am looking forward to it very much.

The list of participants is a kind of international all stars of the people working in the oxygen sensing field. I feel a bit bashful about being asked to be chair—I've been told that this is because I am not doing anything interesting enough to be contentious, and so I can remain dispassionate. I remember years ago my old PhD advisor Tim Biscoe was invited to chair a meeting. He muttered to me, 'I suppose that means I'm an old fart who doesn't do any really useful work any more'.

This field has been almost unique, it seems to me, in the level of disagreement and failure to reach consensus among researchers. I have never understood why. Over the next few days we have a fantastic opportunity to try to understand the origin of these inter-lab differences, to see whether we can resolve them. I believe this will help us to understand the biology, and so I hope that the discussions can be really open and wide ranging so that we can develop strategies to try to iron out these differences.

There are a number of participants here who were invited specifically so they can give us the benefit of their wisdom during the discussion sessions, even though they are not presenting formal papers. I hope you will contribute enthusiastically to the discussion. We begin by looking at the regulation of gene expression by hypoxia, and I'd like to introduce the first paper.

Regulation of gene expression by HIF-1

Gregg L. Semenza, Larissa A. Shimoda* and Nanduri R. Prabhakar†

*Vascular Program, Institute for Cell Engineering; Departments of Pediatrics, Medicine, Oncology, and Radiation Oncology; and McKusick-Nathans Institute of Genetic Medicine, The Johns Hopkins University School of Medicine, Baltimore, Maryland 21205, *Division of Pulmonary and Critical Care Medicine, Department of Medicine, The Johns Hopkins University School of Medicine, Baltimore, Maryland 21224, and †Department of Physiology and Biophysics, Case Western Reserve University, Cleveland, Ohio 22106 USA*

Abstract. Hypoxia-inducible factor 1 (HIF-1) is a critical mediator of physiological responses to acute and chronic hypoxia. First, HIF-1 is required for the development of the systems that mediate these responses, including the heart, blood and blood vessels. Mice with complete HIF-1α deficiency manifest developmental defects that involve all three components of the circulatory system. Second, HIF-1 mediates changes in gene expression that underlie physiological responses to chronic hypoxia, such as increased erythropoiesis and angiogenesis. *Hif1a*$^{+/-}$ mice, which are partially HIF-1α deficient, manifest impaired hypoxia-induced pulmonary vascular remodelling. Smooth muscle cells from pulmonary arteries (PASMCs) of wild-type mice subjected to chronic hypoxia manifest hypertrophy, depolarization, increased $[Ca^{2+}]_i$, and decreased voltage-gated K^+ currents. These responses are impaired in PASMCs from *Hif1a*$^{+/-}$ mice. Carotid bodies isolated from *Hif1a*$^{+/-}$ mice are unresponsive to hypoxia despite normal histology and normal responses to cyanide stimulation. Rat PC12 cells share properties with O_2-sensing glomus cells of the carotid body, including hypoxia-inducible expression of tyrosine hydroxylase, the rate limiting enzyme for catecholamine biosynthesis. In PC12 cells subjected to intermittent hypoxia, Ca^{2+}/calmodulin-dependent kinase activity leads to HIF-1 transcriptional activity and tyrosine hydroxylase mRNA expression. Thus, HIF-1 regulates both acute and chronic responses to continuous and intermittent hypoxia.

2005 Signalling pathways in acute oxygen sensing. Wiley, Chichester (Novartis Foundation Symposium 272) p 2–14

The average adult consumes O_2 at a rate of approximately 250 ml per minute or about 360 litres of O_2 per day. A variety of biochemical reactions require O_2, most notably the process of oxidative phosphorylation, in which electrons are passed from NADH and $FADH_2$ to respiratory cytochromes in the inner mitochondrial membrane, and finally to O_2. The electromotive force that is generated during this process is used to catalyze the formation of ATP, which is utilized as the energy source for most reactions that are required to maintain cellular viability. This consumption of O_2 is dependent upon the activity of the respiratory system, which mediates the intake of 5–6 litres of air per minute or about 8000 litres per day. Once delivered to the pulmonary alveolar air sacs, O_2 diffuses into red blood cells, in which it is bound to haemoglobin, and then transported via the cardiovascular system for delivery to every cell of the body. Through the combined efforts of the respiratory and circulatory systems, every one of the more than 10^{14} cells of a healthy adult obtains sufficient O_2 to maintain metabolic homeostasis. The mechanisms that maintain cellular and systemic homeostasis have been the subject of investigation by physiologists for centuries. However, it has only been within the last decade that a unifying molecular mechanism for the control of oxygen homeostasis within individual cells, in tissues and organs, and within the body as a whole, both during development and in postnatal life, has been elucidated.

Discovery of HIF-1 as a transcriptional regulator of the *EPO* gene

In vertebrates, erythrocytes are specialized for the transport of O_2 from the lungs to body tissue and red cell mass determines the blood O_2-carrying capacity. Specialized cells in the kidney produce erythropoietin (EPO), which is secreted into the bloodstream and binds to receptors on bone marrow erythroid progenitor cells, activating a signal transduction pathway leading to cell survival. When O_2 delivery is reduced, increased levels of EPO are produced, resulting in a compensatory increase in red cell mass. A *cis*-acting regulatory element was identified in the *EPO* gene that is required for hypoxia-induced gene transcription (Beck et al 1991, Pugh et al 1991, Semenza et al 1991). The hypoxia response element (HRE) was used as a molecular probe to identify the binding of a transcription factor, which was designated hypoxia-inducible factor 1 (HIF-1) because it was detected in nuclear extracts of cells exposed to hypoxia and undetectable in nuclear extracts prepared from cells that were cultured under non-hypoxic conditions (Semenza & Wang 1992). HIF-1 was purified by DNA affinity chromatography and shown to be a heterodimer of HIF-1α and HIF-1β subunits (Wang & Semenza 1995). Partial protein sequence analysis provided sufficient information to isolate complete cDNA sequences encoding both subunits (Wang et al 1995). HIF-1α protein levels and transcriptional activity were found to be dramatically regulated by the cellular O_2 concentration (Jiang et al 1996, 1997). O_2-dependent hydroxylation of proline and asparagine residues in HIF-1α represent the mechanism for transducing changes in cellular oxygenation into changes in HIF-1 activity (Epstein et al 2001, Ivan et al 2001, Lando et al 2002, Yu et al 2001). Two additional proteins involved in the negative regulation of HIF-1α protein stability are OS-9, which binds to both the prolyl hydroxylases and to HIF-1α (Baek et al 2005), and ARD1, which acetylates lysine 532 of HIF-1α (Jeong et al 2002). Mitochondrial reactive oxygen species production may also contribute to inactivation of the HIF-1α hydroxylases under hypoxic conditions (Chandel et al 2000).

HIF-1 is required for embryonic survival

Unlike the *EPO* gene, which is expressed only in a limited number of cell types, HIF-1 activity was induced under hypoxic conditions in all cell types tested (Wang & Semenza 1993), which suggested that HIF-1 played a more general role in oxygen homeostasis. Analysis of mice in which the gene encoding either HIF-1α or HIF-1β was inactivated by homologous recombination revealed that HIF-1 was required for embryonic survival (Carmeliet et al 1998, Iyer et al 1998, Maltepe et al 1997, Ryan et al 1998). The absence of HIF-1 activity results in lethality at midgestation with defective development of the heart, blood and vessels, i.e. all three components of the circulatory system.

HIF-1 is a critical regulator of vascularization

In the case of tissue vascularization, each cell insures that it receives adequate perfusion by hypoxia-induced expression of angiogenic growth factors, particularly vascular endothelial growth factor (VEGF) (Shweiki et al 1992), which activates endothelial cells leading to capillary sprouting. Human and rodent *VEGF* genes were shown to contain an HRE in their 5′-flanking region (Levy et al 1995, Liu et al 1995) that was activated by HIF-1 binding (Forsythe et al 1996). More recent gain-of-function and loss-of-function experiments have shown that HIF-1 controls the expression of many of the key angiogenic growth factors including VEGF, placental growth factor, platelet-derived growth factor B, angiopoietin 1 and angiopoietin 2 (Kelly et al 2003), which are produced by hypoxic cells in tissues and bind to receptors on vascular endothelial and smooth muscle cells. In addition, HIF-1 also controls cell-autonomous responses to hypoxia in vascular endothelial cells by regulating the expression of hundreds of genes (Manalo et al 2005). Loss-of-function and gain-of-function studies indicate that HIF-1 plays a critical role in vascularization both during development and in postnatal life (Carmeliet et al 1998, Iyer et al 1998, Kelly et al 2003, Ryan et al 1998).

Involvement of HIF-1 in pulmonary vascular remodelling in response to chronic hypoxia

HIF-1 also controls remodelling of pre-existing vessels in response to hypoxia. When humans and experimental animals are subjected to alveolar hypoxia as a result of chronic obstructive pulmonary disease or exposure to reduced ambient O_2, respectively, pulmonary arterioles undergo a remodelling process involving hypertrophy and hyperplasia of smooth muscle cells in the medial compartment of the vessel wall, which results in a reduction in luminal area, increased resistance to blood flow, and pulmonary hypertension. *Hif1a*$^{+/-}$ mice, which are heterozygous for a null allele at the locus encoding HIF-1α and thus partially HIF-1α deficient, have impaired pulmonary arterial remodelling in response to chronic hypoxia (Yu et al 1999). Electrophysiological studies of pulmonary artery smooth muscle cells (PASMCs) isolated from pulmonary arteries of *Hif1a*$^{+/-}$ mice and wild-type littermates revealed that the hypoxia-induced depolarization and reduction of K_v channel current that were observed in PASMCs from wild-type mice were markedly blunted in the heterozygotes (Shimoda et al 2001). Hypoxia induced hypertrophy of PASMCs isolated from wild-type mice but not their heterozygous littermates. Thus, HIF-1 mediates two of the classic pathological responses to chronic hypoxia: hypertrophy and depolarization of PASMCs.

HIF-1 is required for carotid body responses to acute and chronic hypoxia

Whereas the responses to hypoxia described above occur on a timescale of weeks, acute responses to hypoxia occur within seconds. The classic acute physiological responses to hypoxia are the increase in respiratory and heart rate that occur in response to the stimulation of brainstem centres by neural signals emanating from the carotid body, which is a small organ located at the bifurcation of the carotid artery that functions as the primary chemoreceptor for sensing arterial pO2 (López-Barneo 2003, Prabhakar 2000). Exposure of *Hif1a*$^{+/-}$ mice to acute hypoxia or hypercarbia was associated with increases in respiratory rate (RR), tidal volume, and minute ventilation that were similar to wild-type littermates (Kline et al 2002). However, exposure of wild-type mice to chronic hypobaric hypoxia (three days at 0.4 ATM) resulted in an augmented ventilatory response to a subsequent acute hypoxic exposure, whereas in *Hif1a*$^{+/-}$ mice the acute ventilatory response was actually blunted following chronic hypoxia. The carotid body plays a critical role in ventilatory adaptation to chronic hypoxia. To analyse carotid body function, we performed the Dejours test (Dejours 1962). Exposure of wild-type mice to a brief hyperoxic challenge inhibited RR and minute neural respiration (RR $\times$ integrated phrenic nerve activity), whereas this response was blunted in *Hif1a*$^{+/-}$ mice, providing further evidence for a defect in the carotid body.

When carotid bodies from wild-type mice were exposed to 100% O_2 followed by 12% O_2 there was a dramatic increase in carotid sinus nerve activity. The response was absent in carotid bodies from *Hif1a*$^{+/-}$ mice. However, these carotid bodies responded normally to cyanide administration. Furthermore, immunohistochemistry revealed that glomus cells were present in normal numbers, were of normal morphology and showed normal production of chromogranin A and tyrosine hydroxylase (Kline et al 2002). Thus in mice with only a partial deficiency of

HIF-1α expression the ability of the carotid body to either sense or respond to hypoxia is specifically lost. In these mice, peripheral chemoreceptors have compensated for the loss of carotid body function, similar to the effect of carotid sinus nerve transection, which initially abolishes acute hypoxic ventilatory responses but subsequently leads to a reorganization of the chemoreflex pathway with recovery of the hypoxic ventilatory response (Martin-Body et al 1986). In support of this hypothesis, the ventilatory response to hypoxia was markedly impaired after vagotomy in wild-type mice but not in heterozygotes.

HIF-1 is induced by intermittent hypoxia

In addition to playing an important role in adaptation to chronic hypoxia, the carotid body is required for responses to intermittent hypoxia, which occurs during sleep-disordered breathing, a condition that affects >18 million people in the USA and results in systemic hypertension (Kiley et al 1995). To analyse molecular mechanisms underlying involvement of HIF-1α in carotid body responses to intermittent hypoxia, we have utilized rat PC12 cells, which share many properties with glomus cells of the carotid body, including O_2-regulated neurotransmitter release (Kumar et al 1998) and expression of tyrosine hydroxylase, the rate limiting enzyme for catecholamine biosynthesis (Hui et al 2003). Cells were exposed to alternating cycles of 1.5% O_2 for 30 s followed by 20% O_2 for 4 min (Yuan et al 2004). HIF-1α protein expression and HIF-1 transcriptional activity were induced by exposure of cells to 30, 60 or 120 cycles of intermittent hypoxia. Addition of the intracellular Ca^{2+} chelator BAPTA-AM or the Ca^{2+}/calmodulin-dependent (CaM) kinase inhibitor KN93 blocked the induction of HIF-1 transcriptional activity in response to intermittent hypoxia. CaM kinase activity increased fivefold in cells subjected to intermittent hypoxia. KN93 blocked intermittent hypoxia-induced transcriptional activation mediated by HIF-1α or its coactivator p300, which was phosphorylated by CaM kinase *in vitro*. HIF-1-regulated expression of TH mRNA, encoding tyrosine hydroxylase, was induced by intermittent hypoxia and this effect was blocked by KN93. In contrast, the induction of TH mRNA by continuous hypoxia was not blocked by KN93. HIF-1 transcriptional activity and TH mRNA expression were induced in non-hypoxic cells transfected with a plasmid encoding a constitutively active form of CaM kinase II (Yuan et al 2004). Taken together, these results indicate that intermittent hypoxia induces HIF-1 transcriptional activity and TH mRNA expression via a novel pathway involving CaM kinase, an enzyme that is activated by the increase in intracellular Ca^{2+} levels that occurs during depolarization. Thus, HIF-1 represents a bridge between the acute (depolarization and neurotransmission) and chronic (changes in gene and protein expression) responses to hypoxia.

Acknowledgments

We thank Dr Charles Wiener for helpful comments on the manuscript.

References

Baek JH, Mahon PC, Oh J et al 2005 OS-9 interacts with hypoxia-inducible factor 1α and prolyl hydroxylases to promote oxygen-dependent degradation of HIF-1α. Mol Cell 17:503–512

Beck I, Ramirez S, Weinmann R, Caro J 1991 Enhancer element at the 3'-flanking region controls transcriptional response to hypoxia in the human erythropoietin gene. J Biol Chem 266:15563–15566

Carmeliet P, Dor Y, Herbert JM et al 1998 Role of HIF-1α in hypoxia-mediated apoptosis, cell proliferation and tumour angiogenesis. Nature 394:485–490

Chandel NS, McClintock DS, Feliciano CE et al 2000 Reactive oxygen species generated at mitochondrial complex III stabilize hypoxia-inducible factor-1α during hypoxia: a mechanism of O_2 sensing. J Biol Chem 33:25130–25138

Dejours P 1962 Chemoreceptors in breathing. Physiol Rev 42:335–358

Epstein AC, Gleadle JM, McNeill LA et al 2001 C. elegans EGL-9 and mammalian homologs define a family of dioxygenases that regulate HIF by prolyl hydroxylation. Cell 107:43–54

Forsythe JA, Jiang BH, Iyer NV et al 1996 Activation of vascular endothelial growth factor gene transcription by hypoxia-inducible factor 1. Mol Cell Biol 16:4604–4613

Hui AS, Striet JB, Gudelsky G et al 2003 Regulation of catecholamines by sustained and intermittent hypoxia in neuroendocrine cells and sympathetic neurons. Hypertension 42:1130–1136

Ivan M, Kondo K, Yang H et al 2001 HIFα targeted for VHL-mediated destruction by proline hydroxylation: implications for O_2 sensing. Science 292:464–468

Iyer NV, Kotch LE, Agani F et al 1998 Cellular and developmental control of O_2 homeostasis by hypoxia-inducible factor 1α. Genes Dev 12:149–162

Jeong JW, Bae MK, Ahn MY et al 2002 Regulation and destabilization of HIF-1α by ARD1-mediated acetylation. Cell 111:709–720

Jiang BH, Semenza GL, Bauer C, Marti HH 1996 Hypoxia-inducible factor 1 levels vary exponentially over a physiologically relevant range of O_2 tension. Am J Physiol 271:C1172–C1180

Jiang BH, Zheng JZ, Leung SW, Roe R, Semenza GL 1997 Transactivation and inhibitory domains of hypoxia-inducible factor 1α: modulation of transcriptional activity by oxygen tension. J Biol Chem 272:19253–19260

Kelly BD, Hackett SF, Hirota K et al 2003 Cell type-specific regulation of angiogenic growth factor gene expression and induction of angiogenesis in nonischemic tissue by a constitutively active form of hypoxia-inducible factor 1. Circ Res 93:1074–1081

Kiley JP, Edelman N, Derderian S, Horan M, Littner M 1995 Cardiopulmonary disorders of sleep. In: National Commission on Sleep Disorders Research, Volume Two: Working Group Reports, US Government Printing Office, p 10–75

Kline DD, Peng YJ, Manalo DJ, Semenza GL, Prabhakar NR 2002 Defective carotid body function and impaired ventilatory responses to chronic hypoxia in mice partially deficient for hypoxia-inducible factor 1α. Proc Natl Acad Sci USA 99:821–826

Kumar GK, Overholt JL, Bright GR et al 1998 Release of dopamine and norepinephrine by hypoxia from PC-12 cells. Am J Physiol 274:C1592–C1600

Lando D, Peet DJ, Whelan DA, Gorman JJ, Whitelaw ML 2002 Asparagine hydroxylation of the HIF transactivation domain a hypoxic switch. Science 295:858–861

Levy AP, Levy NS, Wegner S, Goldberg MA 1995 Transcriptional regulation of the rat vascular endothelial growth factor gene by hypoxia. J Biol Chem 270:13333–13340

Liu Y, Cox SR, Morita T, Kourembanas S 1995 Hypoxia regulates vascular endothelial growth factor gene expression in endothelial cells: identification of a 5' enhancer. Circ Res 77:638–643

López-Barneo J 2003 Oxygen and glucose sensing by carotid body glomus cells. Curr Opin Neurobiol 13:493–499
Maltepe E, Schmidt JV, Baunoch D, Bradfield CA, Simon MC 1997 Abnormal angiogenesis and responses to glucose and oxygen deprivation in mice lacking the protein ARNT. Nature 386:403–407
Manalo DJ, Rowan A, Lavoie T et al 2005 Transcriptional regulation of vascular endothelial cell responses to hypoxia by HIF-1. Blood 105:659–669
Martin-Body RL, Robson GJ, Sinclair JD 1986 Restoration of hypoxic respiratory responses in the awake rat after carotid body denervation by sinus nerve section. J Physiol 380:61–73
Prabhakar NR 2000 Oxygen sensing by the carotid body chemoreceptors. J Appl Physiol 88:2287–2295
Pugh CW, Tan CC, Jones RW, Ratcliffe PJ 1991 Functional analysis of an oxygen-regulated transcriptional enhancer lying 3' to the mouse erythropoietin gene. Proc Natl Acad Sci USA 88:10553–10557
Ryan HE, Lo J, Johnson RS 1998 HIF-1α is required for solid tumor formation and embryonic vascularization. EMBO J 17:3005–3015
Semenza GL, Wang GL 1992 A nuclear factor induced by hypoxia via de novo protein synthesis binds to the human erythropoietin gene enhancer at a site required for transcriptional activation. Mol Cell Biol 12:5447–5454
Semenza GL, Nejfelt MK, Chi SM, Antonarakis SE 1991 Hypoxia-inducible nuclear factors bind to an enhancer element located 3' to the human erythropoietin gene. Proc Natl Acad Sci USA 88:5680–5684
Shimoda LA, Manalo DJ, Sham JS, Semenza GL, Sylvester JT 2001 Partial HIF-1α deficiency impairs pulmonary arterial myocyte electrophysiological responses to hypoxia. Am J Physiol 281:L202–208
Shweiki D, Itin A, Soffer D, Keshet E 1992 Vascular endothelial growth factor induced by hypoxia may mediate hypoxia-initiated angiogenesis. Nature 359:843–845
Wang GL, Semenza GL 1993 General involvement of hypoxia-inducible factor 1 in transcriptional response to hypoxia. Proc Natl Acad Sci USA 90:4304–4308
Wang GL, Semenza GL 1995 Purification and characterization of hypoxia-inducible factor 1. J Biol Chem 270:1230–1237
Wang GL, Jiang BH, Rue EA, Semenza GL 1995 Hypoxia-inducible factor 1 is a basic-helix-loop-helix-PAS heterodimer regulated by cellular O_2 tension. Proc Natl Acad Sci USA 92:5510–5514
Yu AY, Shimoda LA, Iyer NV et al Impaired physiological responses to chronic hypoxia in mice partially deficient for hypoxia-inducible factor 1α. J Clin Invest 103:691–696
Yu F, White SB, Zhao Q, Lee FS 2001 HIF-1α binding to VHL is regulated by stimulus-sensitive proline hydroxylation. Proc Natl Acad Sci USA 98:9630–9635
Yuan G, Nanduri J, Bhasker RC, Semenza GL, Prabhakar NR 2004 Ca^{2+}/calmodulin kinase-dependent activation of hypoxia-inducible factor 1 transcriptional activity in cells subjected to intermittent hypoxia. J Biol Chem 280:4321–4328

DISCUSSION

Duchen: Do you see the involvement of HIF-1 in carotid body oxygen sensing as an acute role, or an involvement in the regulation of a channel or some other protein? Do you think that HIF-1 has an acute role to play in oxygen sensing?

Semenza: We don't know what the mechanism is that accounts for the lack of response in the heterozygotes. The most likely explanation is that there are genes that are not being transcribed at sufficient levels to produce critical proteins such as channels or regulators of channels. A more novel role would be if HIF-1 is somehow directly contributing to the response by virtue of the fact that it is being regulated by the oxygen concentration, so the presence or absence of the protein could be used as a signal for things other than transcription. We don't have any evidence for this, though.

Duchen: The actual oxygen sensor in this pathway is the prolyl hydroxylase. Can you manipulate this? It could mediate an oxygen sensing mechanism.

Semenza: The problem is that you are still going to be manipulating HIF-1. It won't answer the question of the mechanism; it will just say that HIF-1 is involved, and we know that already. In terms of the mechanism, unless it led us directly to a protein that was being hydroxylated—which is another possibility—looking at prolyl hydroxylase would tell us little.

Ratcliffe: Wasn't the HIF-1α heterozygote phenotype initially held to be normal, with this phenotype becoming apparent later? The HIF-1β phenotype was also held to be normal initially. What do we now know about this? We might expect that these effects would be seen.

Semenza: They might even be more severe, because HIF-2α has also been implicated.

Ratcliffe: If it was less severe, then this would raise a question.

Semenza: I don't know of anyone who has done experiments with the HIF-1β heterozygote.

Ratcliffe: A related question, then, is with your adenovirus delivery system you could induce a transcriptionally disabled HIF: have you done this yet, and if you have, are there any effects on gene expression?

Semenza: No, we haven't done this yet. We are hoping to isolate the glomus cells and use them. PC12 cells are a useful model, but they have their limitations.

López-Barneo: I have a comment on the role of HIF-1α in acute oxygen sensing in the carotid body. We have done experiments not using the whole carotid body preparation but rather a slice preparation that in our hands mimics what is seen *in vivo* very well. We can see secretory responses to hypoxia that are almost the same as those in *in vivo* preparations. We don't see any change in the hypoxia sensitivity in the HIF-1α heterozygote. We have tried to acutely inhibit prolyl hydroxylases by adding dimethyloxalylglycine (an inhibitor of prolyl hydroxylases) to our slice, and we didn't see any effect on acute oxygen sensing. So, at least in our hands, prolyl hydroxylation doesn't seem to be involved in the acute oxygen sensitivity in the carotid body.

Prabhakar: When you measured the secretory activity by amperometry, I presume that you were measuring the catecholamine secretion. We do not know the role of catecholamines in the sensory excitation by hypoxia. In order to understand the role of transmitters, especially catecholamines in sensory transmission during hypoxia, in addition to secretory activity it is imperative that we measure the sinus nerve activity as an output.

López-Barneo: We are using the carotid body thin slice as a model to study oxygen sensitivity. I am not talking about the whole organ sensitivity: this depends on ATP, acetylcholine release from the terminals, and so on. I am talking about a well established model. I think this catecholamine release in response to hypoxia is a good indication of oxygen sensitivity, regardless of what the catecholamines are doing. We are looking at whether the glomus cell is still oxygen sensitive, and is able to depolarise and induce catecholamine release in a Ca^{2+}-dependent manner. In this preparation we don't see any difference between catecholamine release induced by hypoxia in the normal animal and in the HIF-1α heterozygote.

Semenza: That is the value of the genetic models: they allow us to dissect the various components of the physiological response. They tell us that HIF-1 is not involved in this part of the response, but it must be involved in some other critical aspect of the response to hypoxia, because we see this dramatic effect at the level of the carotid sinus nerve transmission. There has to be another component of this response that is critical, which HIF-1 does control. Your data suggest that it does not control the secretory response. This is interesting, because we have to ask what controls that. Could it be HIF-2α? We are trying to understand at the molecular level how these different responses are controlled.

Duchen: Is there any ultrastructural change in the carotid bodies in these animals?

Prabhakar: We looked only at gross morphology at the light microscopic level and have not done electron microscopy. It's an important question.

Acker: It is well known that *in vivo* organ pO2 distribution is ranging from 0–90 Torr. The mean pO2 is about 20–30 Torr. Most of the cells live under low pO2 conditions. There must be a quite different mechanism for inhibiting the HIF response under these low pO2 conditions. However, cells are able to respond to a change in the whole pO2 field. Most of the cells live under low oxygen conditions but they are not hypoxic. It is difficult to mirror this heterogeneity in tissue culture. Do you have indications that in the organ there is also a distribution of HIF responses?

Semenza: These are complicated questions. It is all relative: the normal oxygen concentration will be different for different cells. Likewise, the level of HIF-1α expression at any particular pO2 is not absolute: it differs from cell to cell. The hydroxylases are not operating under equilibrium conditions. The level of the expression of the enzymes determines the dose–response curve, shifting it in one direction or another. With all the components of the pathway, it seems that they can be decreased or increased and effects are seen in the response: nothing seems to be rate limiting in the pathway. For example, the expression of the hydroxylases may be controlled by many other factors.

Acker: The medulla of the kidney has very low pO2 levels, but we don't see any HIF. The lung has very high pO2 values: perhaps this is the reason why it is so sensitive to changes in pO2. Another example is the liver, which has very drastic mapping of periportal and perivenous distribution of HIF stabilization. There must be a complicated number of control mechanisms.

Chandel: There are some data to suggest that the prolyl hydroxylases (PHDs) come up during hypoxia. This would suggest that chronically, in tissues that are more hypoxic, HIF-1α would be turned off.

Semenza: This raises a new issue: the feedback controls on the system. First of all, there is the complexity of each cell being at a different set point, and then there is the feedback regulation. This has been known for a long time: if someone is made anaemic by removal of large volumes of blood their EPO levels will go up and come back down before the haematocrit has changed at all. This is necessary so that they don't overshoot. If there are too many red blood cells this could be dangerous because of increased blood viscosity. There has to be feedback that is clearly oxygen independent, because it occurs before any change in the oxygen carrying capacity. The up-regulation of the PHD genes is also controlled by HIF-1, and this is one of the mechanisms by which this feedback occurs. Depending on the nature of that response, you could modulate the feedback. That is, the kinetics and the intensity of that feedback relationship can also be modulated. In every cell HIF-1 is going to be induced, but the level of expression of particular downstream genes is modulated by other transcription factors, which creates a whole new level of complexity.

Chandel: Could you comment on HIF-2α? Specifically, it is clear from the phenotypes that HIF-1α and HIF-2α have distinct functions, but there is not a list of distinct target genes. Why?

Semenza: HIF-1α is expressed in all cell types, and coordinates response to hypoxia in all. HIF-2α has a more specialized role, and appears to be much more cell-type specific. It is expressed in a restricted number of cells types and has a more specialized role. Part of the problem is that the experiments haven't been done in the correct cell types. If we did the experiment in a cell type where HIF-2α has a critical physiological function then we might uncover those genes. Doing the experiment in tissue culture cells where it may not play an important role is unlikely to give the answer. The HIF-2α knockout mice have not been pursued to the same degree. Joe Garcia is now doing nice work with them and is posing interesting hypotheses about HIF-2α potentially being involved in the response to oxidative stress that are quite unexpected. I think that analysing physiological responses in the knockout mice will prove key.

Peet: Endothelial cells are one of the cell types where HIF-2α is highly expressed. You had some nice results with endothelial cells. It looked as if many of the target genes you saw changing were similar with hypoxia and over-expression of HIF-1α.

Semenza: That is how the experiment was set up: we were specifically looking for these. This was another case where there was an oversimplification: originally HIF-2α was found in endothelial cells and was hypothesized to play a critical role there whereas HIF-1α was not thought to be active in endothelial cells. This is not correct. There are interesting data emerging about which target genes are regulated by both HIF-1α and HIF-2α and which target genes are distinctly regulated by HIF-1α or HIF-2α.

Duchen: What is the lifetime of HIF?

Semenza: In isolated, perfused, ventilated ferret lung preparations, when we ventilate the lungs with 21% oxygen after ventilating with 0% oxygen the protein is degraded with a half-life of less than one minute. I don't know of any protein that has a shorter half life.

Schumacker: How do you explain the intermittent hypoxia when you give 30 s of hypoxia and 20 min of normoxia, and you see a progressive increase in HIF after just 10–20 s?

Semenza: This is because hypoxia and re-oxygenation, rather than hypoxia, is the signal. We have shown that HIF-1α transactivation is not being mediated through changes in the hydroxylation of HIF-1α, but rather through the phosphorylation of p300 that is mediated by the CaM kinase pathway. Probably, the signal transduction pathways are being activated by reactive oxygen species that arise as a result of hypoxia and then reoxygenation in a repetitive fashion.

Prabhakar: We monitored O_2 profiles near cells during intermittent hypoxia, and found that they drop by about ~25 mmHg with each episode of hypoxia. With sustained hypoxia for 15 minutes, which is equivalent to the 120 cycles of intermittent hypoxia, pO2 drops by about ~60 mmHg. Despite the modest drop in pO2, intermittent hypoxia is more potent a stimulus in activating c-*fos*; whereas 15 min of sustained hypoxia caused a more substantial drop in pO2, it was ineffective in evoking c-*fos*, suggesting that the effectiveness of intermittent hypoxia is due to something more than drop in pO2, which we presume is reactive oxygen species (ROS).

Ward: This would predict that if you have a higher frequency of reoxygenation, you would increase the signal.

Prabhakar: If you change the duration of a single episode of hypoxia, it doesn't have an impact on the magnitude of c-*fos* activation, but if you increase the duration of reoxygenation phase, while keeping the duration of hypoxic episode constant, then you get a proportional increase in c-*fos* activation. It seems that the major stimulus is the reoxygenation, not the absolute fall in pO2 during each episode of hypoxia. We published these observations at the beginning of last year (Yuan et al 2004).

Archer: I have a related question about redox signalling. We found a rat (the Fawn hooded rat; unpublished data) that has a mitochondrial defect leading to impaired hypoxic pulmonary vasoconstriction (HPV) and HIF activation which promotes development of pulmonary hypertension. This seems to be related to a failure to make radicals, and is associated with a loss of HPV in this animal. If this conference is trying to look for unifying themes, what do you think of the argument that the same redox signal that is involved in acute oxygen sensing is involved in the remodelling responses? If this signal is removed, HIF activation may ensue.

Semenza: Another one of the complexities of this system is that under hypoxia it appears that there is actually ROS generation in the mitochondria, which provides a signal that is important for the response as well. I hope Paul Schumacker will tell us later how this impacts on the hydroxylation system in continuous hypoxia. There is some role for ROS in that system as well, which is much harder to understand for me than in intermittent hypoxia, where it is easy to understand how ROS are generated. Obviously, the mitochondria are the main consumers of oxygen in the cell, and that a signal from the mitochondria is also critical for the response to hypoxia shows how well the cell is integrating all these signals at the level of HIF-1 in terms of determining the output.

Ward: Going back to the intermittent bit, there was some work we did a long time ago looking at intermittent hypoxia in rats. We used right ventricular hypertrophy and polycythaemia as a model. One of the things we showed was that the summed length of hypoxia was important. It didn't matter whether we had 12 half-hour pulses through the day or one six hour period: the same level of response was observed. This is slightly different from what you are saying.

Semenza: When your units are hours, I view this as continuous hypoxia, as opposed to 30-second hypoxic exposure, which is what we use in our intermittent hypoxia experiments. Under the conditions that you described, HIF-1α accumulates as a result of decreased hydroxylation during the hypoxic exposure, and the resulting effect on gene expression is cumulative.

Ward: Yes, I suppose from the point of view of the patient with sleep apnoea, the periods of hypoxia are around 45 s at the very most.

Prabhakar: The average duration of apnoea in adult human subjects averages ~12 s, and the range is between 7–30 s.

Ward: The question is whether the tissues actually see significant hypoxia under these conditions, considering the short period and rate of diffusion from the blood to the cells.

Prabhakar: I do not know whether other tissues see hypoxia, but I believe carotid bodies do see hypoxia because of their close proximity to the lungs, and the circulation time from lung to carotid body is ~6 s. I also believe that reoxygenation is more important than hypoxia. In the patient population there appear to be two patterns of sleep apnoea; one monomorphic pattern, wherein the number of apnoeic episodes is far more and associated with a modest drop in arterial O_2 saturation. The other is a polymorphic pattern, where the numbers of apnoeic episodes are relatively less but associated with a greater fall in O_2 saturations.

Semenza: The important point is that the carotid body senses and responds to arterial hypoxemia. The responses are mediated by the carotid body through activation of the sympathetic nervous system. This is why patients with sleep apnoea who are subjected to intermittent hypoxia mainly get systemic hypertension and not pulmonary hypertension, because it is a systemic effect mediated via the carotid body whereas continuous hypoxia has a direct effect on the pulmonary vasculature.

Ward: You still get polycythaemia, which is presumably not mediated via the carotid body and sympathetic system.

Semenza: We have shown that intermittent hypoxia does induce EPO through HIF-1.

Reference

Yuan G, Adhikary G, McCormick AA, Holcroft JJ, Kumar GK, Prabhakar NR 2004 Role of oxidative stress in intermittent hypoxia-induced immediate early gene activation in rat PC12 cells. J Physiol 557:773–783

Regulation of HIF: prolyl hydroxylases

Ineke P. Stolze, David R. Mole and Peter J. Ratcliffe[1]

Henry Wellcome Building for Molecular Physiology, Roosevelt Drive, Oxford, OX3 7BN, UK

Abstract. Hypoxia inducible factor (HIF) is an α/β heterodimeric transcriptional complex that plays a key role in directing cellular responses to hypoxia. Recent studies have defined novel oxygen-sensitive signal pathways that regulate the activity of HIF by post-translational hydroxylation at specific residues within the α subunits. HIF prolyl hydroxylation regulates proteolytic degradation of HIF whereas HIF asparaginyl hydroxylation modulates interaction with transcriptional co-activators. These hydroxylations are catalysed by a set of non-haem Fe(II)- and 2-oxoglutarate (2-OG)-dependent dioxygenases. During catalysis, the splitting of molecular oxygen is coupled to the hydroxylation of HIF and the oxidative decarboxylation of 2-OG to give succinate and CO_2. Hydroxylation at two prolyl residues within the central 'degradation domain' of HIF-α increases the affinity for the von Hippel-Lindau (pVHL) E3 ligase complex by at least three orders of magnitude, thus directing HIF-α polypeptides for proteolytic destruction by the ubiquitin/proteasome pathway. Since the HIF hydroxylases have an absolute requirement for molecular oxygen this process is suppressed in hypoxia allowing the HIF-α to escape destruction and activate transcription. Co-substrate and co-factor requirements for Fe(II), ascorbate, and the Krebs cycle intermediate 2-OG, and inducible changes in the cellular abundance of three closely related HIF prolyl hydroxylases (PHD1–3) provide additional interfaces with cellular oxygen status that may be important in regulating the oxygen-sensitive signal.

2005 Signalling pathways in acute oxygen sensing. Wiley, Chichester (Novartis Foundation Symposium 272) p 15–32

Hypoxia-inducible factor (HIF)

HIF is the central mediator of a large number of adaptive responses to hypoxia involving changes in gene expression. Transcriptional targets include genes involved in angiogenesis, metabolism, cell proliferation, vasomotor control, and erythropoiesis (for review see Semenza 2003, Schofield & Ratcliffe 2004). HIF binds to a core recognition sequence (G/ACGTG) in hypoxia response elements (HRE) of target genes as a heterodimer composed of one α-subunit (HIF-1α, HIF-2α or HIF-3α) and HIF-β, which is also known as the aryl hydrocarbon receptor nuclear

[1] This paper was presented at the symposium by Peter J Ratcliffe to whom correspondence should be addressed.

translocator (ARNT). The HIF-β subunit is constitutively expressed, whereas the stabilization and activity of HIF-α subunits is largely dependent on oxygen and markedly increased by hypoxia. Both subunits belong to the family of basic helix-loop-helix (bHLH) containing PER-ARNT-SIM (PAS) domain proteins. The bHLH and two PAS domains (PAS-A and PAS-B) near the N-terminus of HIF-α are essential for heterodimerization and DNA binding. The C-terminal half of HIF-α consists of an oxygen-dependent degradation domain (ODDD) that contains N- and C-terminal portions (NODDD and CODDD), and two transactivation domains (TADs) separated by an inhibitory domain (ID): an internal transactivation domain that overlaps with the CODDD (NAD) and the C-terminal TAD (CAD) that is distinct from the ODDD (for review see Semenza 2003, Schofield & Ratcliffe 2004).

Oxygen-dependent proteolytic degradation of HIF-α subunits involves polyubiquitination and degradation by the 26S proteasome. This process is dependent on the binding of the von Hippel-Lindau tumour suppressor protein (pVHL) to the ODDD of HIF-α, with pVHL functioning as the recognition component of a multi-component ubiquitin ligase (pVHL-elonginB-elonginC-Cul2-Rbx) (for review see Masson & Ratcliffe 2003). Transactivation studies with HIF regions fused to a heterologous DNA-binding domain, have identified three regions (NODDD, CODDD and CAD) that respond to the classical HIF-activating stimuli such as hypoxia, cobaltous ions and iron chelation.

In contrast to the activation by NODDD and CODDD, whose function is mediated by VHL-dependent proteolysis, the regulation of the transcriptional activation by CAD occurs independently of changes in HIF-α protein expression.

Recent analyses of these processes have shown that HIF is regulated by at least two post-translational hydroxylation steps that regulate protein–protein interactions in an oxygen-dependent manner: prolyl hydroxylation regulates HIF-α(ODDD)-pVHL interaction and subsequently HIF-α degradation (Jaakkola et al 2001),

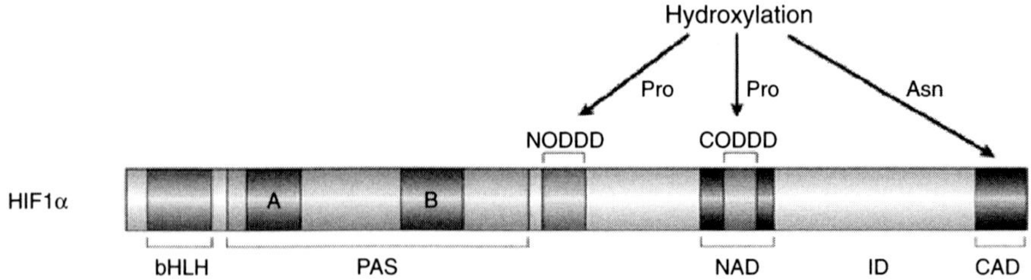

FIG. 1. The domain structure of HIF-1α (adapted from Schofield & Ratcliffe 2004).

whereas asparaginyl hydroxylation modulates HIF-α(CAD)-p300 interaction and transcriptional activity (Lando et al 2002b).

The HIF hydroxylases

The enzymes that catalyse HIF hydroxylation belong to the family of 2-oxoglutarate (2-OG) and iron-dependent dioxygenases that use one atom of molecular oxygen to hydroxylate target prolyl and asparaginyl residues within HIF-α, and the second atom for the oxidative decarboxylation of 2-OG to give succinate and CO_2. Based on precedent for enzymes within this family, it is likely that the hydroxylase/Fe^{2+}/2-OG complex binds to the HIF-α, and substrate binding then primes the enzyme for interaction with molecular oxygen. This reaction requires ascorbate for full catalytic activity, though the precise role of ascorbate remains unclear. Ascorbate may be required for the reduction of the enzyme catalytic iron centre following uncoupled cycles, and/or may act more generally to increase the availability of cellular Fe^{2+}.

HIF prolyl hydroxylases exist in *Caenorhabditis elegans* (Epstein et al 2001) and *Drosophila melanogaster* (Bruick & McKnight 2001). In mammalian cells, three prolyl-4-hydroxylase isoforms, PHD1, PHD2 and PHD3 (for prolyl hydroxylase domain) have been identified (other acronyms are EGLN2, EGLN1 and EGLN3 [for homologues of the EGL-9 protein from *Caenorhabditis elegans*] and HPH3, HPH2 and HPH1 [for HIF prolyl hydroxylase]). These enzymes hydroxylate two independent sites: Pro402 in the NODDD and Pro564 in the CODDD (for human HIF-1α) (Epstein et al 2001). Each site contains a conserved LXXLAP motif that can be independently recognized in hydroxylated form by pVHL and targeted for polyubiquitination and proteasomal degradation (for review see Masson & Ratcliffe 2003).

In vitro analyses have indicated that a different preference of the PHDs for the two hydroxylation sites in HIF-1α exists, with Pro402 not being modified by PHD3 (Epstein et al 2001, Hirsila et al 2003) in contrast to Pro564. Residues lying at a distance from the target prolyl residue, such as Leu574, which has been demonstrated to be important for the action of PHD2 on Pro564 (Kageyama et al 2004), might, at least in part, be responsible for this selectivity. *In vivo* PHD2 appears to contribute more dominantly to the control of HIF-1α than HIF-2α, with PHD3 contributing relatively more to the regulation of HIF-2α than HIF-1α; this may reflect the relative importance of the Pro402 and Pro564 hydroxylations in these two HIF-α isoforms (Appelhoff et al 2004).

In contrast to HIF proteolysis, the transcriptional activity of HIF is regulated by a unique asparaginyl hydroxylase (Lando et al 2002a, Hewitson et al 2002), first identified as factor inhibiting HIF-1 (FIH-1) (Mahon et al 2001). This enzyme uses

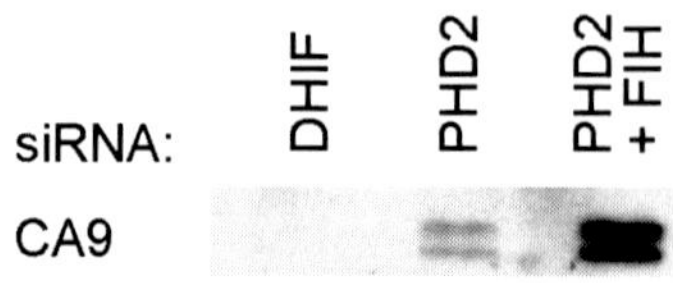

FIG. 2. Synergistic effect of small interfering RNA (siRNA) targeting PHD2 and FIH on the expression of the HIF transcriptional target gene carbonic anhydrase 9 (CA9) in U-2OS cells under normoxic conditions (DHIF siRNA duplex was used as an irrelevant control).

molecular oxygen to hydroxylate Asn803 in the CAD (for human HIF-1α) thus preventing its interaction with the transcriptional co-activator p300 (Hewitson et al 2002, Lando et al 2002a). Under hypoxic conditions both hydroxylation steps are inhibited which allows HIF-α to escape proteasomal degradation and to activate target gene transcription.

Oxygen-dependent activity of HIF hydroxylases

Because of their absolute requirement for molecular oxygen these enzymes have the potential to serve as oxygen sensors. Several reports support this role. *In vitro* assays of enzyme activity using VHL capture by HIF peptides as a measure of hydroxylation show marked oxygen dependency for all three isoforms with reduced hydroxylation activity in modest hypoxia (Epstein et al 2001, Tuckerman et al 2004). Furthermore, PHD overexpression reduces HIF activity in normoxic cells but not in cells cultured at 0.5% oxygen, suggesting that activity is indeed limited in cells by relevant levels of hypoxia (Bruick & McKnight 2001). Additionally, measurements of activity *in vitro* using 2-OG turnover have demonstrated an apparent K_m for oxygen in the range of 230–250 μM for all three PHD enzymes (Hirsila et al 2003). These values are above the concentration of oxygen seen under atmospheric conditions, and high in relation to the physiological oxygen concentrations in tissues. This indicates that at all physiological oxygen tensions, enzyme activity is likely to be limited by oxygen availability, potentially affecting the rate at which HIF-α is hydroxylated, ubiquitinated and degraded. One caveat to these findings is that studies to date have been performed on short HIF-α peptides. Since it is believed that it is the enzyme/substrate complex that binds molecular oxygen, it is possible that results could be different with the full-length HIF-α polypeptide, and might potentially be affected by post-translational modifications of the latter.

Current evidence suggests significant differences between the oxygen dependence of the HIF prolyl and asparaginyl hydroxylases. In contrast to the prolyl hydroxylases, overexpressed FIH is still able to suppress hypoxic induction of CAD and to reduce HRE-mediated reporter gene activity when overexpressed in cells

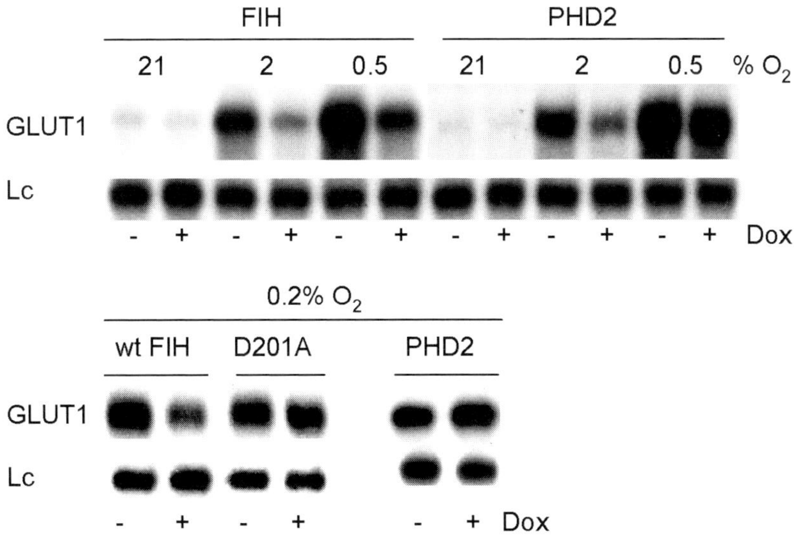

FIG. 3. FIH and PHD2 display a different oxygen dependence in living cells. Stable doxycycline-dependent 'Tet-on' FIH (wild-type and enzymatic inactive mutant D201A) and PHD2 transfectants in the same U-2OS cell background were equilibrated at different oxygen tensions ranging from 21% to 0.2% oxygen for 8 h prior induction with doxycycline for 16 h. To avoid confounding effects from unphysiological actions, enzymes were only modestly over-expressed. RNase protection assay for Glucose transporter-1 (GLUT1) mRNA and U6 SnRNA as loading control (Lc) are shown (adapted from Stolze et al 2004).

cultured at low oxygen concentration (Lando et al 2002a, Mahon et al 2001). Interestingly, the apparent K_m value of FIH for oxygen is substantially lower (90 μM) than that of the PHDs (Koivunen et al 2003). Comparison of modestly over-expressed PHD2 and FIH in cells pre-equilibrated at severe hypoxia supports the physiological relevance of these differences in cells, with FIH showing a distinctly lower oxygen threshold for detectable activity than PHD2 (Stolze et al 2004). Thus one function of FIH appears to be to limit the activity of HIF-α that escapes proteolysis at low oxygen tensions.

Regulation of expression of the HIF prolyl hydroxylases

Because the PHD enzymes do not catalyse the reverse reaction, the abundance of each enzyme will determine the net rate of substrate hydroxylation under any set of conditions. Thus changes in enzyme expression will potentially affect the oxygen sensitive signal. It is therefore of great interest that the expression pattern of prolyl hydroxylase isoforms varies in different tissues and cell types and is also highly responsive to certain inducing stimuli. PHD1 mRNA is widely expressed, but is most abundant in testis. PHD2 and PHD3 mRNAs are also expressed in many tissues with particular high expression of PHD2 mRNA in adipose tissue and heart and PHD3 mRNA in heart and placenta (Oehme et al 2002, Lieb et al 2002).

The expression pattern of PHDs also differs in response to physiological stimuli. In oestrogen receptor-expressing breast carcinoma cell lines, PHD1 gene expression is stimulated in response to oestrogen (Seth et al 2002). PHD3 expression has been reported to be inducible by factors inducing smooth muscle differentiation (Wax et al 1994), by nerve growth factor withdrawal (Lipscomb et al 2001) and by p53 (Madden et al 1996).

Hypoxia increases the expression of PHD2 and PHD3, but not PHD1. The response to hypoxia is at least partly mediated by HIF itself, indicating the existence of a feedback loop potentially limiting HIF induction in hypoxia and promoting destruction following reoxygenation (Appelhoff et al 2004, Berra et al 2003, D'Angelo et al 2003, Epstein et al 2001, Marxsen et al 2004). The *cis*-acting sequences mediating these effects are incompletely characterized but, one functional HRE has been defined in the promoter element of the PHD2 gene (Metzen et al 2005). A further level of control has recently been demonstrated to occur through the action of Siah2 and Siah1a, two members of the E3 ubiquitin ligase family that mediate proteolytic regulation of PHD expression amongst other targets. So far effects on PHD3 have been analysed most thoroughly, and indicate that Siah1a/2 function has the potential to regulate both HIF-α and HIF target gene expression via effects on hydroxylase abundance (Nakayama et al 2004). Since Siah2 transcripts are themselves induced by hypoxia, this process also appears likely to be important in shaping the cellular responses to hypoxia.

As HIF target genes, PHD2 and PHD3 protein expression can be modulated by FIH. The potential of FIH to promote HIF-1α stability by limiting PHD2 and more strikingly PHD3 protein expression in hypoxia (Stolze et al 2004) and possibly reoxygenation represents another regulatory mechanism to adjust prolyl hydroxylase activity and HIF-1α protein expression under hypoxic conditions.

In studies, the most abundant enzyme expressed under normoxic tissue culture conditions is the prolyl hydroxylase PHD2 (Berra et al 2003, Appelhoff et al 2004). When oxygen is not limiting, the amount of endogenous, active PHD2 enzyme will determine the rate of HIF-1α hydroxylation, ubiquitination and degradation. However, in different biological settings such as growth factor stimulation and cellular hypoxia, PHD1 and PHD3, respectively, also contribute to the regulation of HIF-1α and HIF-2α. Thus, following oestrogen stimulation, enhanced PHD1 levels in oestrogen responsive breast carcinoma cells contribute more importantly to both basal and hypoxia inducible levels of HIF-α (Appelhoff et al 2004). PHD3 is particularly strongly induced by hypoxia, and appears to contribute more to limiting HIF-α induction in hypoxia and promoting breakdown following reoxygenation.

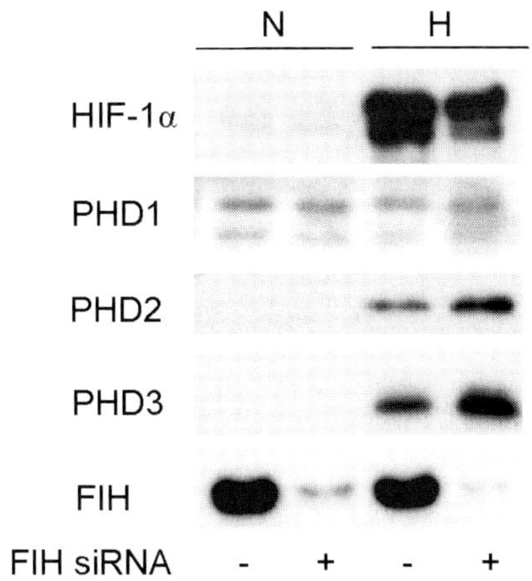

FIG. 4. Small interfering RNA (siRNA) targeting FIH enhances hypoxic induced expression of prolyl hydroxylases PHD2 and PHD3 in U-2OS cells. Cells were transfected with irrelevant control (−) or FIH siRNA duplexes (+) prior normoxic (N) or hypoxic (H, 16 h 1% oxygen) incubation. Immunoblots for HIF-1α, PHD1-3 and FIH are shown (adapted from Stolze et al 2004).

Co-factor requirements of the HIF hydroxylases present additional interfaces with the availability of oxygen

In addition to the use of molecular oxygen as an obligatory co-substrate, requirements for iron and the Krebs cycle intermediate 2-OG for enzyme activity potentially mediate other connections between enzyme activity and the availability of oxygen. Current evidence suggests that both of these processes do contribute to oxygen signalling.

Lability of the HIF hydroxylase catalytic iron centres likely explains the sensitivity of this pathway to activation by iron chelators. In response to oncogenic stimulation, hydroxylation of HIF-1α CODDD has been shown to be often incomplete even in fully oxygenated tissue culture suggesting that factors other than oxygen availability may limit hydroxylase activity (Chan et al 2002). That at least one of these factors is inadequate Fe^{2+}/ascorbate availability has been strongly suggested by the large reduction in normoxic HIF-α levels that is observed when cultures are supplemented by sources of iron and ascorbate (Knowles et al 2003). Interestingly many HIF target genes are themselves involved in iron metabolism and potentially operate to increase cellular iron availability so that this system may contribute to feedback mechanism in iron homeostasis. Though the connections between iron and oxygen homeostasis are incompletely understood, co-ordinate regulation may be teleologically important since the two elements interact to generate cellular toxicity by the generation of radical species. In this respect, it is interesting that a member of the transcription factor activator protein 1 (AP-1) family, the protein oncogene junD, has been proposed to inhibit the HIF system by limiting the production of oxygen radicals that reduce the activity of HIF hydroxylase PHD2 through iron oxidation of the catalytic centre to the Fe^{3+} inactivated state (Gerald et al 2004).

The effect of hypoxia on the mitochondrial electron transport might in theory also affect HIF hydroxylase activity by secondary effects on the availability of Krebs cycle intermediates such as 2-OG. Other potential links could be through mitochondrial radical production and radical based inactivation of the hydroxylase catalytic centre or inactivation of the Krebs cycle aconitase. In addition to effects on 2-OG availability certain Krebs cycle intermediates notably fumarate and succinate can act as competitive analogues of 2-OG for the HIF hydroxylases and are likely involved in promoting increases in HIF activity observed in fumarate hydratase- and succinate dehydrogenase-deficient tumour cells.

Thus it appears that the 2-OG oxygenases are well placed to regulate oxygen sensitive signal pathways. Multiple interfaces with the cellular availability of oxygen potentially allow flexibility in generating precisely tuned physiological responses within the HIF system. Given the predicted existence of a very extensive family of these enzymes in the mammalian genome an important challenge is now to

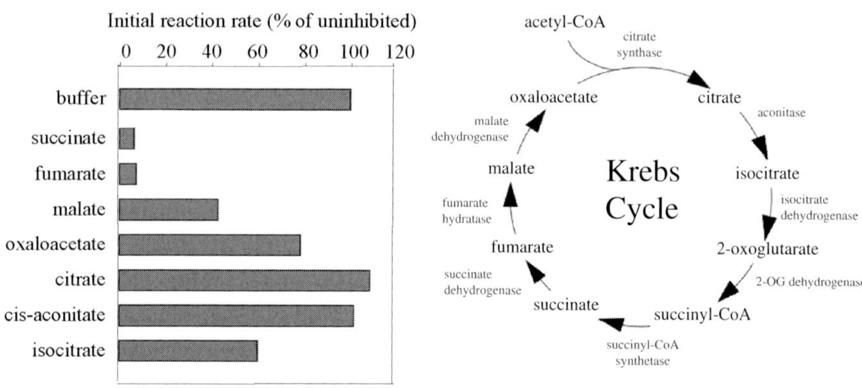

FIG. 5. Inhibition of PHD activity by tricarboxylic acid (TCA) cycle intermediates. VHL capture assays were performed, using purified bacterially expressed PHD2, to modify synthetically produced biotinylated HIF-1α peptide, corresponding to residues 556–574. The peptide was captured by streptavidin beads and the amount of hydroxylation assayed by its ability to capture radiolabelled pVHL, produced in a rabbit reticulocyte lysate IVTT. 150 μM 2-OG, was used in the assay together with 2 mM of the designated TCA cycle intermediate. Initial reaction rates were determined and normalised to the uninhibited reaction. Both succinate and fumarate show profound inhibition of PHD2 activity under these conditions.

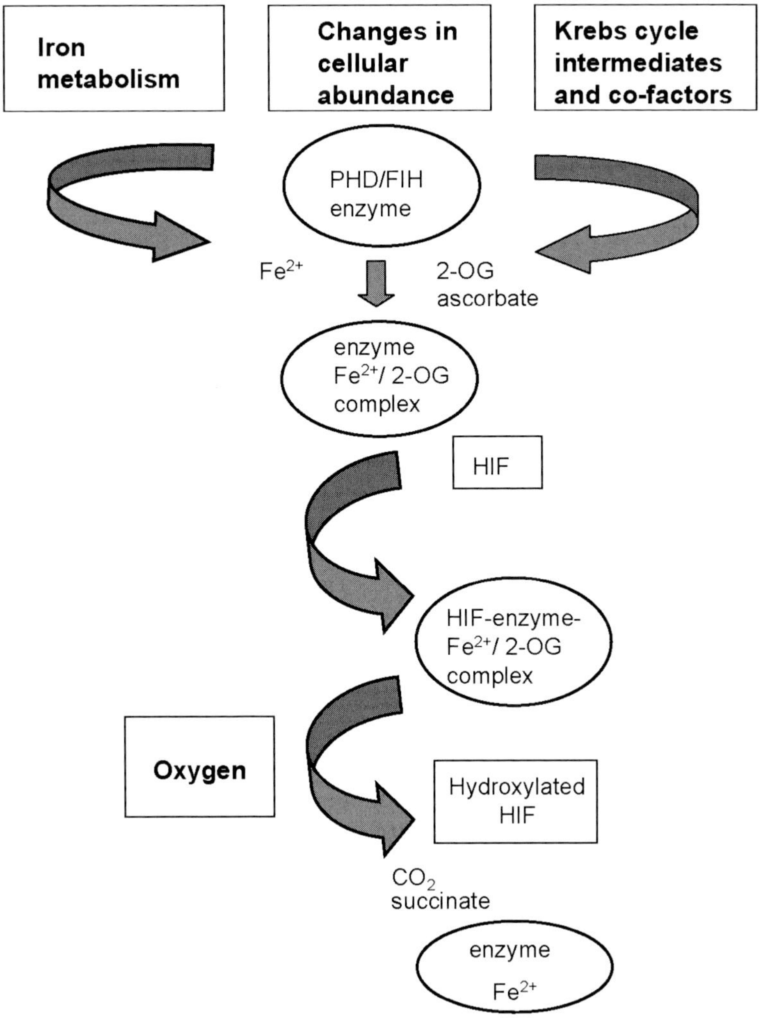

FIG. 6. Co-substrate and co-factor requirements for Fe^{2+}, ascorbate, Krebs cycle intermediates (2-OG) and inducible changes in the cellular abundance provide multiple interfaces with the cellular availability of oxygen and have the potential to provide precisely tuned physiological responses to hypoxia.

determine the existence and nature of other oxygen sensitive pathways that may be regulated by similar processes of protein hydroxylation.

Acknowledgements

This work was supported by The Wellcome Trust, Cancer Research UK and in part by the 6th framework program of the European Commission (EUROXY).

References

Appelhoff RJ, Tian YM, Raval RR et al 2004 Differential function of the prolyl hydroxylases PHD1, PHD2, and PHD3 in the regulation of hypoxia-inducible factor. J Biol Chem 279: 38458–38465

Berra E, Benizri E, Ginouves A et al 2003 HIF prolyl-hydroxylase 2 is the key oxygen sensor setting low steady-state levels of HIF-1α in normoxia. EMBO J 22:4082–4090

Bruick RK, McKnight SL 2001 A conserved family of prolyl-4-hydroxylases that modify HIF. Science 294:1337–1340

Chan DA, Sutphin PD, Denko NC, Giaccia AJ 2002 Role of prolyl hydroxylation in oncogenically stabilized hypoxia-inducible factor-1α. J Biol Chem 277:40112–40117

D'Angelo G, Duplan E, Boyer N, Vigne P, Frelin C 2003 Hypoxia up-regulates prolyl hydroxylase activity. J Biol Chem 278:38183–38187

Epstein ACR, Gleadle JM, McNeill LA et al 2001 *C. elegans* EGL-9 and mammalian homologues define a family of dioxygenases that regulate HIF by prolyl hydroxylation. Cell 107:43–54

Gerald D, Berra E, Frapart YM et al 2004 JunD reduces tumor angiogenesis by protecting cells from oxidative stress. Cell 118:781–794

Hewitson KS, McNeill LA, M.V. R et al 2002 Hypoxia inducible factor (HIF) asparagine hydroxylase is identical to Factor Inhibiting HIF (FIH) and is related to the cupin structural family. J Biol Chem 277:26351–26355

Hirsila M, Koivunen P, Gunzler V, Kivirikko KI, Myllyharju J 2003 Characterization of the human prolyl 4-hydroxylases that modify the hypoxia-inducible factor HIF. J Biol Chem 278: 30772–30780

Jaakkola P, Mole DR, Tian Y-M et al 2001 Targeting of HIF-α to the von Hippel-Lindau ubiquitylation complex by O_2-regulated prolyl hydroxylation. Science 292:468–472

Kageyama Y, Koshiji M, To KKW et al 2004 Leu-574 of human HIF-1alpha is a molecular determinant of prolyl hydroxylation. FASEB J 18:1028–1030

Knowles HJ, Raval RR, Harris AL, Ratcliffe PJ 2003 Effect of ascorbate on the activity of hypoxia inducible factor (HIF) in cancer cells. Cancer Res 63:1764–1768

Koivunen P, Hirsila M, Gunzler V, Kivirikko KI, Myllyharju J 2003 Catalytic properties of the asparaginyl hydroxylase (FIH) in the oxygen sensing pathway are distinct from those of its prolyl-4-hydroxylases. J Biol Chem 279:9899–9904

Lando D, Peet DJ, Gorman JJ et al 2002a FIH-1 is an asparaginyl hydroxylase enzyme that regulates the transcriptional activity of hypoxia-inducible factor. Genes Dev 16:1466–1471

Lando D, Peet DJ, Whelan DA, Gorman JJ, Whitelaw ML 2002b Asparagine hydroxylation of the HIF transactivation domain: a hypoxic switch. Science 295:858–861

Lieb ME, Menzies K, Moschella MC, Ni R, Taubman MB 2002 Mammalian *EGLN* genes have distinct patterns of mRNA expression and regulation. Biochem Cell Biol 80:421–426

Lipscomb EA, Sarmiere PD, Freeman RS 2001 SM-20 is a novel mitochondrial protein that causes caspase-dependent cell death in nerve growth factor-dependent neurons. J Biol Chem 276:5085–5092

Madden SL, Galella EA, Riley D, Bertelsen AH, Beaudry GA 1996 Induction of cell growth regulatory genes by *p53*. Cancer Res 56:5384–5390
Mahon PC, Hirota K, Semenza GL 2001 FIH-1: a novel protein that interacts with HIF-1α and VHL to mediate repression of HIF-1 transcriptional activity. Genes Dev 15:2675–2686
Marxsen JH, Stengel P, Doege K et al 2004 Hypoxia-inducible factor-1 (HIF-1) promotes its degradation by induction of HIF-alpha-prolyl-4-hydroxylases. Biochem J 381:761–767
Masson N, Ratcliffe PJ 2003 HIF prolyl and asparaginyl hydroxylases in the biological response to intracellular O_2 levels. J Cell Sci 116:3941–3049
Metzen E, Stiehl DP, Doege K et al 2005 Regulation of the prolyl hydroxylase domain protein 2 (phd2/egln-1) gene: identification of a functional hypoxia-responsive element. Biochem J 387:711–717
Nakayama K, Frew IJ, Hagensen M et al 2004 Siah2 regulates stability of prolyl-hydroxylases, controls HIF1α abundance, and modulates physiological responses to hypoxia. Cell 117:941–952
Oehme F, Ellinghaus P, Kolkhof P et al 2002 Overexpression of PH-4, a novel putative proline 4-hydroxylase, modulates activity of hypoxia-inducible transcription factors. Biochem Biophys Res Commun 296:343–349
Schofield CJ, Ratcliffe PJ 2004 Oxygen sensing by HIF hydroxylases. Nat Rev Mol Cell Biol 5:343–354
Semenza GL 2003 Targeting HIF-1 for cancer therapy. Nat Rev Cancer 3:721–732
Seth P, I. K, Porter D, Polyak K 2002 Novel estrogen and tamoxifen induced genes identified by SAGE (Serial Analysis of Gene Expression). Oncogene 21:836–843
Stolze IP, Tian YM, Appelhoff RJ et al 2004 Genetic analysis of the role of the asparaginyl hydroxylase FIH in regulating HIF transcriptional target genes. J Biol Chem 279:42719–42725
Tuckerman JR, Zhao Y, Hewitson KS et al 2004 Determination and comparison of specific activity of the HIF-prolyl hydroxylases. FEBS Lett 576:145–150
Wax SD, Rosenfield CL, Taubman MB 1994 Identification of a novel growth factor-responsive gene in vascular smooth muscle cells. J Biol Chem 269:13041–13047

DISCUSSION

Gonzalez: What is the K_m of prolyl hydroxylase for oxygen? Is it high enough as to limit prolyl hydroxylation rate during hypoxia?

Ratcliffe: In our work we didn't measure apparent K_m for oxygen formally, although our data would be concordant with what the Finnish group have published, which is around 200 μM. This is a supraphysiological concentration. It doesn't tell us a great deal, just that oxygen can be limiting over the entire physiological range, which will clearly be less than 200 μM. If the K_m had been extremely low (say less than 2 μM), then clearly that might not be the case.

Gonzalez: In one of the schemes you presented, I saw some regulation of the activity of proline hydroxylase by nitric oxide (NO). Do you have any data for CO?

Ratcliffe: No. One would predict that CO could bind the dioxygen site. I have asked Chris Schofield this question, and he believes that this could happen. However, whether CO ever reaches sufficient concentrations *in vivo* for this to be a physiologically relevant effect is unclear.

González: As you know it has been shown that CO inhibits gene expression triggered by hypoxia. Does CO bind and inhibit proline hydroxylase?

Ratcliffe: I guess so, with the proviso that it might not be a physiological effect.

Acker: In the old days, when we did carotid body research, we thought that the oxygen sensor was likely to be a haem protein with a K_m for oxygen of about 20–30 Torr, just fitting the oxygen distribution curve of the organs. Now it turns out that the PHD oxygen sensor has this unusually high K_m for oxygen of about 160 Torr. This oxygen sensor works under physiological conditions in organs only at about 10% of its maximal activity. This feature makes PHDs very sensitive to other influences. The difficult issue to understand is whether PHDs really sense the oxygen, or are other influences much more important for PHD activity than the oxygen level?

Ratcliffe: That's a good question. As I've indicated, we need to be cautious about K_m values for molecular oxygen derived *in vitro*. It is certainly possible that other processes contribute to the oxygen sensitivity of PHD activity, though this is very difficult to resolve. Changes in the availability of Fe(II) ions in the relevant intracellular microenvironment could influence enzyme activity. However we have difficulty in measuring the Fe(II) ion concentration that the enzyme would see, or, for that matter, the 2-OG concentration that it would see. What we are trying to do is make mutant enzymes, that have altered Fe(II), oxygen, or 2-OG binding *in vitro* and then re-express them in hydroxylase-deficient cells and determine changes to the physiological regulation of HIF. Though difficult, we believe this is about the only way this problem can be resolved. We want to characterize the mutant enzymes in a test-tube, and then complement a deficient cell, and then ask how the physiology is altered.

Acker: There must be some physiological significance to this.

Ratcliffe: Personally, I think you can ignore the K_m. It is unusually high in relation to other enzymes that have been assayed. But these are assays in the test-tube and they use unphysiological amounts of HIF, and indeed a short HIF peptide rather than a HIF molecule.

Archer: There was a comment earlier about the potential link between mitochondria and HIF. On the basis of your α-KG statement, if you had an error, whether it is genetic or acquired in metabolism affecting mitochondria or Krebs cycle, would this result in changes in the activity of prolyl hydroxylase or HIF activity in a pO2-independent manner?

Ratcliffe: That's theoretically possible. I can think of at least three possible links between mitochondrial metabolism, hypoxia and HIF hydroxylation. When electron transport is impaired, there might be a back-up of intermediates in the Krebs cycle. The exact nature of this is difficult to predict. For instance, I don't know in hypoxia at what point the Krebs cycle stops. If it stops by NADH accumulation, as I understand is the case, then this will inhibit several of the enzymes, but how that affects concentrations of metabolites such as 2-OG is unclear. The position might also change with differing availability of glucose. The second possible link is that oxygen radical species might inactivate the HIF hydroxylase. There was a recent paper from the Pasteur Institute that indicates that this might be the case (Gerald et al 2004). The third link is aconitase, another Krebs cycle enzyme that is said to be one of the most sensitive enzymes to radical inactivation. I don't have any problem in thinking of ways in which these effects could influence hydroxylase activity: the question is, do they?

Duchen: How would the α-KG get to the enzyme if the enzyme is cytosolic?

Ratcliffe: That's another good question. There are transporters. Furthermore some of the mitochondrial enzymes have cytosolic forms and it is interesting to speculate as to whether they might have a function in regulating relevant intermediates such as 2-OG in the cytosol. One of the HIF hydroxylase enzymes, PHD3, was said to be mitochondrial: but we don't think this is the case.

Chandel: How low an oxygen level is sufficient to drive the PHDs to hydroxylate HIFs? If you take cells, overexpress PHD2 or 3 and then look at HIF stability as a function of pO_2, how far can you drop the pO_2 and still have the PHDs?

Ratcliffe: The enzyme will continue to turn over at very low oxygen tensions. We know this from test-tube experiments. In cells, for the asparaginyl hydroxylase we found that we could take the oxygen concentration down to 0.2% and still force the hydroxylase to down-regulate HIF activity by overexpression. We can't do this with the prolyl hydroxylases. People think this is inconsistent with the high K_m of the HIF hydroxylases, but it isn't—it just means that the enzyme will work to a degree at low oxygen tensions. In our hands, FIH seems to continue to work at lower pO2s than the PHDs, but these findings are not necessarily connected with the apparently lower K_m for oxygen that has been reported for FIH versus the PHDs.

Buckler: In relation to that, how do you see the interaction between FIH (asparaginyl hydroxylase) and prolyl hydroxylases in controlling transcriptional activity? If FIH works well at low pO2, this might suggest that there is very little HIF-1α around that could bind p300. Why then can you see an additional effect of prolyl hydroxylation on transcription if most of the HIF-1α is already effectively transcriptionally inactive?

Ratcliffe: There are two different questions there. How do FIH and PHD interact to prevent HIF transcriptional activity? What has been said is that FIH could act to inactivate the HIF that escapes prolyl hydroxylation at low oxygen tension. This makes sense; whether it is true or not is uncertain. The other is how can we drive the HIF response with a prolyl hydroxylase inhibitor or indeed genetic inactivation of PHDs in the face of continuing activity of FIH? There must be incomplete asparaginyl hydroxylation or at least incomplete blockage of transcription.

Interestingly when we overexpress FIH in normoxia it also affects the system. This is a situation where there is plenty of oxygen and little HIF. One might expect the normal endogenous level of FIH to have adequate capacity for the full hydroxylation of HIF, but the overexpression data suggest this is not the case. A potential explanation is that the enzyme behaves differently when the substrate concentration is very low. The conditions *in vivo* are certainly likely to be very different from those analysed to date in the test-tube.

Pouysségur: You showed that when you overexpressed FIH, the impact was different dependent on the gene loop. This is something that we have observed also. If FIH is knocked down, some genes are not happy. When FIH is overexpressed some genes will be repressed and others will not. My interpretation is that there are two *trans*-activating domains on HIF-1α that could eventually involve different subsets of genes.

Ratcliffe: It's a good potential explanation.

Pouyssegur: Which genes do you see with a different pattern of induction when you overexpress FIH?

Ratcliffe: We haven't got a list of genes that have that distinct characteristic. We are not working directly on this, although we have noticed that in different cells there are differences. These experiments were simply designed to look at the lowest level of hypoxia under which we could force a reduction in HIF target gene expression by overexpression of FIH and ask whether it was the same or different from the effects of overexpression of the PHDs

Kummer: You have shown inhibition of PHD2 by succinate. Does this lead to altered expression of target genes? We have had a particular interest in succinate dehydrogenase (SDH) and have used succinate and two different blockers of complex II. We made siRNA for each of these subunits. Neither treatment affected the response of these cells to hypoxia.

Ratcliffe: We concentrated on fumarate hydratase deficiency for a reason: the effect of fumarate on the PHDs is much more substantial. In a recent issue of *Cancer Cell* there was an article on SDH showing, broadly speaking, the same phenomenon (Selak et al 2005). They have been able to get positive effects on HIF with suppression of SDH, but in our laboratory the effects were insubstantial. Succinate does behave as a competitive inhibitor, but we didn't see a major induction of HIF. It is possible that these effects are going to be highly sensitive to how the cells are cultured.

Prabhakar: Are the PHDs translocated to the nucleus?

Ratcliffe: It has been reported that PHD1 is predominantly located in the nucleus in overexpression studies. However, we are less certain about the location of the endogenous protein. In cell fractionation studies we find that it is in both the nucleus and the cytoplasm. There is presumably a nuclear export process, and one might be saturating this with overexpression.

Schumacker: You mentioned the *Cancer Cell* paper (Selak et al 2005), and the possibility that Krebs cycle intermediates could be regulating PHDs. One interesting issue that wasn't addressed by that paper was that the phenotypes for many of the Krebs cycle genetic defects are uninteresting, except for some of those associated with complex II. Have you any thoughts on this?

Ratcliffe: I'm not well qualified to answer that. We have certainly looked at human and *C. elegans* mutants and not seen major effects on HIF induction by hypoxia, this included SDH mutants. It is also interesting that, for some reason, mutants of SDH B, C and D have a tumour-promoting predisposition, but not mutants of SDH A. Many of the defects are incomplete and I imagine that there are quantitative and qualitative effects in the way that metabolism is interrupted that we have yet to understand, and may be important in driving these phenotypes.

Schumacker: Most of the Krebs cycle enzymes work at near equilibrium. Wouldn't you expect that a partial defect would cause a back up of substrates, which would increase the levels of intermediates prior to that site?

Ratcliffe: Possibly. However, if you make a mutant you often don't destroy the protein entirely, and there are a whole host of effects that might be generated, for instance, though interfering effects of the abnormal protein. It is easy to oversimplify.

Sylvester: Do you have any basis for predicting what the relative affinity of prolyl hydroxylase would be for NO or CO?

Ratcliffe: No.

Sylvester: Would you predict that NO or CO would be equivalent to O_2 in terms of affinity?

Ratcliffe: I do not want to be drawn on this. NO has been used as a dioxygen analogue in crystallography studies. I do not know the binding affinity.

Duchen: Moncada's group published a paper in *Science* (Hagen et al 2003) suggesting that at low levels NO can inhibit mitochondrial respiration enough to raise pO2 a bit and therefore reduce activation of HIF.

Ratcliffe: There is one problem with this. We and others have examined HIF induction in a number of mitochondrial defective cells which consume little oxygen, and did not see major differences from wild-type cells.

Duchen: There has always been a debate about how much of an oxygen gradient there really is inside cells.

Semenza: In the study where we looked at the dose–response of HIF-1α versus oxygen concentration with Christian Bauer's lab in Zurich, we performed the experiment in the presence and absence of potassium cyanide to eliminate oxygen gradients and oxygen consumption. We didn't see a significant difference in the two response curves (Jiang et al 1996).

Acker: I have one comment on the CO experiments. A classic CO experiment on the carotid body was done by Joels and Neil (1962). They showed that CO excites carotid body chemoreceptor discharge. These experiments could be confirmed by Wilson et al (1994) and our group (Lahiri et al 1999, Streller et al 2002) hinting to mitochondrial complex IV as oxygen sensor in the carotid body. This oxygen sensing property was verified by action spectrum analysis.

Ratcliffe: I would like to avoid confusion over CO. First, I'm not talking about the carotid body. The putative haem protein might or might not be involved in carotid body oxygen sensing. Second, you are correct that there were experiments done with EPO production and CO. They used extremely high concentrations of CO, and the effects weren't that big or all that reproducible. I do not think that CO has a dramatic and reproducible effect on EPO signalling in every system that has been tested.

Acker: It would be a very nice method to identify a haem-based oxygen sensor. This is the attraction of using CO.

Ward: You mentioned the modulation of HIF-1 by oestrogen in the breast cancer cells. Was this through the classic oestrogen nuclear receptors, or is it more like the mode of action of phytooestrogens? There is current interest in phytooestrogens being protective against cancers.

Ratcliffe: I don't know, nor do I know its significance for cancer progression.

Semenza: There is a compound—2-methoxyoestradiol—which is an endogenously produced molecule that has been shown to inhibit HIF-1. Judah Folkman presented data on this recently. He proposed that this compound is a physiological regulator of angiogenesis because it is particularly elevated during pregnancy and seems to be a fairly potent inhibitor of HIF-1.

Ward: Is that working through oestrogen receptors?

Semenza: No, not through oestrogen receptors. Its mechanism of action isn't clear. It is known to affect the microtubule function and may affect the transport of HIF-1α into the nucleus.

López-Barneo: How much succinate do you need to inhibit the prolyl hydroxylases?

Ratcliffe: It is competitive, so it depends how much 2-OG is available.

López-Barneo: Would you expect also to inhibit the hydroxylation of HIF because of mass action, because you have a lot of succinate?

Ratcliffe: Succinate can act competitively with 2-OG presumably competing for binding at the 2-OG binding site. This doesn't exclude other effects of product inhibition.

Harris: Why do you think the enzyme defects in the Krebs cycle give particular patterns of cancer, or cancer at all, for that matter? You can't cause cancer by just giving HIF.

Ratcliffe: It is important to be clear that the data do not necessarily implicate HIF. They tell us that the defect produces enough fumarate accumulation to inhibit an enzyme of the 2-OG family class. That enzyme might have another substrate that causes cancer. It is quite possible that increased HIF is only an associated phenomenon.

Harris: There are many enzymes of that class around. If you look by mass spectrometry or 2D gels you might see other proteins induced by their inhibition in an intact cell.

Ratcliffe: Chris Schofield thinks that there are about 50 enzymes of this class in the mammalian genome. They may be metabolizing small molecules or proteins, but their function is currently largely unknown.

Schumacker: The paper by Gerald et al (2004) recently suggested that the redox state of the iron in the hydroxylases was responsible for regulating the activity. In the *in vitro* assays of prolyl hydroxylase it is necessary to add ascorbate, presumably to reduce the iron to the ferrous state. To what extent do you think redox regulation is playing a role in the cell? In your talk you mentioned that administration of ascorbate to certain cancer cell lines caused a reversal of the normoxic stabilization of HIF. Do you think this is how it is working?

Ratcliffe: There is a very large effect in tissue culture. The implication is that iron or ascorbate availability is limiting for hydroxylase activity under commonly used tissue culture conditions. To what extent does ascorbate and iron availability regulate HIF in the intact organism? I can't tell you.

Sylvester: Would inhibition of the enzyme make it more likely to be a source of reactive oxygen species?

Ratcliffe: I can't tell you. The enzymes do get damaged easily through oxidation and employ a radical mechanism at the catalytic centre. Work with the collagen prolyl hydroxylases indicates that if you alter the collagen substrate you can get an increased number of uncoupled events where the enzyme gets stalled with the iron in an oxidised form. I do not know whether this increases the generation of oxygen radicals.

Archer: In the body we have a specialized oxygen sensing system. I understand why a lot of the initial work was done in cell culture. But if you take the oxygen-sensitive organs and place them *ex vivo*, they rapidly lose their oxygen sensitivity: after a few days it is mostly gone. What advances do you feel could be made by using fresh placenta, fresh ductus and fresh pulmonary artery? Might you then get around some of the problems of studying what may amount to culture-induced artefact in HIF regulation?

Ratcliffe: It's an interesting question. In tissue culture you can look at a range of genes induced by HIF-1 and HIF-2: although there have been some published differences, they all look broadly speaking the same in terms of oxygen sensitivity. But we know that in the whole animal it must be different, since different organs and regions have quite different oxygen environments. I agree with you: following the behaviour of explanted tissue from habitually hypoxic or hyperoxic sites might give some interesting findings. There might well be adaptations that occur in tissue culture and which make the oxygen sensitive poise come to a particular level.

Buckler: As a general principal, are the genes that are regulated by HIF exclusively regulated by HIF, or can other transcription factors also regulate the expression of those same genes?

Ratcliffe: Most commonly other transcription factors regulate the same genes.

Buckler: So the extent to which HIF has an effect on the level of expression of relevant genes would also depend on the levels of activity of other transcription factors.

Ratcliffe: That's absolutely right. It follows that it is very difficult to identify a particular underlying oxygen sensing process by characterising the physiological output. We showed early on that the transcriptional process underlying EPO regulation was widespread. The first gene, other than EPO, that we identified as responding to HIF was phosphoglycerate kinase 1. Initially I found it difficult to believe that glycolytic control and EPO were linked to the same system, but genetically we now have cast iron evidence that this is true—through HIF. However, the physiological characteristics of the responses to hypoxia are quite different.

References

Gerald D, Berra E, Frapart YM et al 2004 JunD reduces tumor angiogenesis by protecting cells from oxidative stress. Cell 118:781–794

Hagen T, Taylor CT, Lam F, Moncada S 2003 Redistribution of intracellular oxygen in hypoxia by nitric oxide: effect on HIF1α. Science 302:1975–1978

Jiang BH, Semenza GL, Bauer C, Marti HH 1996 Hypoxia-inducible factor 1 levels vary exponentially over a physiologically relevant range of O2 tension. Am J Physiol 271:C1172–1180

Joels N, Neil E 1962 The action of high tensions of carbon monoxide on the carotid chemoreceptors. Arch Int Pharmacodyn Ther 139:528–534

Lahiri S, Ehleben W, Acker H 1999 Chemoreceptor discharges and cytochrome redox changes of the rat carotid body: role of heme ligands. Proc Natl Acad Sci USA 96:9427–9432

Selak MA, Armour SM, MacKenzie ED et al 2005 Succinate links TCA cycle dysfunction to oncogenesis by inhibiting HIF-alpha prolyl hydroxylase. Cancer Cell 7:77–85

Streller T, Huckstorf C, Pfeiffer C, Acker H 2002 Unusual cytochrome a592 with low PO2 affinity correlates as putative oxygen sensor with rat carotid body chemoreceptor discharge. FASEB J 16:1277–1279

Wilson DF, Mokashi A, Chugh D, Vinogradov S, Osanai S, Lahiri S 1994 The primary oxygen sensor of the cat carotid body is cytochrome a3 of the mitochondrial respiratory chain. FEBS Lett 351:370–374

General discussion I

Harris: Gregg Semenza, in reoxygenation experiments, in a hypoxia inducible factor-1 (HIF-1) knockout is there still induction of the CaM kinase?

Semenza: We haven't done that experiment yet. However, the HIF-1α heterozygous knockout mice do seem to be defective in response to intermittent hypoxia.

López-Barneo: Can you comment generally on HIF-1α up-regulating and down-regulating gene expression? Is the mechanism for down-regulation of gene expression by HIF clear?

Semenza: There are two potential mechanisms for both up- and down-regulation. One mechanism involves the direct binding of HIF-1 to hypoxia-response elements (HREs) in target genes. The second mechanism involves activation by HIF-1 of a target gene encoding a transcriptional activator or repressor, which would then bind to regulate the expression of other genes—a secondary effect. For gene activation, we know many examples of direct activation. We have identified transcription factors that seem to be targets of HIF-1. This may lead us to secondary targets for activation. In terms of repression, less is known. There are several HIF-1 target genes that do encode repressors. The best studied of these is DEC1, and several pathways have been identified in which induction of DEC1 in response to hypoxia through HIF-1 leads to downstream effects. There has not been a strong experimental demonstration of HIF-1 binding to DNA and having a direct repressive effect on gene expression.

Ratcliffe: We have done some work in collaboration with Adrian Harris that is relevant. We have been interested in the functions of HIF-2, which you might imagine can compete for the same binding site. For some genes we find that they are exclusively HIF-1 targets. If HIF-1 is suppressed by RNAi then there is a major effect on gene expression whereas manipulation of HIF-2 does nothing at all. However, some genes respond positively to HIF-1 and negatively to HIF-2, even though they both bind the same site: an example is the gene encoding BNip3, which is a proapoptotic protein. This is interesting in relation to cancer. VHL-associated cancer has a bias towards expression of HIF-2 rather than HIF-1. Thus it seems that one way the cancer avoids up-regulation of BNip3 as part of the HIF pathway is by enhancing HIF-2 expression over HIF-1. This mechanism is not necessarily used in other cancers which have other ways of disabling this system. However it is a mechanism by which positive and negative transcriptional effects are achieved as part of activation of the HIF system.

Semenza: A complicating factor is the existence of HIF-3α, which is also an oxygen-regulated protein and appears to function as an inhibitor of HIF-1α. HIF-3α is also encoded by a HIF-1 target gene.

Chandel: With regard to the BNip3 data, there is an idea that at least in renal cell carcinomas, HIF-1α is tumour suppressive whereas HIF-2α provides a growth advantage to the tumour. Is this differential effect on BNip3 part of the explanation for the differences?

Ratcliffe: This might well be part of the explanation, but we have not done an intervention experiment to prove this. It is also relevant that in renal cell carcinoma (RCC) lines, cyclin D1, a growth promoter is a specific HIF-2 target just as BNip3 is a specific HIF-1 target. Interestingly, certain other genes appear to change their transcriptional selectivity such that they are HIF-1 targets in most cells but HIF-2 targets in RCC. We believe that this must involve disconnections of the HIF-1 system, as well as possibly new connections with HIF-2. It is interesting that in the cancer that is most associated with HIF activation there is this unusual transcriptional targeting. We believe that in some way the cancer adapts the HIF system to be less antitumorigenic and more protumorigenic.

Chandel: Have there been any more substrates for PHDs other than the HIFs?

Ratcliffe: Nothing that has been published yet.

Archer: In the simple gene profiling of VEGF, you pointed out tissue diversity, showing different responses in four cell lines. Have you done gene profiling to explain the basis for this VEGF tissue heterogeneity?

Semenza: Thus far we have only performed an analysis of hypoxia- and HIF-1-regulated gene expression in arterial endothelial cells.

Archer: We tried that comparing gene expression using microarrays in the Fawn hooded rat (which gets pulmonary hypertension and has a disordered mitochondrial O_2 sensor) versus its consomic control (which does not get pulmonary hypertension). What is interesting is that a lot of the most strongly differentially regulated genes are ESTs (expressed sequence tags). We don't know what these genes are or what their product does.

Kemp: I am intrigued by the intermittent hypoxia issue and the differences between this and chronic hypoxia. I don't think that anyone can doubt that intermittent hypoxia is a strong stimulus, whatever the functional readout. I like the idea that what we are really looking at isn't intermittent hypoxia, but intermittent hyperoxia. This fits very well with that idea that ROS or RNS produced in excess are a strong stimulus. The thing I am having difficulty with is that no one has done the experiment looking at intermittent hyperoxia, which we would expect might be an even stronger stimulus. Also, I believe that chronic hyperoxia produces pathologies such as lung fibrosis and epithelial depopulation, and in the neonate much worse disease. However, those people with sleep apnoea don't suffer this, but suffer with pulmonary hypertension. We don't get the pathologies associated with

ROS toxicity. How can intermittent or chronic hyperoxia cause two different diseases?

Chandel: We have spent a lot of time doing hyperoxia experiments, with the FIO_2s up at around 95% O_2, similar to the ventilation strategies that might be seen in the ICU. We have never been able to elicit an effect on HIF-1α. We have looked from 70–95% O_2 in alveolar epithelial cells, and we have never seen an increase in HIF-1α levels.

Ward: Have you looked at haeme oxygenase?

Chandel: No, just HIF-1α and HIF-2α.

Ward: One of the things we have found was that hyperoxia (95% O_2) causes a large increase in expression of haeme oxygenase 1. However, this only occurred in the presence of a mitogenic stimulus; there appeared to be a synergistic relationship between hyperoxia and growth factors (Pandya et al 2002).

Chandel: Our type 2 cells have 10% serum in them. We have talked about how gene expression can be modulated by a multitude of transcription factors. With respect to HIF-1α, we thought hyperoxia might be a good stimulus.

Nurse: Many years ago you showed that haeme oxygenase was expressed in the carotid body cells. In the HIF-1α heterozygotes is haeme oxygenase expressed in the carotid body?

Prabhakar: We haven't looked into that. The only thing we examined is chromagranin A and tyrosine hydroxylase (TH) expressions in glomus cells in heterozygous mice. Now we are planning a detailed histological investigation of the putative neurotransmitters and synthesizing enzymes in the carotid bodies from HIF-1 heterozygous mice.

Kumar: Like Professor Kemp, I find it hard to understand what is going on with intermittent hypoxia (IH). In Nanduri Prabhakar's experiments, giving a rat or mouse IH with low oxygen intermittent with air is very different to delivering IH to P12 cells. *In vivo*, carotid body cells will have a tissue pO2 of around 40 mmHg, whereas a monolayer of PC12 cells in an incubator will see 160 mmHg. It could be therefore that you are getting more ROS production in your cell cultures than you do in your *in vivo* models. One of the problems in this field is the use of 20% O_2 as normoxia in cell culture experiments.

Evans: I have another question on IH. If you take the carotid body or pulmonary artery and expose them to hyperoxia, carotid body discharge is attenuated (Alcayaga et al 1997, Wang & Fitzgerald 2002) whilst pulmonary artery perfusion pressure remains unaffected for up to 9 h (Hambreaus-Jonzon et al 1997, Jones et al 1984). These effects are not consistent with the effects of hypoxia, yet it is generally accepted that hyperoxia will generate ROS (Freeman et al 1982, Jamieson et al 1986, Sanders et al 1993) and it has been suggested that an increase in ROS may underpin these effects of hyperoxia. If this is the case, then what you are seeing with IH might have nothing to do with the return from hypoxic to normoxic conditions and the generation of ROS by hypoxia. It could equally be due to prolonged and possibly cumulative effects of hypoxia on a discrete, metabolic signalling pathway that may be activated by hypoxia.

Sylvester: In vivo, arterial pO2 can increase during inspiration and decrease during expiration. This oscillation might not occur in tissues, but it probably occurs in the carotid body, which has a high blood flow relative to its oxygen consumption. Would you expect that normal *in vivo* oscillations of pO2 would be seen by the carotid body as IH?

Prabhakar: I don't know, but that is a good point. The question of pO_2 and pCO_2 oscillations in the arterial blood and their impact on the carotid body sensory activity and ensuing ventilatory changes has been extensively studied in the past. In recent years it hasn't had much attention.

Kumar: I have a general question. We get hypoxia when we are sick. Do we have a situation here then, where we are conserving a group of genes to keep us alive for longer while we are ill? Many of the diseases associated with hypoxia occur beyond our optimal reproductive age, so might there be less strong evolutionary pressure to preserve these genes? We also become hypoxic when we ascend to altitude, but not everyone climbs mountains. So what is the evolutionary pressure for conserving these genes? However, we exercise all the time. We have to regulate our breathing precisely for this, and the carotid body is involved. Is the carotid body then primarily a transducer of hypoxia? Just because it can respond to hypoxia, is it necessarily functioning as a hypoxia sensor?

References

Alcayaga J, Iturriaga R, Zapata P 1997 Time structure, temporal correlation and coherence of chemosensory impulses propagated through both carotid nerves in cats. Biol Res 30:125–133

Freeman BA, Topolosky MK, Crapo JD 1982 Hyperoxia increases oxygen radical production in rat lung homogenates. Arch Biochem Biophys 216:477–484

Hambraeus-Jonzon K, Bindslev L, Mellgard AJ, Hedenstierna G 1997 Hypoxic pulmonary vasoconstriction in human lungs. A stimulus–response study. Anesthesiology 86:308–315

Jamieson D, Chance B, Cadenas E, Boveris A 1986 The relation of free radical production to hyperoxia. Annu Rev Physiol 48:703–719

Jones DB, Mihaly GW, Smallwood RA, Webster LK, Morgan DJ, Madsen NP 1984 Differential effects of hypoxia on the disposition of propranolol and sodium taurocholate by the isolated perfused rat liver. Hepatology 4:461–466

Pandya HC, Snetkov VA, Ward JPT, Hirst SJ 2002 Oxygen regulates mitogen stimulated proliferation of fetal airway smooth muscle cells. Am J Physiol 283:L1220–L1230

Sanders SP, Zweier JL, Kuppusamy P et al 1993 Hyperoxic sheep pulmonary microvascular endothelial cells generate free radicals via mitochondrial electron transport. J Clin Invest 91:46–52

Wang HY, Fitzgerald RS 2002 Muscarinic modulation of hypoxia-induced release of catecholamines from the cat carotid body. Brain Res 927:122–137

Regulation of HIF: asparaginyl hydroxylation

Daniel Peet and Sarah Linke

School of Molecular and Biomedical Science, and the ARC Special Research Centre for Molecular Genetics of Development, University of Adelaide, Adelaide, SA 5005, Australia

Abstract. The hypoxia inducible transcription factors (HIFs) are regulated at the level of protein stability and transcriptional activity in an oxygen-dependent manner by prolyl and asparaginyl hydroxylation, respectively. Factor inhibiting HIF (FIH-1) is the only known HIF asparaginyl hydroxylase, and targets a conserved asparaginyl residue within the C-terminal activation domain (CAD) of HIF-α. This represses HIF-mediated transcription by inhibiting the recruitment of p300/CBP coactivators. Recent studies have demonstrated that the function of FIH-1 relative to the HIF prolyl hydroxylases (PHDs) is not redundant, and indicate that FIH-1 is a direct oxygen sensor. This paper will address recent published and unpublished work characterising the role of asparaginyl hydroxylation in the cellular response to hypoxia. The relative oxygen affinities and hypoxic activities of FIH-1 and the PHDs will be discussed. Furthermore, *in vitro* and cell-based assays demonstrating some novel characteristics regarding the substrate specificity of FIH-1, and their potential biological and therapeutic relevance will be presented.

2005 Signalling pathways in acute oxygen sensing. Wiley, Chichester (Novartis Foundation Symposium 272) p 37–53

The recent discovery of hypoxia inducible transcription factor (HIF) hydroxylation has greatly increased our understanding of the cellular hypoxic response and has provided a new perspective on oxygen sensing. Whereas HIF prolyl hydroxylation mediates the oxygen-dependent degradation of the HIF proteins as was discussed in the previous presentation, HIF asparaginyl hydroxylation is important for regulating the oxygen-dependent transcriptional activity of the C-terminal transactivation domains (CADs) of the HIFs. To date only a single HIF asparaginyl hydroxylase has been identified, namely factor inhibiting HIF (FIH-1).

Regulation of the HIF CAD

The HIF-1α and HIF-2α proteins contain both N-terminal and C-terminal transactivation domains (NAD and CAD, respectively). Whereas the activity of the NAD appears to be intimately linked with HIF degradation due to the fact that it over-

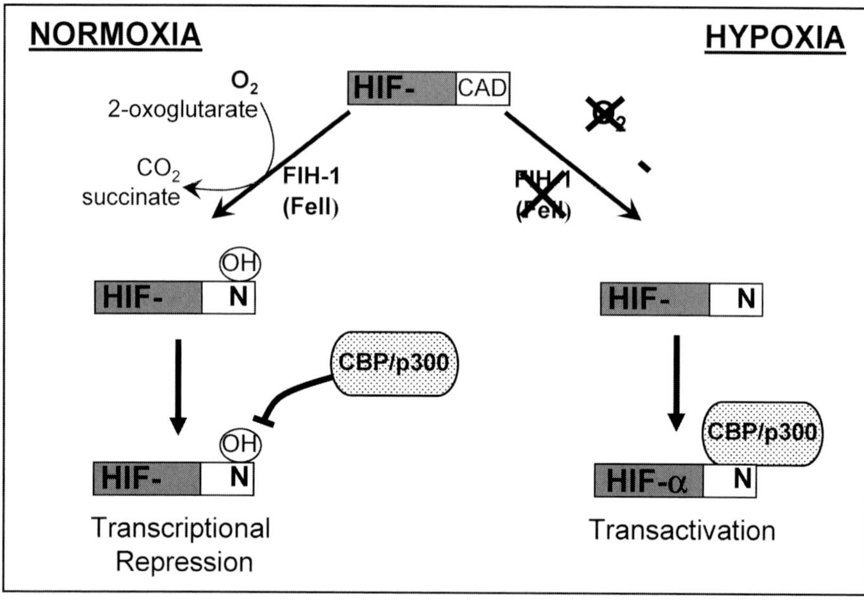

FIG. 1. Schematic representation of regulation of the HIF-α CAD by Asn-hydroxylation. HIF refers to hypoxia inducible transcription factors, CAD is the C-terminal transactivation domain, N indicates asparagine hydroxylation target, and p300/CBP (Creb binding protein) are transcriptional coactivators.

laps with the oxygen-dependent degradation domain, the normoxically repressed transcriptional activity of the CAD is activated by hypoxia independently of protein stability (Fig. 1). This was first revealed by reporter gene experiments showing that the minimal CAD (786–826 hHIF-1α) exhibits constitutive activity as a GAL-DBD chimera whereas the inclusion of an additional 11 residues (775–826 hHIF-1α) results in normoxic repression that is relieved under hypoxia, without a change in protein stability (Pugh et al 1997, Jiang et al 1997, O'Rourke et al 1999).

In an effort to understand the mechanism of CAD regulation, Lando et al (2002a) used mass spectrometry to analyse mHIF-2α 774–874 purified from mammalian cells and demonstrated the oxygen-dependent hydroxylation of a conserved asparaginyl residue (Asn851 mHIF-2α), confirming the involvement of hydroxylation predicted by Sang et al (2002). Subsequent *in vitro* interaction assays (Lando et al 2002a, Hewitson et al 2002) revealed that asparaginyl hydroxylation of HIF-1α and HIF-2α represses CAD activity by preventing its interaction with the essential coactivator p300 (or the homologue Creb binding protein [CBP], reviewed in Lando et al 2003). Structural analyses by NMR of the HIF–CAD/p300 (Freedman et al 2002) and HIF–CAD/CBP (Dames et al 2002) complexes revealed the targeted

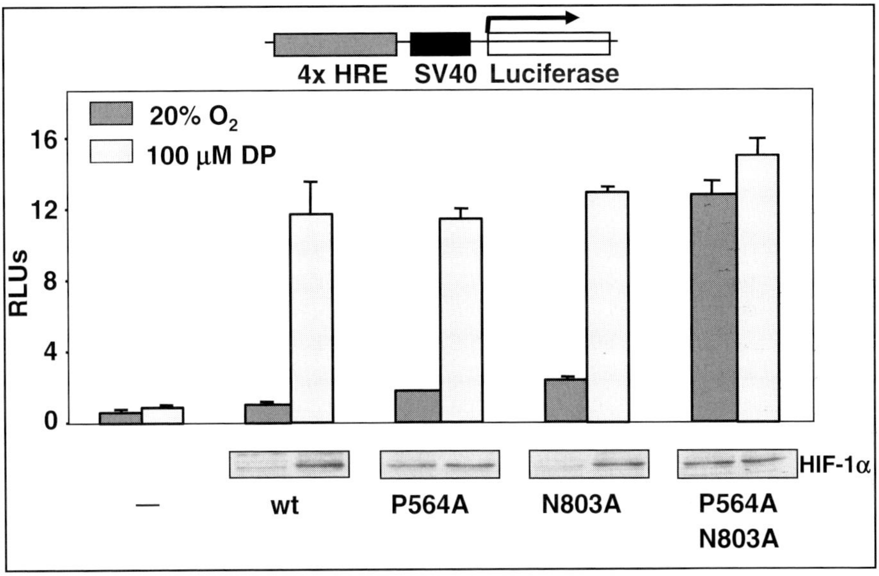

FIG. 2. Full HIF activity requires protein stabilization and CAD derepression. Upper panel shows HRE-driven reporter activity in KA13.5 CHO cells transfected with full length HIF-1α wild-type and mutants as indicated. Experiments performed in triplicate, and illustrated as mean ± SEM ($n = 3$). Lower panels are anti-HIF-1α immunoblots showing HIF-1α protein levels in transfected cells. Results representative of three independent experiments. Methods detailed in Lando et al (2002a).

asparagine buried deep within the interface, and predicted that hydroxylation would disrupt this interaction. This asparaginyl hydroxylation has since been assigned to the pro-S position of the β-carbon, generating the unprecedented *threo* (2S, 3S)-isomer (McNeill et al 2002, Peet et al 2004). The functional significance of asparaginyl hydroxylation in repressing full-length HIF-α protein was confirmed by mutation of P564A (hHIF-1α) or P530A (mHIF-2α), which was sufficient to fully stabilize the proteins at normoxia but only moderately increased HRE-reporter expression (twofold), whereas additional mutation of N803A (hHIF-1α) or N851A (mHIF-2α) increased activity a further four–sixfold, essentially recapitulating induction by hypoxia (Lando et al 2002a) (Fig. 2).

FIH-1

In parallel with identification of asparaginyl hydroxylation mediating CAD repression, the hydroxylase enzyme responsible was discovered. Mahon et al (2001) cloned factor inhibiting HIF-1 (FIH-1) via a yeast two-hybrid assay and demonstrated

			site 1 (LTSYDCEVNAPI) … site 2 (LQGEELLRAL)
HIF-1α	Human	BT009776	LPQLTSYDCEVNAPIQGSRNLLQGEELLRALDQVN
HIF-1α	Bovine	NM_174339	LPQLTSYDCEVNAPIQGSRNLLQGEELLRALDQVN
HIF-1α	Mouse	Q61221	LPQLTSYDCEVNAPIQGSRNLLQGEELLRALDQVN
HIF-1α	Rat	Y09507	LPQLTSYDCEVNAPIQGSRNLLQGEELLRALDQVN
HIF-1α	Rabbit	AY273790	LPQLTSYDCEVNAPIQGSRNLLQGEELLRALDQVN
HIF-1α	Carp	AY450269	LPQLTRYDCEVNAPVQDRHHLLQGEELLRALDQVN
HIF-2α	Human	U81984	LPELTRYDCEVNVPVLGSSTLLQGGDLLRALDQAT
HIF-2α	Bovine	NM_174725	LPELTRYDCEVNVPVPGTSTLLQGGDLLRALDQAT
HIF-2α	Mouse	P97481	LPELTRYDCEVNVPVPGSSTLLQGRDLLRALDQAT
HIF-2α	Rat	NM_023090	LPELTRYDCEVNVPEPGSSTLLQGRDLLRALDQAT
HIF-2α	Quail	AF212989	LPELTRYDCEVNVPVLGSSTLLQGSELLRALDQTT
HIF-2α	Chicken	AF129813	LPELTRYDCEVNVPVLGSSTLLQGSELLRALDQAT
			: ******:* **** :*******
			⊗ ♦♠#

FIG. 3. Clustal W alignment of vertebrate HIF CADs showing 82% amino acid identity within the FIH-1 contact regions. Sites 1 and 2 refer to FIH-1–CAD interaction sites as defined by Ekins et al (2003), 'site 1' (795–806) and 'site 2' (813–822) hHIF-1α; *, complete conservation; conserved substitution; ♠, hydroxylated Asn (803 hHIF-1α); ♦, catalytically important Val (802 hHIF-1α); #, kinetically differentiating Ala/Val (Ala804 hHIF-1α or Val-848 hHIF-2α); ⊗, suggested phosphoacceptor Thr (796 hHIF-1α).

that it functionally repressed CAD activity, particularly under normoxia. FIH-1 was also isolated independently as a novel member of the 2-oxoglutarate (2-OG)-dependent dioxygenase superfamily from a bioinformatic search for putative HIF-α hydroxylases, and was subsequently demonstrated to hydroxylate HIF-α CADs *in vitro* (Lando et al 2002b, Hewitson et al 2002). FIH-1 is strongly conserved between vertebrates, with mouse, rat, human, zebrafish and *Xenopus* orthologous proteins demonstrating approximately 82% amino acid identity (R. Bilton & D. Peet, unpublished work 2003), consistent with the high conservation (82%) of the HIF-α CAD substrate region in vertebrates (Fig. 3). The existence of an invertebrate FIH-1 orthologue is not apparent and no regions within the invertebrate HIF-αs contain discernible similarity to vertebrate CADs.

FIH-1 has been crystallized and its structure determined by three independent groups (Dann et al 2002, Lee et al 2003, Elkins et al 2003), confirming that like other 2-OG-dependent dioxygenases, it utilizes a 'jelly-roll' catalytic core to present a facial triad of His (199), Asp (201) and His (279) which, along with the 2-oxo and 1-carboxylate groups of 2-OG, penta-coordinate the Fe^{2+} cofactor, with water occupying the 6th position. Accordingly, asparaginyl hydroxylation by FIH-1 is expected to occur as is indicated for other 2-OG-dependent dioxygenases, whereby the Fe^{2+}–enzyme complex binds substrates sequentially, with 2-OG followed by the peptide followed by oxygen. A resulting ferric-superoxo species (Fe^{3+}-$O_2^{\bullet -}$) attacks 2-OG, generating CO_2, succinate and a reactive ferryl intermediate [Fe (IV) = O $\leftrightarrow$ Fe (III)-O$^\bullet$] that elicits irreversible hydroxylation of the asparaginyl β-carbon (reviewed in Schofield & Ratcliffe 2004). Similar to other 2-OG-dependent dioxygenases, we have found ascorbate to be required for FIH-1 activity *in vitro*.

The structures highlight some potentially exploitable differences between FIH-1 and other 2-OG-dependent dioxygenases. Typically these enzymes employ an RXS/T motif located on the 8th β strand to hydrogen bond with the 5-carboxylate of 2-OG, whereas FIH-1 utilizes Lys214, Thr196 and Tyr145 derived from unprecedented β-strands. This unusual 2-OG binding strategy may explain some marked differences in K_i values of 2-OG analogues observed for FIH-1 versus the PHDs (HIF prolyl hydroxylases) and the collagen-4-prolyl hydroxylases (Elkins et al 2003, Koivunen et al 2004), although it may not be completely unique to FIH-1. The FIH-1 'jelly-roll' has primary sequence homology to the cupin-like 'JmjC domain' found within the jumonji transcription factor family (Hewitson et al 2002 and references therein), some members of which contain equivalent Fe^{2+}/2-OG-binding residues. This suggests that many JmjC-domain-containing proteins may be 2-OG dependent dioxygenases that with FIH-1 comprise a separate subclass (Elkins et al 2003). More interestingly, FIH-1 exhibits distinctive quaternary structure, existing as a homodimer with an extensive interface (Dann et al 2002, Lee et al 2003, Elkins et al 2003), which is obligate for CAD hydroxylation (Dames et al 2002, Lancaster et al 2004a).

FIH-1 peptide substrate specificity

Structural analysis of the HIF-1α CAD/FIH-1 complex has provided extensive insight into the substrate binding mechanism (Elkins et al 2003). The primary amide of HIF-1α Asn803 forms hydrogen bonds with FIH-1 Gln239 and Arg238 side chains, explaining the preference for asparaginyl over aspartyl residues (Lando et al 2002b, Hewitson et al 2002). The CAD/FIH-1 interaction involves two separate contact regions with the CAD, the dominant 'site 1' (795–806) which includes the hydroxylated asparagine and contributes 10 hydrogen bonds, and 'site 2' (813–822) which contributes only two hydrogen bonds (Fig. 3). Accordingly, *in vitro* hydroxylation experiments with purified FIH-1 and synthetic peptides have shown that hydroxylation still occurs for peptides omitting 'site 2', albeit only 9–40% as efficiently as for peptides containing both sites. (Elkins et al 2003, Koivunen et al 2004). Maximum catalysis, however, requires longer peptide substrates (35-mer) including both sites and additional flanking residues (Koivunen et al 2004), similar to the PHDs where maximum catalysis requires a 17-mer peptide (Hirsila et al 2003).

Comparison of CAD-p300/CBP versus CAD/FIH-1 structures confirms the mutual exclusivity of these two CAD interactions (p300/CBP minimally requires hHIF-1α CAD 786–826), and highlights the diversity of structures the CADs can adopt (Kung et al 2000, Dames et al 2002, Freedman et al 2002). The CAD, which in isolation is not structured (Dames et al 2002), is capable of diverse 'induced fit' configurations. In complex with FIH-1, 'site 1' residues lie extended within a groove and adopt a tight turn involving Val802 to Ala804 that directs the Asn into the catalytic core, whereas the same region adopts an alpha helix in complex with p300/CBP (Dames et al 2002, Freedman et al 2002, Elkins et al 2003). 'Site 2' however, adopts an analogous alpha helix in complex with both FIH-1 and the coactivators. Importantly, the FIH-1/CAD complex entails mutual ordering, including Val-802 CAD causing a marked rearrangement of Trp296 FIH-1.

Results and discussion

In addition to influencing the FIH-1 configuration, CAD Val802 has an integral role in asparaginyl positioning (Elkins et al 2003). In agreement with the structural predictions, we have demonstrated that the HIF-1α CAD V802A mutant exhibits a four–fivefold lower V_{max} than wild-type, without a change in K_m, and is constitutively active in cell-based reporter assays (Linke et al 2004). Molecular modelling provides an explanation for the catalytic impairment, predicting major repositioning of the Asn β-carbon. Given the similarly central role of Ala804 (HIF-1α), it is of interest that this residue is fully conserved for HIF-1α orthologues whereas HIF-2α orthologues contain a fully conserved valine (Fig. 3). A difference in hydroxylation efficiency between the paralogues has been predicted (Hewitson & Schofield 2004) and finds support in the demonstration by Koivunen et al 2004 that HIF-2α peptides are hydroxylated at 1–7% the rate of comparable HIF-1α peptides. In agreement, we find that HIF-2α exhibits a two fold lower V_{max} than HIF-1α *in vitro* and significantly higher normoxic activity in cell-based reporter assays, a change that is reversible by the exchange of these residues. This difference in hydroxylation rate and subsequently in transcriptional activity may ultimately contribute to the functional differences observed between HIF-1α and HIF-2α *in vivo* (C. Bracken, W. Balrak & D. Peet, unpublished work 2004).

Interestingly, despite strict conservation between HIF-1α and HIF-2α orthologues, our kinetic comparisons of the 'site 1' mutants Y798A, D799A, C800A, E801A, P805A and the intervening G808A reveal similar values for K_m, V_{max}, and transcriptional inducibility compared to wild-type. This indicates functional compensation by neighbouring residues within the extended 'site 1' region. Furthermore, since most CAD residues that contact FIH-1 also form direct interactions with p300/CBP, residue conservation is not a specific predictor of importance for FIH-1 recognition (Linke et al 2004). Our results are reminiscent of the PHDs, whereby the conserved LXXLAP motif tolerates broad substitutions to all positions except the adjacent alanine, which displays narrow tolerance to substitution (Huang et al 2002, Li et al 2004). Given the large recognition sites involving much functional redundancy between contributing residues for FIH-1 and the PHDs, no consensus hydroxylation sequence for either has been established, precluding bioinformatic substrate prediction and favouring the use of methods such as yeast two-hybrid screens for identifying other substrates. Beyond the HIF-αs, no other substrates have been reported for FIH-1 or the PHDs in the literature to date.

Oxygen sensing by FIH-1

It is highly likely that FIH-1 directly senses cellular oxygen concentration and conveys this to HIF-α. Clearly HIF-α Asn-hydroxylation is correlated with oxygen tension, being detectable under normoxia (21% atmospheric oxygen, generating ~225 μM O_2 in physiological solution), and undetectable during hypoxia (<1% atmospheric oxygen, generating ~11 μM O_2 in physiological solution) (Lando et al 2002a, Vanderkooi et al 1991). To assign oxygen-sensitivity directly to FIH-1 requires a K_m for oxygen close to the normal cellular oxygen concentration such that physiological hypoxia has a profound impact on catalytic rate. Furthermore, this hydroxylation should be below saturation, even during normoxia, such that reductions in oxygen result in significant amounts of unhydroxylated HIF-α. Kinetic analyses performed using FIH-1 purified from insect cells with a synthetic CAD peptide have indeed demonstrated a K_m for oxygen of 90 μM (equivalent to a physiological solution equilibrated at 8% oxygen) (Koivunen et al 2004). Given that the partial pressure of oxygen measured within a range of vertebrate tissues is just 10–35 μM (Massabuau 2003), FIH-1 catalytic rate should directly reflect physiological oxygen fluctuations. Endogenous FIH-1 activity is also evidently limiting in cells. Overexpression represses HIF-1 activity (Mahon et al 2001, Lando et al 2002b, Metzen et al 2003) and reduces endogenous HIF-1 target gene expression (Stolze et al 2004). Thus FIH-1 has the credentials of a direct oxygen sensor. *In vivo*, this prediction holds. FIH-1 is active within cells during normoxia and moderate to severe hypoxia, demonstrated by both knockdown (RNAi) in cells equilibrated at 21% to 0.2% oxygen, causing HIF-1 target gene expression to increase, and also by FIH-1 overexpression, which represses HIF-1 target gene expression at 0.2% oxygen but importantly, is ineffectual under total anoxia (Stolze et al 2004). The PHDs also appear to have the capacity to sense oxygen, with K_ms in the range of 230–250 μM O_2 *in vitro* (Hirsila et al 2003), non-saturating activity and sensitivity to hypoxia *in vivo* (Huang et al 2002, Metzen et al 2003, Appelhoff et al 2004, Stolze et al 2004).

The higher oxygen affinity of FIH-1 compared with the PHDs *in vitro* is suggestive of an oxygen-sensing cascade in which HIF-α becomes stabilized under moderate hypoxia but is still subject to CAD repression, whereas under severe hypoxia both forms of repression should be relieved. In support of a cascade, overexpression of PHD2 is only moderately repressive to HIF-1 target gene expression at 0.5% oxygen and is ineffectual at 0.2%, whereas FIH-1 represses HIF-1 target gene expression at 0.2% (Stolze et al 2004). Of interest, we have examined endogenous HIF-α expression in parallel with CAD activity under conditions of 21 to 0.5% oxygen and have uncovered cell-line specificity regarding the apparent sensitivity of FIH-1 to hypoxia, whereas prolyl hydroxylation seems to exhibit similar oxygen sensitivity across the cell lines examined. In certain cell types, the CAD is activated under hypoxia an order of magnitude more severe than is needed for detectable protein stabilization, whereas in other cell lines oxygen decrements cause CAD up-regulation before the protein becomes detectable (C. Bracken & D. Peet, unpublished work 2004).

The HIF hydroxylases appear to directly sense oxygen with respect to HIF-α, and it will be of interest to see if they regulate other oxygen-responsive cellular events. In addition, oxygen-sensing might be substrate-directed. The FIH-1/CAD structure reveals a hydrogen bond between HIF-1α Asn803 and the Fe^{2+}-binding Asp201 in FIH-1 which oxygen-binding is predicted to disrupt, thus energetically constraining oxygen affinity (Elkins et al 2003). Given that CAD binding, which is prerequisite to dioxygen binding, induces significant structural changes in FIH-1, it may actually determine oxygen affinity, and this may vary with different peptide substrates.

Functional redundancy

Despite three (differentially operative) PHDs, FIH-1 activity is not redundant. Stolze et al (2004) depleted endogenous FIH-1 by RNAi in a range of cell lines, causing an increase in HIF-1 target gene expression under oxygen concentrations in which the PHDs are active. Thus low levels of stable HIF-α exist in normoxia and are repressed by FIH-1 in a functionally relevant manner. FIH-1 expression is broad according to both EST data (Mahon et al 2001) and immunoblotting of a range of cells (Stolze et al 2004) and mouse tissues (K. Lisy & D. Peet, unpublished work 2004), suggesting a ubiquitous indispensable function. The non-redundancy of FIH-1 is unsurprising given the lack of close structural homologues, with the only other asparaginyl hydroxylase known to date being the highly divergent aspartyl/asparaginyl-β-hydroxylase (BAH), which hydroxylates the β-carbon of both Asn and Asp residues, generating the alternative diastereoisomer (erythro-β-hydroxyasparaginyl/aspartyl) within CXN/DY/FXCXC sequences found in EGF-like domains of various proteins (Stenflo et al 1989 and references therein). The major mechanism of FIH-1 as an HIF-α repressor is clearly catalytic, given that non-catalytic mutant proteins, or wild-type enzymes treated with inhibitors such as dimethyloxalylglycine (DMOG) and anoxia fail to exert their effects (Lando et al 2002b, Hewitson et al 2002, Stolze et al 2004). However, additional functions cannot yet be ruled out. FIH-1 reportedly interacts with pVHL (Mahon et al 2001), although CAD repression remains intact in pVHL-deficient cells (Sang et al 2002). FIH-1 also interacts with HDACs although active repression by HIF-α has yet to be demonstrated (Mahon et al 2001).

Regulation of FIH-1/CAD

Growth factors and oncogenic signalling are capable of both stabilizing and activating HIF-α (reviewed by Bilton & Booker 2003), the latter of which may involve prevention of FIH-1-mediated CAD repression. Experiments by Datta et al (2004) suggest that PKCζ-mediated reduction of FIH-1 levels may maximise HIF-1 activity in various pVHL-deficient and non-deficient tumour cells. Enzymatic reversal or further modification of Asn-hydroxylation would provide additional positions for regulation of the HIF system, but while various chemically feasible strategies have been proposed, such phenomena have not yet been demonstrated (Schofield & Ratcliffe 2004, Lancaster et al 2004b). Direct regulation of the CAD is also suggested; evidence implies that the threonine within the CAD 'site 1' (Thr796 in hHIF-1α and Thr844 in mHIF-2α, Fig. 3) is a phosphoacceptor that increases CAD activity under normoxia (Gradin et al 2002, Elkins et al 2003). Accordingly, a synthetic phospho peptide was not detectably hydroxylated *in vitro* whereas the comparable unmodified peptide was an efficient substrate. Molecular modelling predicted impaired binding to FIH-1 as the cause (Lancaster et al 2004a), which we have recently produced data to support. Using purified HIF-1α CAD T796D versus wild-type substrates we find a sixfold higher K_m for the T796D mutant, without a change in V_{max} *in vitro*. Similarly, in cell-based reporter assays, the T796D CAD shows constitutive activity compared with wild-type, in agreement with Gradin et al (2002), indicating that a negative charge at this position does impair Asn hydroxylation (Fig. 4). Finally FIH-1 itself may be regulated post-translationally.

FIH-1 as a therapeutic target

Both FIH-1 and the PHDs are functionally non-redundant HIF-α repressors that are candidates for pharmaceutical perturbation by small molecule inhibitors or allosteric activators. However, structural information for the PHDs is lacking, whereas the information gained from the CAD/FIH-1 structure can support a

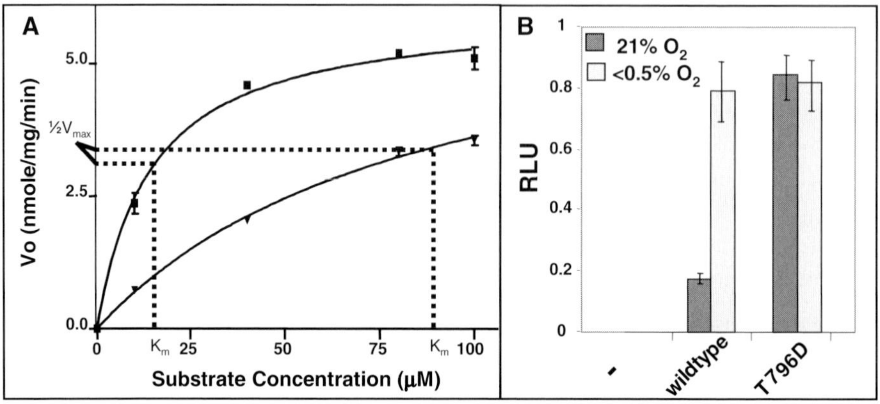

FIG. 4. HIF-1α CAD T796D shows reduced hydroxylation efficiency *in vitro* and is constitutively active in cell-based reporter assays. (A) *In vitro* hydroxylation assay of HIF-1α CAD T796D shows a sixfold increase in K_m (89 μM) compared with wild-type (15 μM) without a change in V_{max} (wild-type = 6, T796D = 6.9 nmol/mg/min). (B) Gal-DBD reporter gene experiments in HEK 293T cells show constitutive normoxic activity of Gal HIF-1α CAD T796D compared with wild-type. Experiments performed in triplicate, and illustrated as mean ± SEM (n = 3), with results representative of three independent experiments. Methods detailed in Linke et al (2004).

rational predictive approach to small molecule design and the distinct structural and functional features of FIH-1 should provide opportunity for high specificity. In particular inhibitors that disrupt the peptide interface between FIH-1 monomers or a mimetic of the tight turn involving CAD Val802 to Ala/Val804 would reduce non-specific effects that might arise from 2-OG analogues, given the potential similarity in 2-OG-binding between FIH-1 and many of the JmjC-domain containing proteins (Hewitson et al 2002, Elkins et al 2003). Proof of principle that FIH-1 inhibition or activation affects HIF activity is provided by knockdown and over-expression experiments that increase or reduce HIF target gene expression, respectively (Stolze et al 2004). Further work must therefore address the structural and functional suitability of FIH-1 as a therapeutic target. For example, inhibition of FIH-1 will activate HIF-1, which may facilitate blood vessel formation within ischaemic tissues, whereas activation of FIH-1 will repress HIF-1 activity and thus reduce tumour vascularization and growth (reviewed by Hewitson & Schofield 2004).

It is envisaged that the information we and others have generated regarding HIF asparaginyl hydroxylation, FIH-1 characterization and substrate specificity will continue to enhance our understanding of the cellular hypoxic response, its role in numerous human diseases, and ultimately facilitate rational drug design.

Acknowledgements

We acknowledge the support of the National Health and Medical Research Council of Australia and the ARC Special Research Centre for the Molecular Genetics of Development, and thank the Peet lab and Murray Whitelaw for their assistance in preparing this paper.

References

Appelhoff RJ, Tian YM, Raval RR et al 2004 Differential function of the prolyl hydroxylases PHD1, PHD2, and PHD3 in the regulation of hypoxia-inducible factor. J Biol Chem 279:38458–38465

Bilton RL, Booker GW 2003 The subtle side to hypoxia inducible factor (HIFalpha) regulation. Eur J Biochem 270:791–798

Dames SA, Martinez-Yamout M, De Guzman RN, Dyson HJ, Wright PE 2002 Structural basis for Hif-1 alpha/CBP recognition in the cellular hypoxic response. Proc Natl Acad Sci USA 99:5271–5276

Dann CE, 3rd, Bruick RK, Deisenhofer J 2002 Structure of factor-inhibiting hypoxia-inducible factor 1: An asparaginyl hydroxylase involved in the hypoxic response pathway. Proc Natl Acad Sci USA 99:15351–15356

Datta K, Li J, Bhattacharya R, Gasparian L, Wang E, Mukhopadhyay D 2004 Protein kinase C zeta transactivates hypoxia-inducible factor alpha by promoting its association with p300 in renal cancer. Cancer Res 64:456–462

Elkins JM, Hewitson KS, McNeill LA, et al 2003 Structure of factor-inhibiting hypoxia-inducible factor (HIF) reveals mechanism of oxidative modification of HIF-1 alpha. J Biol Chem 278:1802–1806

Freedman SJ, Sun ZY, Poy F, et al 2002 Structural basis for recruitment of CBP/p300 by hypoxia-inducible factor-1 alpha. Proc Natl Acad Sci USA 99:5367–5372.

Gradin K, Takasaki C, Fujii-Kuriyama Y, Sogawa K 2002 The transcriptional activation function of the HIF-like factor requires phosphorylation at a conserved threonine. J Biol Chem 277:23508–23514

Hewitson KS, McNeill LA, Riordan MV, et al 2002 Hypoxia-inducible factor (HIF) asparagine hydroxylase is identical to factor inhibiting HIF (FIH) and is related to the cupin structural family. J Biol Chem 277:26351–26355

Hewitson KS, Schofield CJ 2004 The HIF pathway as a therapeutic target. Drug Discov Today 9:704–711

Hirsila M, Koivunen P, Gunzler V, Kivirikko KI, Myllyharju J 2003 Characterization of the human prolyl 4-hydroxylases that modify the hypoxia-inducible factor. J Biol Chem 278:30772–30780

Huang J, Zhao Q, Mooney SM, Lee FS 2002 Sequence determinants in hypoxia-inducible factor-1alpha for hydroxylation by the prolyl hydroxylases PHD1, PHD2, and PHD3. J Biol Chem 277:39792–39800

Jiang BH, Zheng JZ, Leung SW, Roe R, Semenza GL 1997 Transactivation and inhibitory domains of hypoxia-inducible factor 1alpha. Modulation of transcriptional activity by oxygen tension. J Biol Chem 272:19253–19260

Koivunen P, Hirsila M, Gunzler V, Kivirikko KI, Myllyharju J 2004 Catalytic properties of the asparaginyl hydroxylase (FIH) in the oxygen sensing pathway are distinct from those of its prolyl 4-hydroxylases. J Biol Chem 279:9899–9904

Kung AL, Wang S, Klco JM, Kaelin WG, Livingston DM 2000 Suppression of tumor growth through disruption of hypoxia-inducible transcription. Nat Med 6:1335–1340

Lancaster DE, McNeill LA, McDonough MA et al 2004a Disruption of dimerization and substrate phosphorylation inhibit factor inhibiting hypoxia-inducible factor (FIH) activity. Biochem J 383:429–437

Lancaster DE, McDonough MA, Schofield CJ 2004b Factor inhibiting hypoxia-inducible factor (FIH) and other asparaginyl hydroxylases. Biochem Soc Trans 32:943–945

Lando D, Peet DJ, Whelan DA, Gorman JJ, Whitelaw ML 2002a Asparagine hydroxylation of the HIF transactivation domain a hypoxic switch. Science 295:858–861

Lando D, Peet DJ, Gorman JJ, Whelan DA, Whitelaw ML, Bruick RK 2002b FIH-1 is an asparaginyl hydroxylase enzyme that regulates the transcriptional activity of hypoxia-inducible factor. Genes Dev 16:1466–1471

Lando D, Gorman JJ, Whitelaw ML, Peet DJ 2003 Oxygen-dependent regulation of hypoxia-inducible factors by prolyl and asparaginyl hydroxylation. Eur J Biochem 270:781–790

Lee C, Kim SJ, Jeong DG, Lee SM, Ryu SE 2003 Structure of human FIH-1 reveals a unique active site pocket and interaction sites for HIF-1 and von Hippel-Lindau. J Biol Chem 278:7558–7563

Li D, Hirsila M, Koivunen P et al 2004 Many amino acid substitutions in a HIF-1alpha-like peptide cause only minor changes in its hydroxylation by the HIF prolyl 4-hydroxylases: Substitution of 3,4-dehydroproline or azetidine-2-carboxylic acid for the proline leads to a high rate of uncoupled 2-oxoglutarate decarboxylation. J Biol Chem 279:55051–55059

Linke S, Stojkoski C, Kewley RJ, Booker GW, Whitelaw ML, Peet DJ 2004 Substrate requirements of the oxygen-sensing asparaginyl hydroxylase factor-inhibiting hypoxia-inducible factor. J Biol Chem 279:14391–14397

Mahon PC, Hirota K, Semenza GL 2001 FIH-1: a novel protein that interacts with HIF-1alpha and VHL to mediate repression of HIF-1 transcriptional activity. Genes Dev 15:2675–2686

Massabuau JC 2003 Primitive, and protective, our cellular oxygenation status? Mech Ageing Dev 124:857–863

McNeill LA, Hewitson KS, Claridge TD, Seibel JF, Horsfall LE, Schofield CJ 2002 Hypoxia-inducible factor asparaginyl hydroxylase (FIH-1) catalyses hydroxylation at the beta-carbon of asparagine-803. Biochem J 367:571–575

Metzen E, Berchner-Pfannschmidt U, Stengel P et al 2003 Intracellular localisation of human HIF-1 alpha hydroxylases: implications for oxygen sensing. J Cell Sci 116:1319–1326

O'Rourke JF, Tian YM, Ratcliffe PJ, Pugh CW 1999 Oxygen-regulated and transactivating domains in endothelial PAS protein 1: comparison with hypoxia-inducible factor-1alpha. J Biol Chem 274:2060–2071

Peet DJ, Lando D, Whelan DA, Whitelaw ML, Gorman JJ 2004 Oxygen-dependent asparagine hydroxylation. Methods Enzymol 381:467–487

Pugh CW, O'Rourke JF, Nagao M, Gleadle JM, Ratcliffe PJ 1997 Activation of hypoxia-inducible factor-1; definition of regulatory domains within the alpha subunit. J Biol Chem 272:11205–11214

Sang N, Fang J, Srinivas V, Leshchinsky I, Caro J 2002 Carboxyl-terminal transactivation activity of hypoxia-inducible factor 1 alpha is governed by a von Hippel-Lindau protein-independent, hydroxylation-regulated association with p300/CBP. Mol Cell Biol 22:2984–2992

Schofield CJ, Ratcliffe PJ 2004 Oxygen sensing by HIF hydroxylases. Nat Rev Mol Cell Biol 5:343–354

Stenflo J, Holme E, Lindstedt S, et al 1989 Hydroxylation of aspartic acid in domains homologous to the epidermal growth factor precursor is catalyzed by a 2-oxoglutarate-dependent dioxygenase. Proc Natl Acad Sci USA 86:444–447

Stolze IP, Tian YM, Appelhoff RJ, et al 2004 Genetic analysis of the role of the asparaginyl hydroxylase factor inhibiting hypoxia-inducible factor (HIF) in regulating HIF transcriptional target genes. J Biol Chem 279:42719–42725

Vanderkooi JM, Erecinska M, Silver IA 1991 Oxygen in mammalian tissue: methods of measurement and affinities of various reactions. Am J Physiol 260:C1131–1150

DISCUSSION

Ratcliffe: In your experiments on HIF-2, did you look at the other PHD enzymes? Your result is the same as the one we showed—you don't see the dominance of PHD2 on HIF-2. However, if we inhibit all the PHDs in combination, our experience is that we then see an induction of HIF-2. Have you knocked the other enzymes out?

Peet: We haven't done all three together. We have preliminary results on 1 and 3, and we don't see any significant effects if we knock them out.

Ratcliffe: We think the iron binding does vary from enzyme to enzyme. It also may be cell-type specific. It is unlikely that iron is drifting in and out of the enzyme without some more sophisticated mechanism. I suspect the effects of chelators are indirect, not taking the iron directly out of the catalytic centre.

Peet: We have done overexpression experiments that support the idea that overexpression of the PHDs 1, 2 and 3 can work on HIF-1α. These aren't good experiments because they involved overexpression, but it seems that there is this difference where HIF-2α doesn't appear to be regulated by any of the PHD isoforms.

Chandel: I was surprised that in your mutations where you changed the proline to the alanine this wasn't sufficient to drive gene expression. Several people have reported that just the mutations are sufficient to drive HRE-driven luciferase. What is the difference in your system?

Peet: I think the main difference is overexpression. In our hands, if we overexpress enough wild-type you could add any mutation and get good induction of the reporter. The implication of this is that we are probably saturating the prolyl hydroxylases and are getting stabilized transcription factors. So in our experiments we aimed at moderate levels of expression. The proteins we were transfecting in, were still being stabilized by hypoxia. This is when we saw that sort of induction.

Chandel: So why do VHL mutations still allow gene expression?

Peet: It seems that in those cell types there are other effects such that the function of FIH1 is compromised.

Semenza: Decreased FIH-1 mRNA expression has been linked to the activation of a PKC-zeta pathway in 786-O renal carcinoma cells.

Ratcliffe: Ineke Stolze has some data indicating that expression is preserved in VHL-defective cells such as 786-0.

Stolze: Yes, the protein is expressed.

Chandel: If it is expressed, why does the loss of VHL allow you to get gene expression? Based on Dan Peet's model, you wouldn't.

Ratcliffe: For the same reason you get HIF target gene expression when you overexpress HIF. A functional FIH does not completely suppress HIF transcription for whatever reason. For some reason this enzyme does not work maximally under a number of VHL-defective and VHL-competent circumstances.

Pouysségur: A splice variant of HIF-1α lacking the C-terminus is still fully active in *trans*-activation. Whatever the level of FIH we still think that the N-transactivating domain could have a potential action, even if FIH is not knocked down.

Ratcliffe: One of the puzzles in the FIH field is that when the C-terminal activation domain was first defined, all of us found that if N-terminal pieces of sequence were added, we got more and more repression in the presence of oxygen. This would imply that this sequence is needed for FIH action. However this is wholly inconsistent with the *in vitro* data we have. I'm looking for an explanation: do you have one?

Peet: No. There is a level of complexity there that we don't understand.

Ratcliffe: We wondered whether this was due to steric hindrance. FIH is a dimer, and all the experiments delineating the regulation of the HIF C-terminal activation domain were done on Gal-HIF fusion proteins. We thought that fusion to Gal might introduce artificial steric constraints. Has anyone done deletion experiments on the HIF native molecule?

Peet: Our suggestion is that there probably is another level of complexity here. There are other factors involved. FIH-1 and the C-terminus alone don't fit with the data.

Ratcliffe: In vitro we could see no difference in the substrate efficacy between a 51-mer containing residues 775–826 and a 40-mer containing amino acids 786–826, whereas *in vivo* residues 786–826 are not oxygen sensitive.

Peet: We have tried a 40-mer, and it is not as good as a 90-mer or 100-mer *in vitro*. There is a difference in hydroxylation efficiency, but it is not black and white. If you see a subtle difference *in vitro* and then translate that into the cell based system, the difference is usually magnified. I don't think this is the whole explanation, but with many of the experiments where we have made point mutations, such as with a mutation to valine adjacent to the asparagine, we might see a fourfold decrease in hydroxylation efficiency *in vitro*, but in a cell-based system we see that the valine mutation is constitutively active. Rather than being a fourfold difference, it is not hydroxylated at all. We see a significant amount of hydroxylation of the last 40 amino acids *in vitro*. We couldn't predict this because it is constitutively active in our cell-based assay.

Duchen: Gregg Semenza, earlier on you raised the issue of the oxygen-insensitive regulation of HIF by other signalling pathways. How do the oxygen sensitive and insensitive pathways interact? I had the impression that the insensitive pathway would override the sensitive pathway.

Semenza: That isn't necessarily true. Dan Peet described a mechanism for counteracting the negative regulation of transactivation through phosphorylation. Immediately you can imagine kinases that would phosphorylate that site. Now the requirement for FIH-1 is completely lost. I'd be interested to know what kinase is phosphorylating that residue and what signals it is responding to. This provides a beautiful mechanism for oxygen-independent regulation of transactivation, just as we have shown that the synthesis of HIF-1α protein can be increased through oxygen-independent means.

Ward: In effect, succinate is an end product of the process. If more succinate is being produced by mitochondrial respiration, then this will inhibit that pathway if succinate production is high in the mitochondria. This would act as a limiting effect. Are we talking the same ballpark figures in terms of concentrations? There could be a negative interaction here, it seems.

Semenza: I have no idea what those concentrations are.

Ratcliffe: It is true that these succinate enzymes could mediate product inhibition. As ever, the problem is being able to measure it in the right compartment, to know whether it does indeed do that. In theory it could. It is easy to see how metabolism might interface with hypoxia.

Duchen: One of the interesting things that came across this morning is that as we start looking at alterations in succinate dehydrogenase activity, for example, the temptation is to start thinking about changes in free radical production or some form of respiratory change. But your data suggest that the presence of the end product may be enough.

Ratcliffe: It might be enough.

Ward: It is the location of the succinate that matters, whether it is in the mitochondria or cytosol, and in particular what the relationship between the two is.

Murphy: Related to that, you might expect hypoxia to prevent respiration and thereby back up electron flow through the ubiquinone pool and thereby inhibit succinate dehydrogenase (SDH) that way. This could be another mechanism of altering the activity of SDH apart from mutations.

Duchen: If you do that is there much of a change in succinate concentration in the mitochondria? Will that get out into the cytosol?

Murphy: In principle succinate transport across the mitochondrial inner membrane should be reversible. This would occur by the action of the dicarboxylate carrier enabling succinate to come out of the mitochondria.

Acker: The mitochondria have a very high oxygen affinity. They consume oxygen constantly down to very low pO2 values under normal physiological conditions.

Ward: It depends on what the perimitochondrial oxygen concentration is, which will be determined by rate of O2 consumption by the mitochondria compared to the rate of diffusion across the cell. Many studies, including those of Mike Duchen and ourselves, have shown that 'physiological' hypoxia increases mitochondrial NADH in intact cells and tissues, suggesting reduction of the electron transport chain.

Acker: I doubt that there is a large sink to the mitochondria. Even if there is one, the critical pO2 of isolated mitochondria is below 1 Torr. We think for the carotid body that there is a special low affinity component of the complex IV to get the mitochondrial oxygen sensing property adapted to the tissue pO2 levels.

Buckler: Not necessarily. You can have competitive inhibitors. CO and NO are competitive inhibitors that can markedly shift the apparent K_m for oxygen of cytochrome a3.

Acker: There may be even more other effectors influencing the pO2 affinity.

Schumacker: I have a general observation. We have heard today discussions of hypoxia, where the hypoxia level was in the range of 2–5%, and others where it was 0.2% oxygen. The latter is a level of oxygenation that would normally limit cellular respiration. We need to be careful about what we call hypoxia, and what we are calling near anoxic conditions.

Semenza: The cells are not seeing 0.2% oxygen: that is what is being delivered into the chamber. The level is much higher in the medium.

Chandel: Pre-equilibrated media can be used.

Semenza: Most people don't do that experiment.

Ratcliffe: How long do you preequilibrate for?

Schumacker: For the media, the night before. The cells go in and we change the media with preequilibrated media, and the cells sit there for four hours.

Ratcliffe: People working on the oxygen gradient issue tell me that there are two effects, which we have so far not taken into account. First, the equilibration with the medium, which will occur within a number of hours. The other is the equilibration with the plastic. Most of the plasticware that we use contains oxygen which will leach out over a matter of days.

Schumacker: I am only stepping back to reflect on the fact that we are putting numbers on this abscissa so as to generate a dose–response curve. My point is that we need to define what we are calling hypoxia, and to be careful that the numbers we assign to the dose–response curves correspond with the cellular oxygen levels that exist under the experimental conditions.

Duchen: I had expected that the question of what constitutes hypoxia would be a recurrent theme through this meeting. It is obviously important to establish the borderline between the physiological and the pathological. Gregg Semenza, you showed a dose–response curve for oxygen tension showing the range of oxygen tensions required to activate HIF. Can you comment here?

Semenza: The experiment was performed using an instrument called a tonometer, which allows the cells to pass through a very thin film of medium that is equilibrated with a gas mixture containing the desired oxygen concentration. The experiment was also performed in the presence of cyanide in order to block oxygen consumption by the cells. Under these conditions, the intracellular oxygen concentration was equal to the concentration of oxygen delivered to the medium.

Duchen: Then you also have to refer those values back to what the real oxygen tension is in the tissue.

Semenza: As has been mentioned, there is tremendous variation both between tissues and within tissues with regard to *in vivo* oxygen concentrations.

Oxygen-sensing by ion channels and mitochondrial function in carotid body glomus cells

José López-Barneo, Patricia Ortega-Sáenz, José I. Piruat and María García-Fernández

Laboratorio de Investigaciones Biomédicas, Departamento de Fisiología and Hospital Universitario Virgen del Rocío, Universidad de Sevilla, Sevilla, Spain

Abstract. Carotid body glomus cells release transmitters in response to hypoxia due to the increase of excitability resulting from inhibition of O_2-regulated K^+ channels. The mechanisms involved in the detection of changes of O_2 tension are unknown. Inhibition of the mitochondrial electron transport chain (ETC) at proximal and distal complexes induces external Ca^{2+}-dependent catecholamine secretion. At saturating concentration of the ETC inhibitors, the cellular response to hypoxia is maintained. However, rotenone, a complex I blocker, selectively occludes the responsiveness to hypoxia of glomus cells in a dose-dependent manner. The effect of rotenone is not mimicked by complex I inhibitors acting on different sites. We have also generated a knock-out mouse lacking SDHD, the small membrane-anchoring protein of the succinate dehydrogenase (complex II) of the mitochondrial electron transport chain. Homozygous $Sdhd^{-/-}$ animals die at early embryonic stages. Heterozygous $Sdhd^{+/-}$ mice show a general, non-compensated, deficiency of complex II activity, and abnormal enhancement of resting carotid body secretion rate due to decrease of K^+ conductance and persistent Ca^{2+} influx into glomus cells. However, responsiveness to hypoxia of carotid bodies from $Sdhd^{+/-}$ mice remains intact. These data strongly suggest that sensitivity to hypoxia of carotid body glomus cells is not linked in a simple way to mitochondrial electron flow. Nevertheless, it is possible that a rotenone-sensitive molecule critically participates in acute carotid body oxygen sensing.

2005 Signalling pathways in acute oxygen sensing. Wiley, Chichester (Novartis Foundation Symposium 272) p 54–72

The carotid body (CB) is a major arterial chemoreceptor that senses blood O_2 to evoke hyperventilation and sympathetic activation in response to hypoxemia (for a review see López-Barneo et al 2001). The O_2-sensing elements in the CB are neurosecretory glomus, or type I, cells, which are organized in innervated clusters in close contact with blood vessels. Glomus cells are electrically excitable and have several types of O_2-sensitive K^+ channels in their membranes (López-Barneo et al 1988, Delpiano & Hescheler 1989, Peers 1990, Stea & Nurse 1991, Buckler 1997). Although the O_2-sensitive K^+ channel type can change in the different animal species, it is broadly accepted that inhibition of the membrane K^+ conductance by low O_2 tension (pO2) is the critical event leading to depolarization, external Ca^{2+} influx and activation of neurotransmitter release, which, in turn, stimulates the afferent sensory fibres. This 'membrane model' of chemotransduction, suggested by the electrophysiological experiments, has been confirmed by monitoring cytosolic $[Ca^{2+}]$ and quantal catecholamine secretion in single dye-loaded cells (López-Barneo et al 1993, Buckler & Vaughan-Jones 1994, Ureña et al 1994, Montoro et al 1996, Carpenter et al 2000, Pardal et al 2000).

Despite the progress in glomus cell physiology the nature of the mechanisms coupling changes of pO2 to K^+ channel activity remain unknown. CB research is hampered by the small size of the organ and the difficulty to obtain a preparation with reproducible sensitivity to hypoxia, since this is a property easily lost by the cells. Because modulation of some K^+ channels by pO2 is seen in excised patches (Ganfornina & López-Barneo 1991, Riesco-Fagundo et al 2001) it was postulated that O_2 sensing in glomus cells depends on membrane-delimited mechanisms. In fact, membrane bound haeme oxygenase 2 has recently been suggested to mediate O_2 sensitivity of heterologously expressed and glomus cell maxi-K^+ channels (Williams et al 2004). On the other hand, several investigators have traditionally considered mitochondria as the major players in CB O_2 sensing because, similarly to hypoxia, inhibitors of the electron transport chain (ETC) or mitochondrial uncouplers increase the afferent activity of the sinus nerve (see, for instance, Mills & Jöbsis 1972). It has also been reported that cyanide and anoxia release Ca^{2+} from mitochondria in isolated rabbit glomus cells (Biscoe & Duchen 1990, Duchen & Biscoe 1992). Although the 'membrane model' of CB chemostransduction is unquestioned, the interest in mitochondria has resurged in recent years since mitochondrial uncouplers raise cytosolic Ca^{2+} and reduce background K^+ permeability in rat CB cells (Buckler & Vaughan-Jones 1998). In addition, mutations in the small membrane-anchoring protein of succinate dehydrogenase (SDHD) in mitochondrial complex II are the main cause of familial hereditary paraganglioma (PGL) a mostly benign, highly vascularized tumour of the CB (Baysal et al 2000). Similar to CB of individuals exposed chronically to hypoxia, PGL tumours display cellular hyperplasia/anaplasia. Moreover, the prevalence of PGL in individuals with *SDHD* mutations increases in high-altitude living populations (Astrom et al 2003). Thus, it has been proposed that SDHD participates in O_2 sensing and that PGL tumours are induced by defects in the detection of blood O_2 levels (Baysal et al 2000, Gimenez-Roqueplo et al 2001). It is, therefore, plausible that mitochondria are the site for glomus cell O_2 sensing and that in response to hypoxia they signal the membrane to modulate K^+ channel activity.

Herein, we summarize recent work in our laboratory designed to investigate whether sensitivity of intact glomus cells to hypoxia is altered by mitochondrial dysfunction induced either pharmacologically (Ortega-Sáenz et al 2003) or by deletion of the *SDHD* gene (Piruat et al 2004). We describe the effect of mitochondrial ETC inhibition on glomus cell K^+ conductances and responsiveness to low pO2 as well as the physiological features of CB glomus cells from *Sdhd* knockout mice.

Experimental procedures

Most of the experiments were performed using the carotid body thin slice preparation in which the response of glomus cells to low pO2 can be studied in almost optimal physiological conditions (Pardal et al 2000, Pardal & López-Barneo 2002). Amperometric recordings of catecholamine released from rat or mouse CB slices as well as patch clamp studies on isolated glomus cells were carried out following the procedures previously described (Ortega-Sáenz et al 2003, Piruat et al 2004). Changes in the solution composition or experimental conditions not previously published are indicated in the figure legends. Generation and genetic characterization of the *Sdhd* knockout mice are described in Piruat et al (2004).

Results and discussion

Glomus cell sensitivity to hypoxia and pharmacological ETC inhibition

In rat CB slices hypoxia (switching from a solution with a pO2 of 150 mmHg to another with ≈15 mmHg) induced catecholamine release in an external Ca^{2+}-dependent manner (Fig. 1A). Similar to hypoxia, exposure of glomus cells to ETC inhibitors also produced an external Ca^{2+}-dependent secretory response. The effect is shown for rotenone and cyanide (inhibitors of complex I and IV, respectively) in Fig. 1B & C, but it was also observed with blockers of complex II (thenoyltrifluoroacetone; TTFA) and III (myxothiazol and antimycin A) (Ortega-Sáenz et al 2003). The interaction between hypoxia and the mitochondrial electron flow was studied in cells exposed to low pO2 before and during application of ETC inhibitors. The rationale behind these experiments was that if hypoxia exerts its effect through alteration of the mitochondrial electron flow, preincubation with saturating concentrations of ETC blockers would prevent any further effect of low pO2. In contrast, the effects of hypoxia and ETC inhibition would be additive, at least partially, if they were acting through separate pathways. With the exception of rotenone, that occluded sensitivity to hypoxia (Fig. 1E), the ETC tested evoked secretory responses that were additive with that elicited by hypoxia (Fig. 1D–F). The differential effect of rotenone is particularly evident when the cumulative secretion signals recorded during concomitant exposure to the ETC inhibitors and hypoxia are shown at an expanded time base (right panels in Fig. 1 D & E). The effect of rotenone was observed in all the cells studied ($n = 28$) with concentrations of the drug between

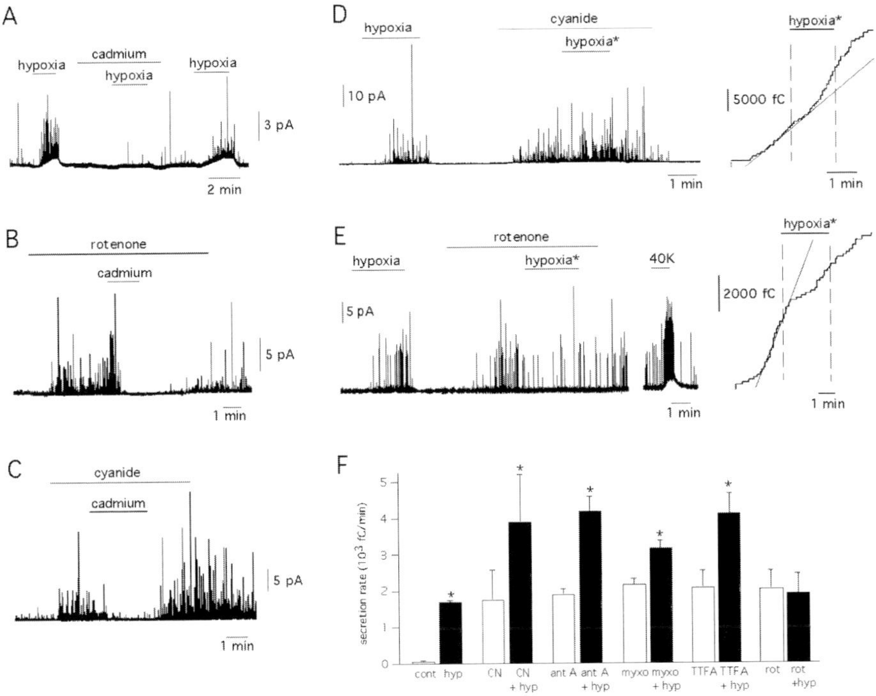

FIG. 1. Secretory responses of glomus cells to hypoxia and to the inhibition of the mitochondrial electron transport. (A–C) Amperometric signals showing catecholamine release from a glomus cell exposed to either hypoxia ($\approx$ 20 mmHg) or to electron transport inhibitors (5 μM rotenone and 100 μM cyanide). Each spike represents an exocytotic event. In all cases secretion is reversibly abolished by external application of 0.3 mM cadmium. (D, E) Secretory responses of glomus cells exposed concomitantly to hypoxia ($pO_2 \approx 20$ mmHg) and to cyanide (100 μM) or rotenone (5 μM). Left panels in D & E are the amperometric recordings, right panels are the cumulative secretion signals (in femtocoulombs) before, during, and after the exposure to hypoxia in the presence of the ETC inhibitors. The exposures to hypoxia are marked with an asterisk in the left panels. The straight lines represent the slopes (secretion rates) of the cumulative secretion signals immediately before the exposure to hypoxia. (F) Average secretion rate (ordinate in femtocoulombs/min, mean $\pm$ SE) measured in cells in various experimental conditions. From left to right: Control ($pO_2 \approx 150$ mmHg, $n = 17$ cells) and hypoxia ($pO_2 \approx 20$ mmHg, $n = 17$ cells). Cyanide and cyanide plus hypoxia (100 μM, $n = 4$ cells). Antimycin A (0.1 to 1 μg/ml, $n = 13$ cells) and antimycin A plus hypoxia ($n = 7$ cells). Myxothiazol (0.1 to 1 μg/ml, $n = 6$ cells) and myxothiazol plus hypoxia ($n = 6$ cells). TTFA (0.1 to 0.3 μM, $n = 5$ cells) and TTFA plus hypoxia ($n = 5$ cells). Rotenone (0.1 to 5 μM, $n = 14$ cells), rotenone plus hypoxia ($n = 12$ cells). Asterisks indicate statistically significant difference ($P < 0.05$) between each pair of samples.

0.1 and 5 μM and was mimicked by 1-methyl-4-phenylpyridinium ion (MPP+) (which binds to the same site as rotenone) but not by other drugs (as rhein or diophenyleneiodonium) acting on different sites in mitochondrial complex I (Ortega-Sáenz et al 2003). These data suggested that ETC inhibition and hypoxia might act through separate signalling pathways. However, rotenone, at concentrations that fully block complex I (Degli Esposti 1998, Vaux et al 2001) or saturate rotenone binding sites (Higgins & Greenamyre 1996) selectively occludes responsiveness to hypoxia in CB cells. Therefore, it seems that a rotenone (and MPP+)-inhibitable molecule participates critically in CB O_2 sensing.

To further investigate the effect of ETC inhibitors on glomus cells, we recorded ionic currents from patch clamped cells using the perforated patch-clamp technique. Although the action of the ETC inhibitors tested in this set of experiments was rather variable, their most common qualitative effects are summarized on Fig. 2. Rotenone, antimycin A and cyanide all produced a marked reduction in the amplitude of the K^+ currents that, particularly in the case of antimycin A, was frequently irreversible (Fig. 2A). Application of depolarizing ramps demonstrated that besides

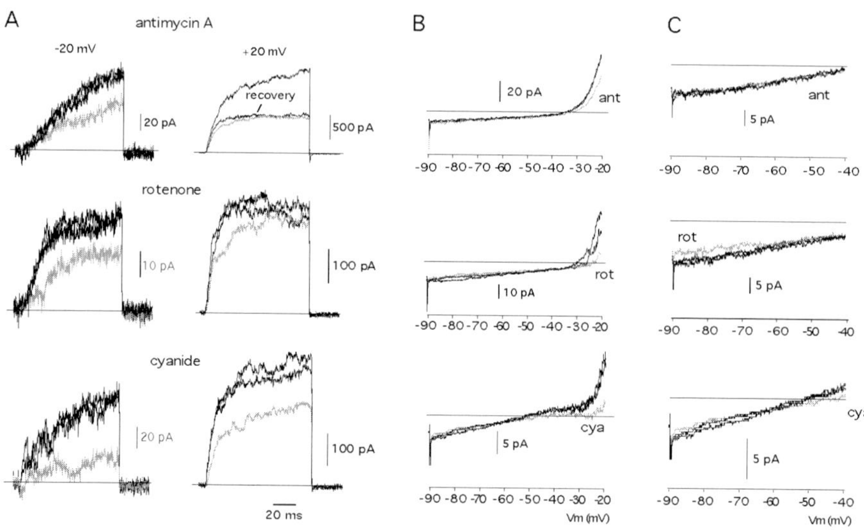

FIG. 2. Electrophysiological effects of mitochondrial inhibitors on dispersed rat glomus cells. (A) Macroscopic K^+ currents recorded from perforated patch clamped cells subjected to step depolarization to the indicated membrane potentials (holding potential, −80 mV). Control and recovery traces are in black. Traces recorded in the presence of antimycin A (1 μg/ml), rotenone (3 μM) or cyanide (100 μM) are in grey. (B, C) Currents recorded during the application of depolarizing voltage ramps in perforated patch clamped glomus cells. Experimental conditions as in A.

the effects on voltage-dependent outward current amplitude, rotenone and cyanide also reduced membrane resting conductance and displaced to more positive potentials the zero current level (Fig. 2B & C). These observations suggested that, as indicated by Wyatt & Buckler (2004), the two agents decrease background K^+ conductance. In contrast, antimycin A (1 μg/ml) had no affect on the resting conductance of glomus cells (Fig. 2C). Hence, antimycin A and cyanide, which have similar effects on secretion, have different actions on the resting K^+ conductance. Rotenone and cyanide, with almost identical electrophysiological effects influenced in different ways the O_2 sensitivity of the glomus cells. These data suggest that glomus cell secretion induced by the ETC inhibitors is probably the result of the pharmacological effect of these drugs on K^+ channels, unrelated to their effect on the mitochondrial electron transport. In fact, rotenone reversibly inhibits the macroscopic K^+ currents in cells devoid of mitochondria (Searle et al 2002).

Persistent carotid body activation with normal glomus cell responsiveness to hypoxia in partially Sdhd-deficient mice

To evaluate the participation of mitochondrial complex II in CB O_2 sensing, we have generated a mouse model lacking SDHD, a component of succinate dehydrogenase (Piruat et al 2004). SDHD deficiency is the most frequent manifestation of the hereditary paraganglioma (PGL, Baysal 2003), a catecholamine-secreting tumour in the neck affecting the carotid body (CB). PGL tumours display cellular hyperplasia/anaplasia similar to the CB of individuals exposed to chronic hypoxia (Lahiri et al 2000, Wang & Bisgard 2002), thus it has been proposed that the ultimate cause of PGL tumours is a defect in sensing environmental O_2 levels (Baysal et al 2000, Gimenez-Roqueplo et al 2001, Rustin et al 2002, Baysal 2003). We have tested in young adult mice whether partial *Sdhd* deficit and inhibition of mitochondrial complex II activity alter CB function by monitoring single-cell secretory activity in slices of the whole organ. Similar to rat CB slices (Pardal et al 2000, Ortega-Sáenz et al 2003), the response to hypoxia of glomus cells in mouse CB slices was characterized by a sharp and reversible burst of secretory events (Fig. 3A). This response was maintained, or even augmented, in *Sdhd*$^{+/-}$ mice (Fig. 3B–C), indicating that partial deficiency of complex II activity does not seem to alter glomus cell responsiveness to hypoxia. However, spontaneous CB activity in normoxic conditions was increased by ~2.4-fold (statistical significance, $P < 0.05$) in *Sdhd*$^{+/-}$ animals as compared with wild-type littermates (Fig. 3D). The higher resting excitability of CB cells in *Sdhd*$^{+/-}$ animals also explained the slight increase of the hypoxic response. The spontaneous secretory activity of *Sdhd*$^{+/-}$ glomus cells was reversibly abolished by blockade of Ca^{2+} channels with Cd^{2+} (Fig. 3E), thus suggesting that it was due to persistent extracellular Ca^{2+} influx through membrane channels (Pardal et al 2000, Ortega-Sáenz et al 2003).

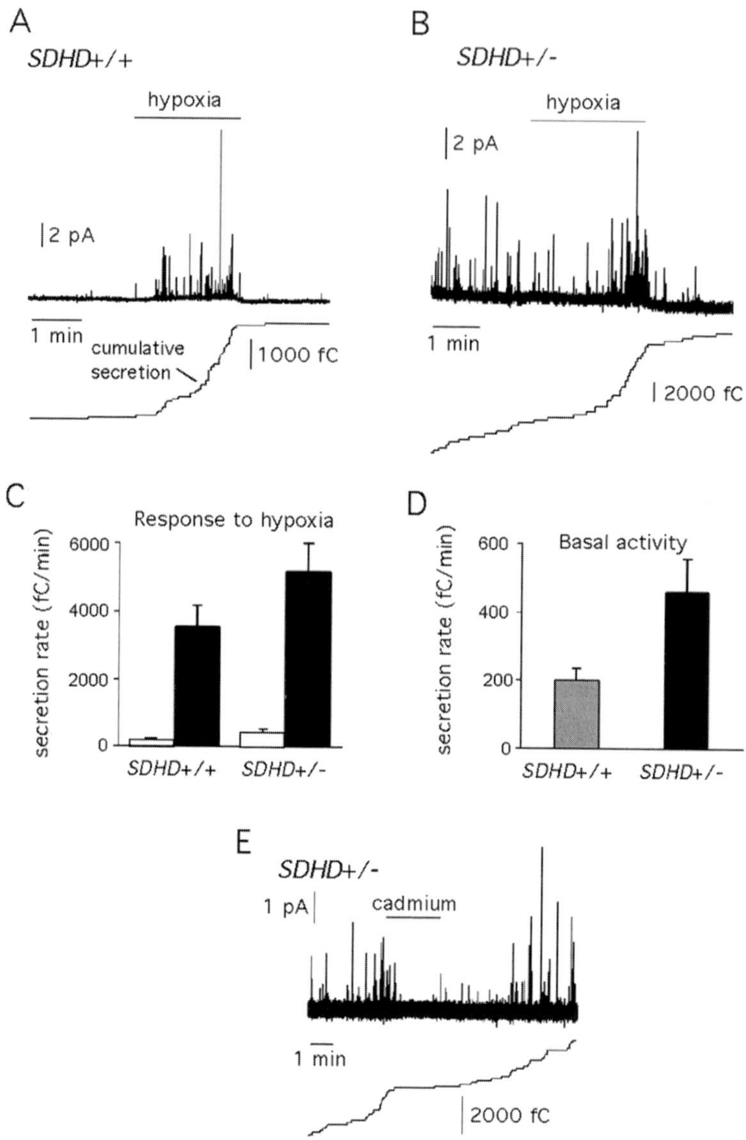

FIG. 3. Secretory activity and responsiveness to hypoxia of carotid body glomus cells. (A, B) Secretory response to hypoxia of glomus cells of *Sdhd*$^{+/+}$ and *Sdhd*$^{+/-}$ mice. Cumulative secretion (in fC) for each experiment are shown. (C) Quantification of the secretory response to low O_2 tension (in fC/last min hypoxia) of SDHD deficient (mean ± SE, $n = 16$) and wild-type ($n = 18$) mice. Anova test, $P = 0.12$. (D) Quantification of the spontaneous secretory activity in a carotid body slice from *Sdhd*$^{+/+}$ (mean ± SE, $n = 16$) and wild-type ($n = 18$) mice. Kruskal-Wallis test, $P = 0.01$. (E) Spontaneous secretory activity in a *Sdhd*$^{+/-}$ glomus cell and reversible blockade by application of 0.3 mM extracellular cadmium.

The constitutive functional activation of CB glomus cells observed in *Sdhd*-deficient mice occurred with only minor morphological changes in CB structure, without obvious organ enlargement or gross histological modifications. We have only observed a slight, although significant, increase in the percentage of glomus cells, identified by their immunoreactivity to tyrosine hydroxylase (TH), the rate-limiting enzyme for catecholamine synthesis (Piruat et al 2004). These subtle glomus cell hyperplasia probably precede CB tumour transformation. In humans, loss of heterozygosity (LOH) of the wild-type allele is required for PGL to occur (Baysal et al 2000, Baysal 2003). The absence of tumours in the *Sdhd*$^{+/-}$ mice suggests that the induction of LOH in CB glomus cells depends on different factors in humans and rodents.

Decrease of glomus cell K^+ current density in Sdhd$^{+/-}$ mice

We have studied the K^+ currents in voltage clamped glomus cells, as it is known that in this preparation K^+ channels regulate membrane potential and cellular excitability (see Pardal et al 2000 and references therein). Representative families of K^+ currents recorded from *Sdhd*$^{+/+}$ and *Sdhd*$^{+/-}$ glomus cells using the perforated patch clamp technique are illustrated in Fig. 4A, B. The average K^+ current density-voltage relationships are shown in Fig. 4C. Although other changes in the K^+ currents have been reported before (see Piruat et al 2004), the most obvious and consistent modification induced by partial deletion of the *Sdhd* gene was a reduction in total K^+ current density from $4.6 \pm 1.5\,pA/\mu m^2$, $n = 6$ in *Sdhd*$^{+/+}$ to $2.5 \pm 0.8\,pA/\mu m^2$, $n = 6$ for *Sdhd*$^{+/-}$, at +20 mV. It is known that Ca^{2+}-activated K^+ channels contribute to the macroscopic voltage-dependent K^+ current and the resting potential of rodent glomus cells (Wyatt & Peers 1995, Pardal et al 2000), and that in most tissues these channels act as counter-regulatory devices that prevent excessive cell depolarization. Therefore, the biophysical changes observed in patch clamped *Sdhd*$^{+/-}$ glomus cells could account for the persistent secretory activity detected in intact cells by amperometry.

Conclusions

Our data indicate that, as hypoxia, ETC inhibitors induce a secretory response in carotid body glomus cells, which is dependent on extracellular Ca^{2+}. They also suggest that sensitivity to hypoxia of carotid body glomus cells is not linked in a simple way to mitochondrial electron flow. In fact, all the ETC inhibitors tested have a non-specific pharmacological action as blockers of voltage-gated K^+ channels. However, a rotenone (and MPP^+)-sensitive molecule, located either inside or outside mitochondria, appears to participate in acute oxygen sensing in the carotid body. Partial genetic deficiency of mitochondrial complex II due to *Sdhd* deletion

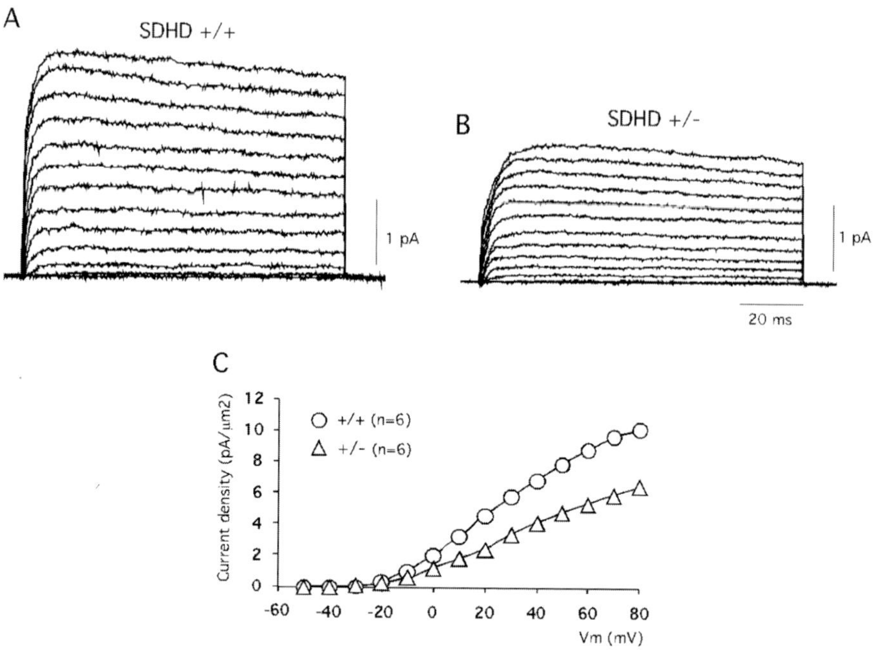

FIG. 4. Macroscopic K^+ currents in patch clamped dispersed glomus cells of wild-type and SDHD-partially deficient mice. (A, B) Families of representative outward K^+ currents recorded in $Sdhd^{+/+}$ and $Sdhd^{+/-}$ glomus cells during 100 ms depolarizing pulses reaching membrane potentials between −40 and +80 mV in steps of 10 mV. C. K^+ current density (ordinate) versus voltage (abscissa) relationship in wild-type (circles) and $Sdhd^{+/-}$ (triangles) glomus cells. Each point is the average of six different experiments.

produces persistent carotid body glomus cell activation and a decrease of K^+ channel current density. Sensitivity to hypoxia of these cells was, however, unaltered.

Acknowledgements

J. López-Barneo received the 'Ayuda a la Investigación 2000' of the Juan March Foundation. Research was also supported by The Lilly Foundation and by grants of the Andalusian Government, Spanish Ministries of Health and of Science and Technology.

References

Astrom K, Cohen JE, Willett-Brozick JE et al 2003 Altitude is a phenotypic modifier in hereditary paraganglioma type 1: evidence for an oxygen-sensing defect. Hum Genet 113:228–237

Baysal BE 2003 On the association of succinate dehydrogenase mutations with hereditary paraganglioma. Trends Endocrinol Metab 14:453–459

Baysal BE, Ferrell RE, Willett-Brozick JE et al 2000 Mutations in *SDHD*, a mitochondrial complex II gene in hereditary paraganglioma. Science 287:848–851

Biscoe TJ, Duchen MR 1990 Responses of type I cells dissociated from the rabbit carotid body to hypoxia. J Physiol 428:39–59

Buckler KJ 1997 A novel oxygen-sensitive potassium current in rat carotid body type I cells. J Physiol 498:649–662

Buckler KJ, Vaughan-Jones RD 1994 Effects of hypoxia on membrane potential and intracellular calcium in rat neonatal carotid body type I cells. J Physiol 476:423–428

Buckler KJ, Vaughan-Jones RD 1998 Effects of mitochondrial uncouplers on intracellular calcium, pH and membrane potential in rat carotid body type I cells. J Physiol 513:819–833

Carpenter E, Hatton CJ, Peers C 2000 Effects of hypoxia and dithionite on catecholamine release from isolated type I cells of the rat carotid body. J Physiol 523:719–729

Degli Esposti M 1998 Inhibitors of the NADH-ubiquinone reductase: an overview. Biochim Biophys Acta 1364:222–235

Delpiano MA, Hescheler J 1989 Evidence for a pO_2-sensitive K^+ channel in the type-I cell of the rabbit carotid body. FEBS Lett 249:195–198

Duchen MR, Biscoe TJ 1992 Relative mitochondrial membrane potential and $[Ca^{2+}]_i$ in type I cells isolated from the rabbit carotid body. J Physiol 450:33–61

Ganfornina MD, López-Barneo J 1991 Single K^+ channels in membrane patches of arterial chemoreceptor cells are modulated by O_2 tension. Proc Natl Acad Sci USA 88:2927–2930

Gimenez-Roqueplo AP, Favier J, Rustin P et al 2001 The R22X mutation of the *SDHD* gene in hereditary paraganglioma abolishes the enzymatic activity of complex II in the mitochondrial respiratory chain and activates the hypoxia pathway. Am J Hum Genet 69:1186–1197

Higgins DS, Greenamyre JT 1996 [^{3}H]dihydrorotenone binding to NADH: Ubiquinone reductase (complex I) of the electron transport chain: An autoradiographic study. J Neurosci 16:3807–3816

Lahiri S, Prabhakar NR, Forster RE 2000 Oxygen sensing. Molecule to man, Kluwer Academic/Plenum Publishers, New York, NY, USA

López-Barneo J, López-López JR, Ureña J et al 1988 Chemotransduction in the carotid body: K^+ current modulated by pO_2 in type I chemoreceptor cells. Science 241:580–582

López-Barneo J, Benot AR, Ureña J 1993 Oxygen sensing and the electrophysiology of arterial chemoreceptor cells. News Physiol Sci 8:191–195

López-Barneo J, Pardal R, Ortega-Sáenz P 2001 Cellular mechanism of oxygen sensing. Annu Rev Physiol 63:259–287

Mills E, Jöbsis FF 1972 Mitochondrial respiratory chain of carotid body and chemoreceptor response to changes in oxygen tension. J Neurophysiol 35:405–428

Montoro RJ, Ureña J, Fernández-Chacón R et al 1996 Oxygen sensing by ion channels and chemotransduction in single glomus cells. J Gen Physiol 107:133–143

Ortega-Sáenz P, Pardal R, García-Fernández M et al 2003 Rotenone selectively occludes sensitivity to hypoxia in rat carotid body glomus cells. J Physiol 548:789–800

Pardal R, López-Barneo J 2002 Low glucose-sensing cells in the carotid body. Nature Neurosci 5:197–198

Pardal R, Luwewig U, García-Hirschfeld J et al 2000 Secretory responses of intact glomus cells in thin slices of rat carotid body to hypoxia and tetraethylammonium. Proc Natl Acad Sci USA 97:2361–2366

Peers C 1990 Hypoxic suppression of K^+ currents in type I carotid body cells: selective effect on the Ca^{2+}-activated K^+ current. Neurosci Lett 119:253–256

Piruat JI, Pintado CO, Ortega-Saenz P et al 2004 The mitochondrial SDHD gene is required for early embryogenesis, and its partial deficiency results in persistent carotid body glomus cells activation with full responsiveness to hypoxia. Mol Cell Biol 24:10933–10940

Riesco-Fagundo AM, Pérez-García MT, González C et al 2001 O_2 modulates large conductance Ca^{2+}-dependent K^+ channels of rat chemoreceptor cells by a membrane-restricted and CO-sensitive mechanism. Circ Res 89:430–436

Rustin P, Munnich A, Rötig A 2002 Succinate dehydrogenase and human diseases: new insights into a well-known enzyme. Eur J Hum Genet 10:289–291

Searle GJ, Hartness ME, Hoareau R et al 2002 Lack of contribution of mitochondrial electron transport to acute O_2 sensing in model airway chemoreceptors. Biochem Biophys Res Comm 291:332–337

Stea A, Nurse CA 1991 Whole-cell and perforated-patch recordings from O_2-sensitive rat carotid body cells grown in short- and long-term culture. Pflügers Arch Eur J Physiol 418:93–101

Ureña J, Fernández-Chacón R, Benot AR et al 1994 Hypoxia induces voltage-dependent Ca^{2+} entry and quantal dopamine secretion in carotid body glomus cells. Proc Natl Acad Sci USA 91:10208–10211

Vaux EC, Metzen E, Yeates KM et al 2001 Regulation of hypoxia inducible factor is preserved in the absence of a functioning mitochondrial respiratory chain. Blood 98:296–302

Wang ZY, Bisgard GE 2002 Chronic hypoxia-induced morphological and neurochemical changes in the carotid body. Microsc Res Tech 59:168–177

Williams SE, Wootton P, Mason HS et al 2004 Hemoxygenase-2 is an oxygen sensor for a calcium sensitive potassium channel. Science 306:2093–2097

Wyatt CN, Peers C 1995 Ca^{2+}-activated K^+ channels in isolated type I cells of the neonatal rat carotid body. J Physiol 483:559–565

Wyatt CN, Buckler KJ 2004 The effect of mitochondrial inhibitors on membrane currents in isolated neonatal rat carotid body type I cells. J Physiol 556:175–191

DISCUSSION

Weir: What temperature were your experiments done at, and does it matter whether you do them at room temperature?

López-Barneo: All the experiments on carotid body slices are done at between 34 and 36 °C. The experiments in dispersed cells were done at room temperature. At room temperature the slices also respond to hypoxia, but the effect is slower and less potent.

Weir: Was the membrane potential depolarised with any of the mitochondrial inhibitors?

López-Barneo: I can't remember. But I suspect that those decreasing K^+ conductance would depolarise the membranes. I am not sure about antimycin A. We haven't measured directly the membrane potential in response to metabolic activity.

Weir: Is the additional rotenone effect in ablating the hypoxic effect related to its ability to cause depolarization?

López-Barneo: I can't answer that, since we have not performed this type of experiment.

Buckler: Your ion channel experiments were all in perforated patch. So how do you know that the effects of rotenone or antimycin on your channels are non-selective and non-specific, as opposed to being mediated via an effect upon on the mitochondria?

López-Barneo: We did whole-cell recording in perforated patches and measured the K^+ currents. Antimycin A, rotenone and cyanide all produce, in addition, to mitochondrial inhibition blockade of K^+ channels. We don't see any major difference between them. However, when we do the experiment using the same concentration in the slices we see a differential effect with each. All three activate the carotid body, but only rotenone will block the sensitivity to hypoxia. The way we interpret this is that what we are seeing in the whole cell recording is a pharmacological action. I am concerned about the use of these drugs without testing what else they do. When you say a molecule is a mitochondrial inhibitor, it can do many other things. One thing that interests us is the differential effect of antimycin A, rotenone and cyanide.

Buckler: Did you do a dose–response for antimycin A or cyanide with respect to the hypoxia effect? With rotenone, whether you got additivity (with hypoxia) or not depended on the concentration of rotenone you used. Did you do the same for cyanide or antimycin A? If you push the concentration of antimycin or cyanide up higher can you saturate the response?

López-Barneo: We haven't done this. In our hands, if we go to high concentrations of cyanide we see a response due to release of calcium from internal stores and therefore you can't do the experiment. With antimycin A, the concentration we are using is already non-reversible.

Buckler: Just because you get Ca^{2+} release from internal stores doesn't mean that you can't do the experiment. You are looking for additivity between hypoxia and metabolic inhibitors.

López-Barneo: In our hands, if we go to higher concentrations of cyanide we get Ca^{2+} release from the stores, massive secretion and irreversible damage to the cells.

Buckler: If you say that a higher concentration of antimycin is toxic, then this suggests to me that the concentration of antimycin you are using routinely is not toxic because it is probably insufficient to fully inhibit mitochondrial function. So how then can you take additivity between a sub maximal level of antimycin and hypoxia as evidence that hypoxia acts via a pathway independent of the mitochondrion?

López-Barneo: The assumption I made is that the drugs are used at saturating concentrations for electron chain inhibition. The doses we are using are at least five times higher than others have reported will inhibit the electron transport chain.

Murphy: I know nothing about carotid bodies, but the parallel between glucose-stimulated insulin release in pancreatic β cells and the response of the carotid body to hypoxia must be one that is familiar to you. In β cells mitochondrial function affects the ATP/ADP ratio and thereby alters the K^+ channel and consequently activates the Ca^{2+} channel. In the carotid body an effect of oxygen on the

ATP/ADP ratio could also act on the K^+ channel in a similar way. The critical thing is that the ATP/ADP ratio is the link, a mitochondrial way of signalling through a K^+ channel and then to a Ca^{2+} channel. If you could measure the ATP/ADP ratio in these cells during hypoxia that might be quite informative. This is easy to do even in very small numbers of cells because the assay is very sensitive. With respect to mitochondrial respiratory chain inhibitors, there is considerable evidence that many of these have secondary effects, in addition to inhibition of respiration: usually at concentrations considerably above the effect of the inhibitors on respiration. It seems important to me to first show that you have added just enough inhibitor to completely inhibit respiration and then determine whether the inhibitor is having an effect on the endpoint of interest. Then you can increase the concentration of the inhibitor to see if the inhibitor has secondary effects in addition to inhibition of respiration. For example, at low concentrations rotenone is known to inhibit complex I, but at higher concentrations it is thought to have a secondary effect and increase ROS production. Another example is antimycin which inhibits the cytochrome/Bcl1 complex, but at higher concentrations it also affects Bcl2 family proteins. To summarize, it seems critical to me to first be certain that the effects of the inhibitors are due solely to the inhibition of respiration.

López-Barneo: It is difficult to do this.

Murphy: I accept that it is technically tricky to measure respiration in small amounts of tissue, but you could achieve the same end by measuring a surrogate of respiration such as the ATP/ADP ratio, or cytochrome reduction levels, which are far easier to measure in small numbers of cells. Similarly, you also assess the membrane potential using a fluorescent probe such as JC1.

Buckler: Determining whether electron transport is fully blocked or not is actually quite easy in isolated cells. You use a fluorescent indicator of mitochondrial potential, fully depolarize the mitochondria with a suitable uncoupler and then see whether they can repolarize when you withdraw the uncoupler in the presence of whichever inhibitor of electron transport you are interested in (see e.g. Wyatt & Buckler 2004).

López-Barneo: I don't think you can do a good measurement. We do have a drug that in addition to inhibiting mitochondrial electron transport will produce a blockade of the K^+ channel. I don't think you can measure mitochondrial membrane potential in glomus cells.

Rich: I have a comment concerning the protocol for using these mitochondrial inhibitors. Do you have a continuous flow of a buffer with the inhibitor in it, in your experiments?

López-Barneo: Yes.

Rich: In this case it raises a difficult technical question that you have to address. The majority of these inhibitors, such as rotenone and antimycin A, have partition coefficients between the aqueous solution and the targeted membranes of 10^8 or more for partitioning into the membrane compared with the liquid. Hence, if you are doing this addition in a cuvette and add 1 μM rotenone, essentially all of the inhibitor will go into the membrane and almost none will be left free in solution. However, if you have a continuous flow, and you are continuously replenishing the rotenone, the inhibitor will continue to build up in the membranes until its concentration becomes astronomical. In fact, you can think of it as replacing a lipid membrane by a rotenone membrane! Secondary effects are therefore inevitable, which are not of any physiological relevance.

López-Barneo: That is my point. We are getting a high concentration of the drug. This is the basis of all our assumptions. But even in this situation there can be further sensitivity to hypoxia in the case of myxothiazol, antimycin A and cyanide but not in the case of rotenone.

Rich: Getting back to the technical aspects, you can look in the literature and see that rotenone has a binding constant of something like 10^{-9} M for the complex I site. You might assume that you could add just 10 times that concentration to the cells and get 90% inhibition, which is true, but only if the inhibitor has unhindered access it its binding sites. But in practice, if you only add 10 times the K_m to cells, the rotenone firstly encounters the cell membrane and because the partition coefficient is 10^8 or more towards the membrane, the rate constant of entry into the membrane is very fast but the rate constant out towards the mitochondrial sites can be hours. It enters the cell membrane, and you don't get any inhibition of mitochondrial respiration because of difficulty in equilibrating to the mitochondria. Hence, when you add 10 nM rotenone to cells there is no effect on mitochondrial respiration even though this concentration is above the binding constant. It is not because there is a different affinity; it is because of the difficulty of delivering the drug to the site through the other membranes where it resides for a long time. All this needs to be taken into account when we are looking at these chemicals with a complex system such as a whole cell. There are great technical difficulties in titrating because any inhibition will be extremely time dependent.

López-Barneo: Of course, we care about these technical issues, but in fact the complications are working in our favour in the experiments. For example, a relatively large concentration of antimycin A put in the bath increases a lot of secretory activity because it is probably doing bad things to the cell, but it still isn't reducing sensitivity to hypoxia, whereas rotenone is having an effect.

Rich: But these are all artefactual effects from flooding the membranes with inhibitor.

López-Barneo: The sensitivity to hypoxia is a very sudden phenomenon in the cell. It is irreversible with rotenone. With the other drugs hypoxia will increase secretory activity, and when this is removed the effect of hypoxia disappears even when the drug remains. It could be that rotenone is not binding anywhere, of course. Hypoxia could be having some non-specific effect on the cell. We saw this a couple of years ago and since then we have been trying hard to get rotenone-binding sites outside mitochondria. It could be that this is a non-specific effect of hypoxia, though.

Rich: There are hundreds, perhaps thousands, of inhibitors that hit the rotenone site. It might be worth trying some of these.

López-Barneo: Other blockers of mitochondrial complex I didn't do this, but MTPT, which binds to the same site as rotenone on mitochondria, did.

Rich: Those two you mention are complicated ones that are non-specific, but you might consider some of the ultraspecific tight binding inhibitors. Some of them have unmeasurably tight binding constants.

Gurney: I have a question concerning the link between the K^+ channel, the Ca^{2+} channel and secretion. Do you just see a lack of reproducibility in the effect of hypoxia on K^+ current in the isolated cells, or is it also variable in the slices?

López-Barneo: More in the isolated cells, but it is still there in the slices.

Gurney: That makes me wonder whether K^+ channels actually mediate the secretory response, which seems robust. I wonder if the cadmium effect is due to blockage of non-selective cation channels rather than voltage-dependent Ca^{2+} channels. The ATP effect on the cells could be relevant here: although cytoplasmic ATP seemed to have a big effect on the Ca^{2+} current, you indicated that it doesn't seem to affect the secretory response. This would also tend to dissociate the Ca^{2+} channels from the secretory response. Could it be that voltage-gated Ca^{2+} channels are not necessarily a mediator?

Lopez-Barneo: We didn't measure the secretory response related to ATP. We measured the response to hypoxia in cells dialysed with and without ATP. We looked at the Ca^{2+} concentration.

Gurney: This wasn't affected, which makes me think that the rise in Ca^{2+} concentration is not due to the L-type Ca^{2+} channels. Could it be that non-selective cation channels are responsible?

López-Barneo: We have looked at the effect of hypoxia on cation channels. It doesn't affect them. People working on rat carotid body cells have shown that the macroscopic K^+ channels are inhibited by hypoxia or the resting potential background K^+ channels are inhibited by hypoxia, but no one has reported an increase in the cation conductance in response to hypoxia. We are now seeing that hypoglycaemia is producing an increase in this cation conductance.

Gurney: The effects on K^+ currents are really quite small. Is this enough to cause depolarization? They are small at quite positive potentials, so at the resting potential what would happen?

López-Barneo: This is an ongoing question. The resting potential in the perforated patches is 40–50 mV. When we look carefully at the ramps to see where the voltage-gated K channels are activated, we can see that some channels are activated around −40 mV. If we use TEA in these preparation this will produce depolarisation and increase secretory activity.

Gurney: Are the ramps you are looking at purely activating K^+ current?

López-Barneo: It could also be another cation current. The nature of the channel that is responsible for the receptor potential for the depolarization in carotid body cells is a question that has been debated over the last decade. I haven't been working on the rabbit carotid body for a decade now, but I find it very different from the rat carotid body. The rabbit carotid body contains a lot of Na^+ channels and carotid body cells *in situ* and at the oxygen concentration in the arterial blood probably fire action potentials repetitively. Voltage-dependent K^+ channels are probably more important in the rabbit than in the rat where resting potential K^+ channels are needed to initiate a depolarizing receptor potential.

Evans: I was wondering about the range of experiments you were presenting. If we disregard non-selective actions of the various mitochondrial inhibitors that you have used, you could have a varying degree of inhibition of mitochondrial function. This could underpin the variable effects on carotid body glomus cells of different mitochondrial inhibitors and the variability seen with respect to their ability to attenuate or not the response to hypoxia. Where the effect of glucose is concerned, the other issue that I'd like to raise is that rotenone may well block the response to hypoxia by cutting out the mitochondria. However, this doesn't necessarily mean that the signal transduction pathway utilized by glucose removal is different from that activated by hypoxia. By blocking the mitochondria you will increase a cell's reliance on glycolysis for ATP production. Any effect on cells of altering the glucose concentration could be mediated by the modulation of a metabolic signalling pathway sensitive, in some way, to changes in ATP supply. Such a pathway could be regulated by changes in the metabolic status of a cell by changes in glycolytic and/or mitochondrial ATP supply.

López-Barneo: We know that low glucose in carotid body cells produces an increase in secretory activity. This is dependent on the rise of internal Ca^{2+} concentration due to influx from the outside. In our preliminary voltage clamp experiment we showed that low glucose also inhibited the macroscopic K^+ current. We have found that in addition to this the more reproducible physiological effect of low glucose is a depolarizing receptor potential that is due to activation of resting non-selective cation conductance. This can be seen in the presence of TEA and iberiotoxin. This conductance is not affected by low pO2. It seems that a major mechanism for signalling low glucose is this cationic conductance. We know that this isn't related to glucose itself, because when glucose is replaced by O-methyl glucose (which is imported but which isn't phosphorylated or metabolized), there is no effect. We don't know what the signal generated as a consequence of glucose metabolism is that is modulating the non-selective conductance in glomus cells. We have data showing that the non-selective cation conductance is activated by low glucose and produces a depolarization of the carotid body cell.

Gonzalez: Do you think that the difference between rotenone and the other mitochondrial inhibitors could be because rotenone inhibits the production of reactive oxygen species?

López-Barneo: We haven't looked at this. Although the experiments in which we load the cells with new internal solutions are very complicated, we could investigate the point you are raising by loading the cells with a large amount of DTT or another reducing agent. Of course, it could be that reactive oxygen species produced in mitochondria are signalling the membrane. However, it is interesting that rotenone is the only mitochondrial complex I inhibitor that in our preparation can occlude sensitivity to hypoxia. We cannot discard that somehow the mitochondrial complex I acts as an oxygen sensor. I would expect that both antimycin A and cyanide would increase the radical production at the rotenone site. However they behave differently with respect to the background K^+ channels (cyanide inhibits them and antimycin A appears to have no effect).

Gonzalez: I was disappointed to see that the carotid bodies on your knockout mice were not hypertrophic. This is surprising in view of your findings that the carotid bodies are permanently activated and that their chemoreceptor cells have permanently high levels of intracellular Ca^{2+}. It is constant activation of chemoreceptors cells that leads to hypertrophy of the carotid bodies in permanent residents at high altitude.

López-Barneo: To have a tumour you probably need to lose the normal allele. We are changing the genetic background of this animal model to see whether in a different background we can get carotid body tumours.

Archer: This is an elegant mouse model, but the problem with the BL6 background is that it doesn't undergo hypertrophy very well. This is a recurring problem with BL6 backgrounds: they don't have a robust remodelling response to vascular injury. If you are failing to see hypertrophy when you are expecting it this might well be the mouse strain-related phenomenon. Although Paul and I don't agree on the effects of hypoxia on pulmonary vascular radical production, we do agree that mitochondria are important as sensors. What is surprising is that when you gave a variety of mitochondrial inhibitors to the carotid body, they caused carotid body discharge, thereby mimicking what hypoxia does, and one of the inhibitors actually ablated subsequent hypoxic responses. Wouldn't the simplest interpretation of these observations be that mitochondria are important in the carotid body's hypoxic response? You attribute the effects to non-specific effects. When you get all those mitochondrial inhibitors and they all cause discharge in the carotid body, wouldn't that point at the mitochondria? I am surprised that your interpretation was that it was not mitochondria.

López-Barneo: I don't mind interpreting the data a different way, the problem is that I don't know how to do it. The way I interpret these data is that it is unlikely that acute oxygen sensing in the carotid body is associated with mitochondria.

Duchen: With rotenone, one of the things that surprises me is that it doesn't ablate the response to hypoxia, it stops it. Your slope was cumulative and secretion was repressed by rotenone. I don't understand how this works. If rotenone was doing what it does, you'd expect it to continue irrespective of whether you apply hypoxia or not. Why do you get a decrease in secretion? How do you interpret this?

López-Barneo: On average, hypoxia wouldn't decrease secretory activity in rotenone. This occasionally happened but not always. When we used large concentrations of rotenone (say above 1 μM) the effect of rotenone was rather irreversible and after removal of the drug the cells were still insensitive to hypoxia. In these cases hypoxia did not increase secretory activity of the rotenone-treated cells but it did not produce inhibition of secretion either.

Duchen: We did the same experiment many years ago, measuring [Ca^{2+}] as an endpoint, and found that all the mitochondrial inhibitors blocked the [Ca^{2+}] response to hypoxia.

Chandel: Does FCCP do the same thing?

Duchen: Yes.

Chandel: So why not add FCCP first and then use the mitochondrial inhibitors? This would get rid of the effect the mitochondrial membrane potential might have on Ca^{2+}.

López-Barneo: The problem with glomus cells is that this would probably kill them.

Gonzalez: If you use uncouplers you would have maximal activation of the cells.

Chandel: But you would still see hypoxic activation on top of this.

Gonzalez: I doubt it, particularly if you use the high concentration of uncouplers needed to fully abolish the mitochondrial potential.

Buckler: There have been a number of studies looking at chemoreceptor discharge with antimycin and oligomycin. More recently there has been a study looking at FCCP (Mulligan et al 1981, Mulligan & Lahiri 1982, Mosqueria & Iturriaga 2002). They all seem to find occlusion, i.e. there is little or no effect of hypoxia in the presence of mitochondrial inhibitors.

Duchen: The main difference is the endpoint that is measured.

Schumacker: Your rationale for the SDHD knockout was very clear, but to be fair, the mechanisms that cause tumour formation in cells and those that regulate the oxygen sensing capacity might be very different. It would seem that the appropriate controls would be to look at the homozygous SDHA knockout. You ought to see the same phenotype with the SDHA as with the SDHD if succinate was the mechanism, or if the mechanism was related to a general slow down of the Krebs cycle.

López-Barneo: Has the SDHA knockout been done?

Schumacker: No, but there are ES cells.

Gonzalez: A quick comment. I think that most familial chemodectomas are produced by fibroblast growth in the carotid bifurcation. Very few chemodectomas in humans are functional, that is, very few chemodectomas are due to proliferation of chemoreceptor cells.

López-Barneo: This might be true. Nevertheless, it is also well known that paragangliomas contain a large number of catecholamine-secreting glomus cells. In fact, they can produce systemic hypertension.

References

Mosqueria M, Iturriaga R 2002 Carotid body chemosensory excitation induced by nitric oxide: involvement of oxidative metabolism. Respir Physiolo Neurobiol 131:175–187

Mulligan E, Lahiri S 1982 Separation of carotid body chemoreceptor responses to O_2 and CO_2 by oligomycin and by antimycin A. Am J Physiol 242:C200–206

Mulligan E, Lahiri S, Storey BT 1981 Carotid body O_2 chemoreception and mitochondrial oxidative phosphorylation. J Appl Physiol 51:438–446

Wyatt CN, Buckler KJ 2004 The effect of mitochondrial inhibitors on membrane currents in isolated neonatal rat carotid body type I cells. J Physiol 556:175–191

The role of TASK-like K^+ channels in oxygen sensing in the carotid body

Keith J. Buckler, Beatrice A. Williams*, Rodrigo Varas Orozco and Christopher N. Wyatt†

*University Laboratory of Physiology. Parks Road, Oxford, OX1 3PT, UK, *The University of Western Ontario, London, Ontario, N6A 5B9, Canada, and †Division of Biomedical sciences, School of biology, St Andrews, Fife KY16 9TS, UK*

Abstract. The carotid body plays an important role in initiating protective responses to hypoxemia. The primary oxygen sensing cells are the glomus or type 1 cells. Hypoxia evokes the secretion of neurotransmitters from these cells which then excite afferent nerves. This response is mediated via membrane depolarization and voltage-gated Ca^{2+} entry. Studies from this laboratory have revealed that membrane depolarization in response to hypoxia is primarily the result of inhibition of background K^+ channels which show strong similarities to the acid sensitive tandem-P-domain K^+ channels TASK-1 and TASK-3. The background K^+ channels of type-1 cells are also very sensitive to inhibition of mitochondrial energy metabolism and, in excised patches, appear to be directly activated by ATP. Thus these TASK-like background channels would appear to confer the ability to sense changes in oxygen levels, pH and metabolism upon the type 1 cell. The key issue of whether the effects of hypoxia are mediated through changes in metabolism remains unanswered but the effects of inhibition of mitochondrial energy metabolism and of hypoxia upon background K^+ channels is mutually exclusive suggesting that there is a close link between metabolism and oxygen sensing in the type 1 cell.

2005 Signalling pathways in acute oxygen sensing. Wiley, Chichester (Novartis Foundation Symposium 272) p 73–94

Peripheral chemoreceptors play a vital role in initiating protective responses to hypoxemia including an increase in ventilation and modulation of regional blood flow, cardiac output and catecholamine secretion. The principle arterial chemoreceptor is the carotid body. This organ is comprised of primary receptive cells, type 1 cells, which synapse with afferent neurons that project to the brainstem. Type 1 cells respond to hypoxia with neurosecretion that leads to excitation of afferent nerves via purinergic receptors (Zhang et al 2000). The neurosecretory response of the type 1 cell is promoted by voltage-gated Ca^{2+} entry in response to hypoxia-induced depolarization (Buckler & Vaughan Jones 1994a, Montoro et al 1996). This paper documents our attempts to elucidate the causes of this hypoxia induced membrane depolarization, to identify the ionic channels responsible and to identify potential oxygen sensing mechanisms.

Methods

We have utilized an enzymatically isolated neonatal rat type 1 cell preparation as this permits the use of fluorescent probes to investigate Ca^{2+} signalling and voltage clamp to investigate electrical signalling. The majority of our studies have been conducted using bicarbonate buffers (Buckler et al 1991) at 35–36° C (the carotid body is notably temperature sensitive, Gallego et al 1979). For whole cell electrophysiology we use the perforated patch technique in order not to interfere with normal cellular biochemistry. Type 1 cells are very small, typically 10 μm diameter, such that conventional whole cell recording will lead to very rapid intracellular dialysis and loss of many cellular constituents. Electrophysiological studies at the single channel level were conducted using conventional tight seal cell attached or inside out patch clamp techniques. (Further details on methodology can be found in Dasso et al 1997, Buckler 1997, Williams & Buckler 2004.)

Results and discussion

Electrical signalling and Ca^{2+} influx

Studies conducted upon freshly isolated type 1 cells have shown that they respond to a variety of chemoreceptor stimuli including hypoxia, acidosis and metabolic poisons with rapid membrane depolarization and voltage-gated Ca^{2+} entry (Buckler & Vaughan Jones 1994a,b, 1998, Wyatt & Buckler 2004). This response is believed to be central to the chemotransduction process in the intact carotid body. In order to elucidate the cause of this stimulus evoked depolarization we used the whole cell voltage-clamp technique to apply voltage ramps across a limited range of membrane potentials spanning the normal resting potential of type 1 cells. Current vs. voltage (I/V) plots constructed from these experiments revealed a marked reduction in membrane conductance (slope of the I/V plot) and a consequential inward, depolarizing, shift in membrane currents at the resting membrane potential in response to lowering oxygen levels (Fig. 1B). This effect was graded with the decline in pO2 with a $K_{1/2}$ of 12–13 Torr (Fig. 1D). By subtracting the membrane current recorded under low (hypoxia) or zero (anoxia) oxygen conditions from that recorded under normoxic conditions we constructed an I/V plot for the oxygen sensitive currents that contribute to the control of resting membrane potential. This O_2-sensitive current is shown in Fig. 1C (line labelled 4.5 mM $[K^+]_o$). Note that this current reverses at about −90 mV (i.e. at potentials positive to −90 mV oxygen sensitive current is positive and at potentials negative to −90 mV current is negative). The reversal potential of an ionic current is determined by the valency and the transmembrane concentration gradient of the ions carrying that current. In a normal physiological buffer, K^+ current would be expected to have a reversal potential close to −90 mV. Confirmation that the oxygen sensitive current is indeed carried primarily by K^+ ions was obtained by raising the external potassium ion concentration (to 20 mM, Fig. 1D) and observing that the reversal potential for the oxygen sensitive current shifted to a less negative potential in close agreement with that predicted by the Nernst equation (Buckler 1997). Since the O_2-sensitive currents presented in Fig. 1C were derived by subtracting currents measured under anoxic conditions from those recorded under control conditions they reflect the I/V relationship of a K^+ current that is *inhibited* by hypoxia. It is therefore primarily the inhibition of a background K^+ current that initiates membrane depolarization.

Properties of background K^+ channels

The O_2-sensitive background K^+ currents in type 1 cells are resistant to the classical K^+ channel inhibitors tetraethylammonium (TEA) and 4-amino pyridine (4-AP) (Buckler 1997). This has facilitated the study of background K^+-channels in isolation from many other voltage-gated and Ca^{2+}-activated K^+ channels. Thus the oxygen sensitive background K^+ current has been shown to exhibit only weak outward rectification in the presence of normal levels of extracellular K^+ and a linear dependence of membrane current upon voltage in symmetrical $[K^+]$ (Fig. 1E,F).

Using cell-attached patch clamp recording we have identified K^+ channels open at negative membrane potentials whose activity is inhibited by hypoxia (Fig. 2A–C). With K^+ as the main cation in the pipette these ion channels exhibited a relatively low conductance (14–16 pS; Fig 2E), were highly K^+ selective, insensitive to TEA and 4-AP and active over a wide range of membrane potentials (Fig. 1D). These channels are therefore most likely to be responsible for generating the oxygen-sensitive background K^+ current described above. We have found these channels to be present in the majority of cell attached patches (indeed many patches contain more than one channel) and therefore presume that they must be of relatively high abundance in the cell membrane.

In addition to being O_2 sensitive, background K^+ currents are also notably pH sensitive (Fig. 3A) and therefore probably play an important role in acid sensing as well as O_2 sensing (Buckler & Vaughan Jones 1994b). Oxygen-sensitive background K^+-current is also inhibited by the local anaesthetic bupivicaine, by barium and by quinidine (Buckler et al 2000) and is strongly stimulated by the gaseous general anaesthetic halothane (Fig. 3B). The biophysical and pharmacological properties of these background K^+ channels/currents are therefore remarkably similar to those

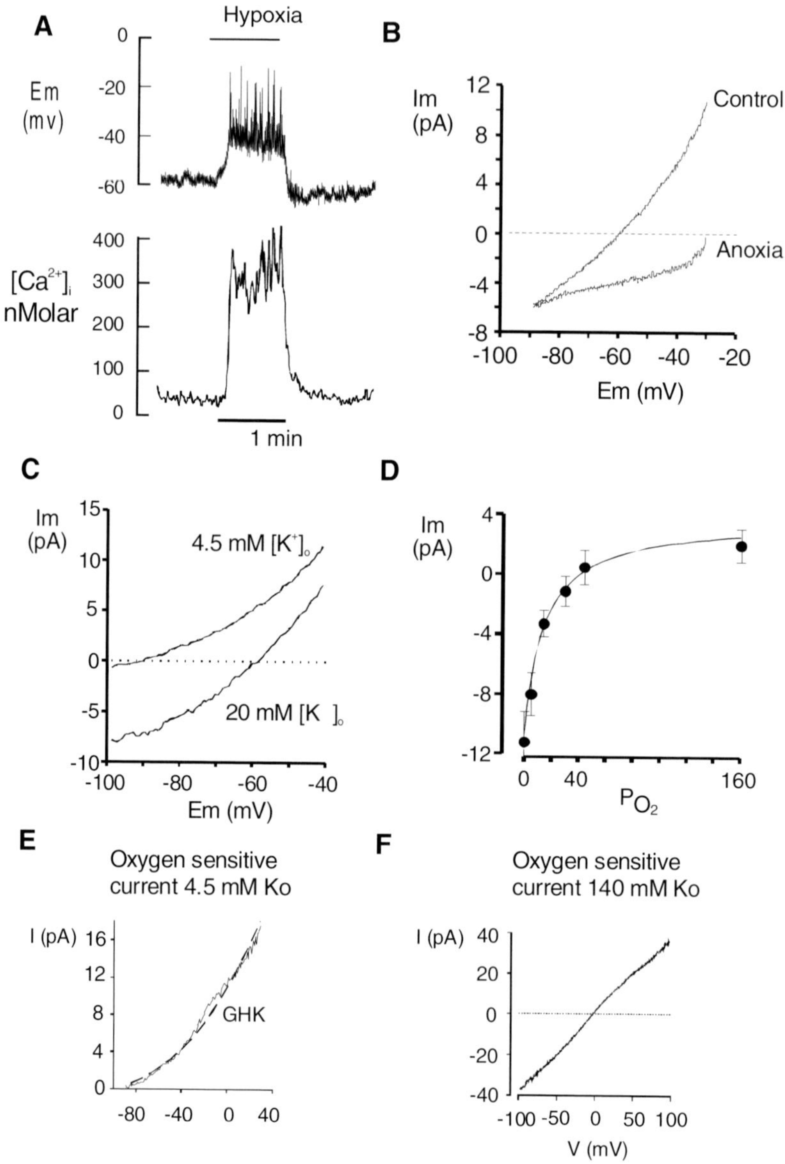
A
Hypoxia
Em (mv)
0
-20
-40
-60
[Ca2+]i nMolar
400
300
200
100
0
1 min
B
Im (pA)
12
8
4
0
-4
-8
Control
Anoxia
-100 -80 -60 -40 -20
Em (mV)
C
Im (pA)
15
10
5
0
-5
-10
4.5 mM [K+]o
20 mM [K]o
-100 -80 -60 -40
Em (mV)
D
Im (pA)
4
0
-4
-8
-12
0 40 160
PO2
E
Oxygen sensitive current 4.5 mM Ko
I (pA)
16
12
8
4
0
GHK
-80 -40 0 40
F
Oxygen sensitive current 140 mM Ko
I (pA)
40
20
0
-20
-40
-100 -50 0 50 100
V (mV)

of TASK channels (members of the tandem P-domain K^+ channel family; Reyes et al 1998, Kim et al 1999, 2000, Rajan et al 2000). We have also recently observed that zinc ions inhibit the oxygen-sensitive background K^+ current and that Mg^{2+} significantly reduces single channel conductance (Fig. 3C&D; Williams & Buckler 2004). Mg^{2+} and Zn^{2+} have been reported to have similar effects upon TASK-3 but not on TASK-1 (Rajan et al 2000, Lopes et al 2000, Clarke et al 2004). Thus, pharmacologically the oxygen-sensitive background channels of type 1 cells closely resemble TASK-3; the single channel conductance of this native channel is however closer to that of TASK-1 than TASK-3.

In summary the molecular identity of the oxygen-sensitive background K^+ channel has not yet been fully resolved, it is clearly a TASK-like channel but its pharmacological and biophysical properties do not exactly correspond to either TASK-1 or TASK-3 channels. There is however some evidence that TASK-1 and TASK-3 channels may form heteromultimers (Kang et al 2004, Czirjak & Enyedi 2001) and they may also associate with other proteins/subunits, all of which could subtly alter their characteristics.

Modulation of background K^+ channels: role of metabolism

Almost all inhibitors of oxidative phosphorylation, including uncouplers, numerous electron transport inhibitors (including cyanide, carbon monoxide,

FIG. 1. Background K^+ channels and oxygen sensing in type 1 cells. (A) Simultaneous recording of membrane potential and intracellular calcium in an isolated rat type 1 cell exposed to a hypoxic solution (pO2 approx 5 Torr). (B) Whole-cell current–voltage (I/V) relationship for a type 1 cell determined using the voltage-clamp technique to apply voltage ramps from −90 to −30 mV. Note that the control I/V relationship transects the zero current axis at about −60 mV: this corresponds to the cells normal resting membrane potential. Under anoxic conditions (N_2 equilibrated +0.5 mM $Na_2S_2O_4$, pO_2 = 0 Torr) there is a substantial decrease in the cell's resting membrane conductance (decrease in the slope of the I/V relationship) and the anoxic I/V does not transect the zero current axis, i.e. there is no stable resting membrane potential in this voltage range (this cell would therefore have depolarized in response to anoxia). (C) Current–voltage relationship of oxygen-sensitive currents in type 1 cells. Oxygen-sensitive current is determined using voltage ramps (as above) under both normoxic and anoxic or hypoxic conditions. The whole-cell I/V obtained under anoxic or hypoxic conditions is then subtracted from that obtained under control (normoxic) conditions to reveal the oxygen-sensitive component. This experiment was performed under conditions of both normal (4.5 mM) and elevated (20 mM) extracellular K^+ using *anoxia*. (D) Oxygen sensitivity of resting membrane currents. Membrane current was measured at −50 mV as pO2 was lowered from 160 to 0 Torr. Data were fitted by a rectangular hyperbola with a $K_{1/2}$ of 13 Torr. (E) Whole-cell I/V relation for oxygen-sensitive current (control, *hypoxia*; pO2 approx 5 Torr) recorded at normal extracellular K^+ in the presence of 5 mM 4-AP and 10 mM TEA. Broken line, Goldman Hodgkin Katz constant field equation fit to data. (F) Whole-cell I/V relation for oxygen-sensitive current (control, *hypoxia*; pO2 approx 5 Torr) recorded at 140 mM extracellular K^+ in the presence of 5 mM 4-AP and 10 mM TEA.

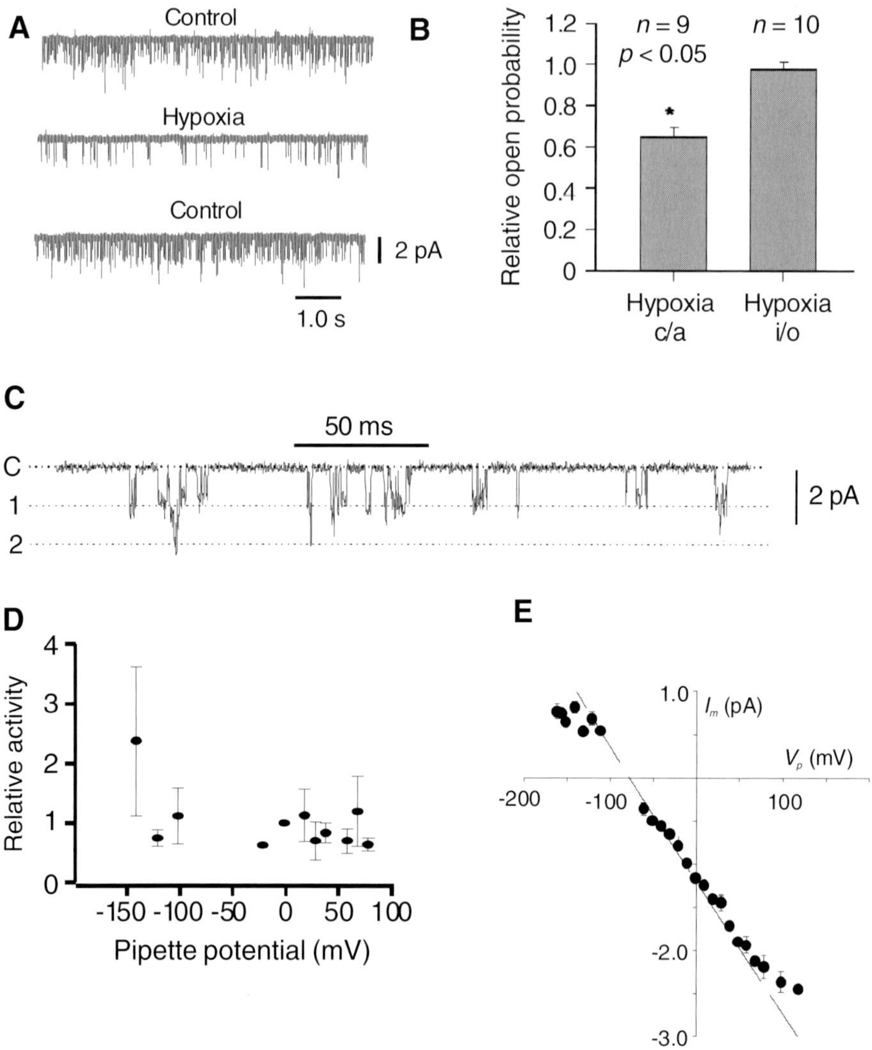

FIG. 2. Biophysics of background K^+ channels. (A) Single channel recording of background K^+ channels in a cell attached patch (at an approximate membrane potential of −70 mV). Note marked reduction in channel opening under hypoxic conditions (pO_2 approx 5 Torr). (B) Effects of hypoxia on single channel activity of background K^+ currents in cell attached patches (showing inhibition by hypoxia) and in excised inside out patches (resistant to hypoxia). (C) Cell-attached patch recording of background K^+ channel activity. (D) Open probability of background K^+ channel activity as a function of pipette voltage in cell attached patch. Patch electrode contained 5 mM 4-AP + 10 mM TEA. (E) Single channel current as a function of voltage for background K^+ channels in the cell attached patch. Pipette [K^+] was 146 mM.

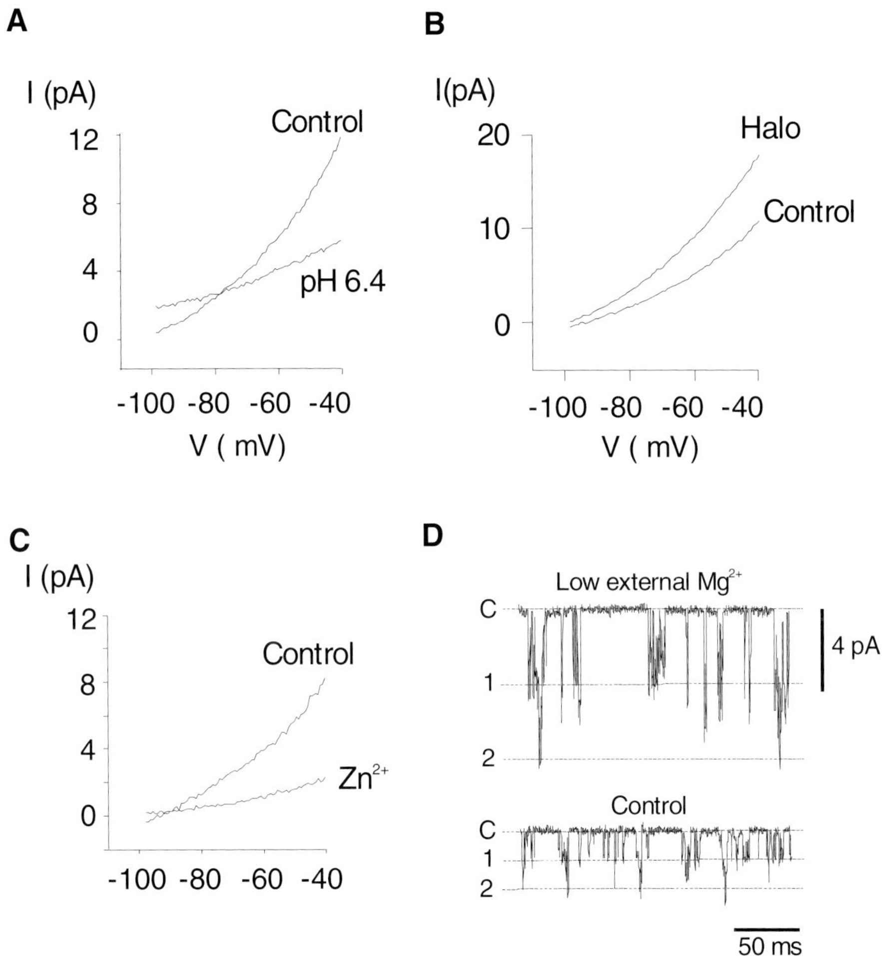

FIG. 3. Pharmacology of oxygen-sensitive background K^+ currents. (A) Mean (n = 5) oxygen-sensitive current (normoxia-hypoxia; pO2 approx 5 Torr) recorded at a normal extracellular pH (7.4; control) and at reduced extracellular pH (6.4) showing acid inhibition of oxygen-sensitive background K^+ current. (B) Mean (n = 9) oxygen-sensitive current (normoxia-hypoxia; pO2 approx 5 Torr) recorded in the absence (control) and presence of halothane (Halo). Halothane was introduced into solution by including it in the gas mixture used to bubble solutions (at 1.5%). (C) Mean (n = 6) oxygen sensitive current (normoxia-hypoxia; pO2 approx 5 Torr) recorded in the absence (control) and presence of 200 μM zinc. (D) Cell-attached patch recordings of background K^+ channels at pipette potential of +60 mV (approx −120 mV membrane potential) in the presence of 3.7 mM external (pipette)-free magnesium (control) and in the nominal absence of magnesium (low external Mg^{2+} approx 0.9 μM). Note that in the nominal absence of external magnesium, single channel current is approximately doubled.

myxothiazol, antimycin A and rotenone) and inhibitors of ATP synthase (oligomycin) are potent stimulants of the carotid body (Ortega Saenz et al 2003, Wilson et al 1994, Anichkov & Belen'kii 1963, Mulligan et al 1981). A number of recent studies have shown that metabolic inhibitors act in a manner remarkably similar to that of hypoxia in that they also inhibit background K^+ channels leading to membrane depolarization, voltage-gated Ca^{2+} entry and an abrupt rise in intracellular Ca^{2+} (Buckler & Vaughan Jones 1998, Wyatt & Buckler 2004). For example, Fig. 4 shows the effects of cyanide.

The observation that a very diverse range of inhibitors of mitochondrial energy metabolism including uncouplers (DNP and FCCP), inhibitors of electron transport at complex 1 (rotenone), complex 3 (myxothiazol) and complex 4 (CN and carbon monoxide) and an inhibitor of ATP synthase (oligomycin) all produce the same effect, i.e. inhibition of background K^+ channel activity (Wyatt & Buckler 2004, Buckler & Vaughan Jones 1998, Barbe et al 2002) suggests that these effects are most likely to be due to inhibition of ATP production and not a consequence of other events occurring in mitochondria (e.g. O_2^- production, see Wyatt & Buckler [2004] for full discussion). Although the precise link between ATP synthesis and background K^+ channel activity has not yet been determined, we have observed that ATP has a direct effect upon channel activity in excised patches. Following patch excision, background K^+ channel activity runs down dramatically suggesting that these channels are strongly modulated by cytosolic constituents. Following rundown addition of millimolar levels of ATP to the intracellular aspect of excised patches results in a rapid increase of channel activity (Fig. 5; Williams & Buckler 2004). Thus

FIG. 4. Effects of metabolic inhibitors on background K^+ currents. (A) Simultaneous recording of membrane potential and intracellular Ca^{2+} in a type 1 cell. Note that application of 2 mM cyanide (CN^-) leads to an abrupt membrane depolarization and rise in intracellular Ca^{2+} in current clamp but not when the cell is voltage-clamped (middle section of trace). (B) Effects of cyanide on membrane currents in type 1 cells. Figure shows mean ($n = 6$) whole-cell membrane current under control conditions, in the presence of 2 mM cyanide and in the presence of 2 mM cyanide + hypoxia (pO_2 approx 6 Torr). Note that cyanide causes a marked decline in membrane conductance but that in the presence of cyanide hypoxia has no effect upon membrane current. (C) Inhibition of background K^+ channel activity in cell-attached patches by 2 mM cyanide. (D) Comparison of effects of 2 mM cyanide and 250 μM 2,4-dinitrophenol (DNP) on the single-channel activity of background K^+ currents in cell-attached (black bars) and inside–out patches (white bars). Note that both CN^- and DNP cause a 60% inhibition of channel activity in the cell-attached patch only. (E) Effects of hypoxia (pO_2 approx 5 Torr) on whole cell membrane currents in a control group ($n = 46$) of type 1 cells. Compare with lack of effect of hypoxia in the presence of cyanide (part B this figure). (F) Summary of the effects of metabolic inhibitors on oxygen sensitive current at −50 and −70 mV. Control, control—hypoxia (pO_2 approx 5 Torr, $n = 46$); Rot, rotenone—rotenone and hypoxia ($n = 6$); CN, cyanide—cyanide and hypoxia ($n = 6$); and FCCP, FCCP—FCCP and hypoxia ($n = 4$). Cyanide was used at a nominal concentration of 2 mM (see Wyatt & Buckler 2004), and rotenone and FCCP at 1 μM. Note that rotenone, cyanide and FCCP all abolish the oxygen-sensitive current.

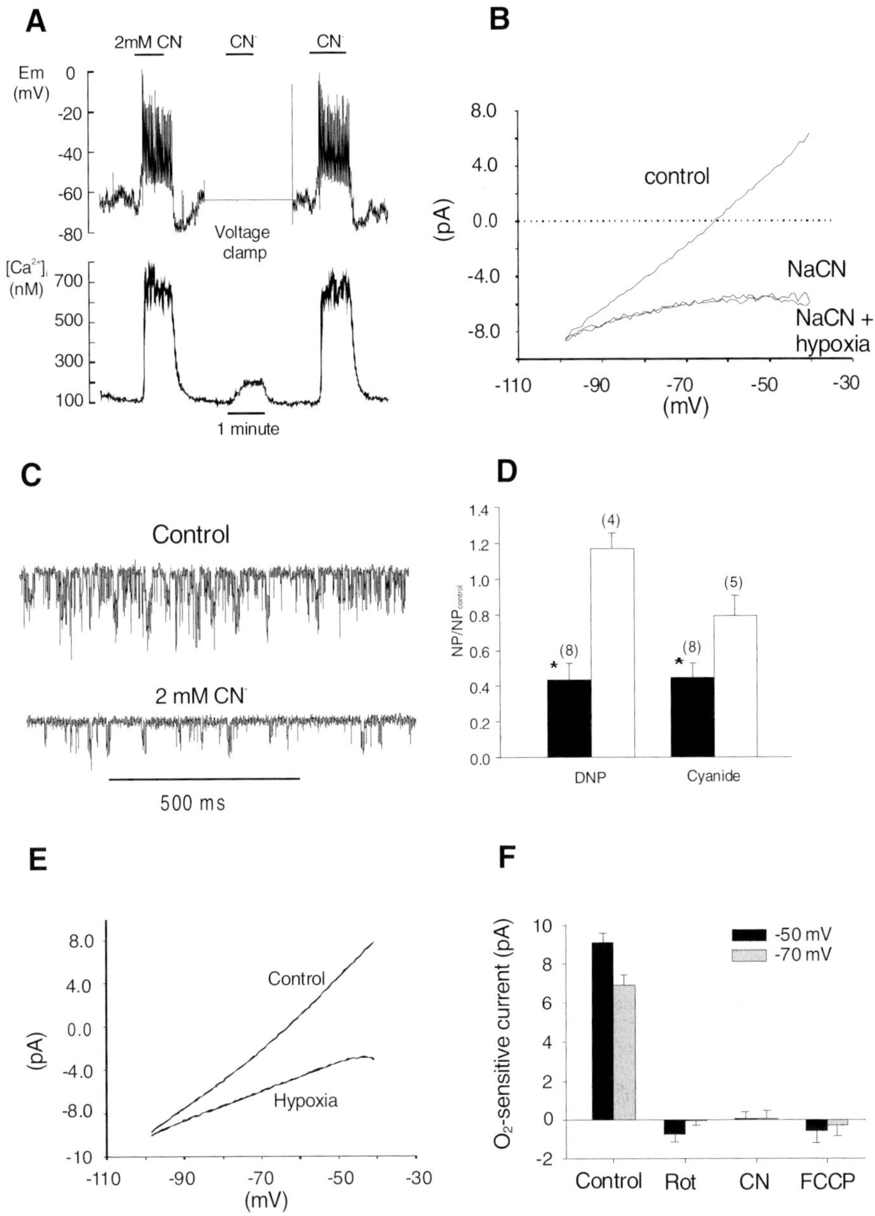
A
2mM CN-
CN-
CN-
Em (mV)
0
-20
-40
-60
-80
Voltage clamp
[Ca2+]i (nM)
700
500
300
100
1 minute
B
8.0
4.0
0.0
-4.0
-8.0
(pA)
control
NaCN
NaCN + hypoxia
-110
-90
-70
-50
-30
(mV)
C
Control
2 mM CN-
500 ms
D
NP/NPcontrol
1.4
1.2
1.0
0.8
0.6
0.4
0.2
0.0
(4)
(5)
(8)
(8)
*
*
DNP
Cyanide
E
8.0
4.0
0.0
-4.0
-8.0
-10
(pA)
Control
Hypoxia
-110
-90
-70
-50
-30
(mV)
F
O2-sensitive current (pA)
10
8
6
4
2
0
-2
-50 mV
-70 mV
Control
Rot
CN
FCCP

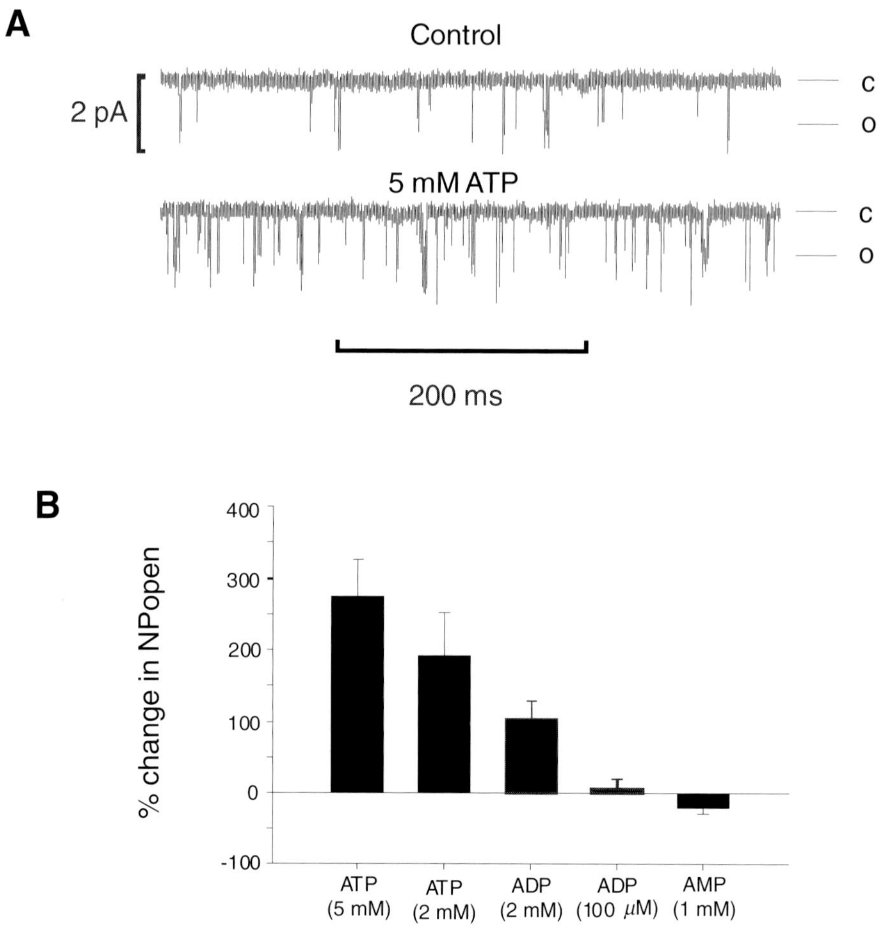

FIG. 5. Effects of ATP on background K^+ channel activity. (A) Excised inside–out patch recording of background K^+ channel activity in the absence and presence of 5 mM MgATP (−70 mV membrane potential). (B) Summary of effects of various adenine nucleotides upon background K^+ channel activity in excised patches.

changes in cytosolic ATP could provide a direct link between mitochondrial function and background K^+ channel activity.

Metabolism and oxygen sensing

Neither hypoxia nor metabolic poisons fully inhibit background channel activity, on average metabolic inhibition reduces channel activity, or whole cell resting K^+ conductance, by about 60% (Fig. 4D; Buckler & Vaughan Jones 1998, Wyatt & Buckler 2004). The effects of metabolic inhibition and hypoxia upon background K^+ currents are not however additive but mutually exclusive. In the presence of CN, FCCP or rotenone, hypoxia has no detectable effect upon residual background K^+ current (Fig. 4E,F; Wyatt & Buckler 2004). This observation suggests that not only do hypoxia and metabolic inhibitors modulate the same channels but that the effects of hypoxia are dependent upon metabolism. It has similarly been observed in intact carotid body preparations that inhibitors of oxidative phosphorylation appear to selectively abolish sensitivity to hypoxia whilst leaving responses to acidic stimuli or to nicotinic agonists intact (Mulligan et al 1981, Mulligan & Lahiri 1982, Mosqueria & Iturriaga 2002).

Hypoxia and metabolism

One of the oldest theories as to how hypoxia is sensed by the carotid body postulates that it is mediated by the mitochondrion. The major criticism of this hypothesis is that in most tissues pO2 needs to be reduced to very low levels before O_2 availability significantly limits mitochondrial function. Measurements using fluorescent oxygen sensors have shown that chemoreceptor activity increases sharply when microvascular pO2 falls from a normoxic level of around 50 Torr to below 20–30 Torr (Lahiri et al 1993). This level of O_2 sensitivity is comparable to that of background K^+ currents, Ca^{2+} signalling and neurosecretion in isolated type 1 cells (Biscoe & Duchen 1990, Perez Garcia et al 1992, Buckler & Vaughan Jones 1994a, Montoro et al 1996), but is considerably higher than that of mitochondrial O_2 consumption for which, in most tissues, the P_{50} is less than 0.5 Torr (Gnaiger et al 1998). Indirect studies of mitochondrial function, utilizing rhodamine 123 to monitor mitochondrial membrane potential or cellular autofluorescence as an index of NADH levels, have however shown that type 1 cell mitochondria depolarize and NADH levels increase when pO2 is reduced below 40–20 Torr (Duchen & Biscoe 1992a,b). These observations lend support to the original hypothesis that mitochondrial function is unusually sensitive to oxygen in the carotid body (Mills & Jobsis 1970). Clearly there is a need for further confirmation of this hypothesis as it not obvious how the apparent O_2 affinity of mitochondrial function could vary so greatly between type 1 cells and other cells when there is only one mitochondrial gene encoding cytochrome a_3.

Conclusions

In summary, TASK-like background K^+ channels play a key role in mediating the response of type 1 cells to hypoxic stimuli. These channels also appear to be markedly sensitive to metabolic inhibition, an effect which may be mediated, at least in part, through direct sensitivity to ATP. There also appears to be a close link between oxygen sensing and energy metabolism. At present however, we cannot differentiate between the mitochondrion serving in the capacity of 'oxygen sensor' or simply playing an accessory role in providing the conditions necessary for oxygen sensing to occur (e.g. by provision of ATP to maintain normal K^+-channel function and/or biochemical processes in the oxygen signalling pathway). It is, however, intriguing that the type 1 cell seems to be exceptionally well equipped with a rapid signalling pathway linking mitochondrial metabolism to electrical signalling.

References

Anichkov S, Belen'kii M 1963 Pharmacology of the carotid body chemoreceptors. Pergamon Press, Oxford, England

Barbe C, Al-Hashem F, Conway AF et al 2002 A possible dual site of action for carbon monoxide-mediated chemoexcitation in the rat carotid body. J Physiol 543:933–945

Biscoe TJ, Duchen MR 1990 Responses of type I cells dissociated from the rabbit carotid body to hypoxia. J Physiol 428:39–59

Buckler KJ 1997 A novel oxygen-sensitive potassium current in rat carotid body type I cells. J Physiol 498:649–662

Buckler KJ, Vaughan Jones RD 1994a Effects of hypoxia on membrane potential and intracellular calcium in rat neonatal carotid body type I cells. J Physiol 476:423–428

Buckler KJ, Vaughan Jones RD 1994b Effects of hypercapnia on membrane potential and intracellular calcium in rat carotid body type I cells. J Physiol 478:157–171

Buckler KJ, Vaughan Jones RD 1998 Effects of mitochondrial uncouplers on intracellular calcium, pH and membrane potential in rat carotid body type I cells. J Physiol 513:819–833

Buckler KJ, Vaughan Jones RD, Peers C, Nye PC 1991 Intracellular pH and its regulation in isolated type I carotid body cells of the neonatal rat. J Physiol 436:107–129

Buckler K, Williams B, Honoré E 2000 An oxygen-, acid- and anaesthetic-sensitive TASK-like background potassium channel in rat arterial chemoreceptor cells. J Physiol 525:135–142

Clarke CE, Veale EL, Green PJ, Meadows HJ, Mathie A 2004 Selective block of the human 2-P domain potassium channel, TASK-3, and the native leak potassium current, IKSO, by zinc. J Physiol 560:51–62

Czirjak G, Enyedi P 2001 Formation of functional heterodimers between the TASK-1 and TASK-3 two pore domain potassium channel subunits. J Biol Chem 277:5426–5432

Dasso LL, Buckler KJ, Vaughan Jones RD 1997 Muscarinic and nicotinic receptors raise intracellular Ca^{2+} levels in rat carotid body type I cells. J Physiol 498:327–338

Duchen MR, Biscoe TJ 1992a Mitochondrial function in type I cells isolated from rabbit arterial chemoreceptors. J Physiol 450:13–31

Duchen MR, Biscoe TJ 1992b Relative mitochondrial membrane potential and $[Ca^{2+}]_i$ in type I cells isolated from the rabbit carotid body. J Physiol 450:33–61

Gallego R, Eyzaguirre C, Monti Bloch L 1979 Thermal and osmotic responses of arterial receptors. J Neurophysiol 42:665–680

Gnaiger E, Lassnig B, Kuznetsov A, Rieger G, Margreiter R 1998 Mitochondrial oxygen affinity, respiratory flux control and excess capacity of cytochrome c oxidase. J Exp Biol 201:1129–1139

Kang D, Han J, Talley EM, Bayliss DA, Kim D 2004 Functional expression of TASK-1/TASK-3 heteromers in cerebellar granule cells. J Physiol 554:64–77

Kim Y, Bang H, Kim D 1999 TBAK-1 and TASK-1, two-pore K^+ channel subunits: kinetic properties and expression in rat heart. Am J Physiol 277:H1669–1678

Kim Y, Bang H, Kim D 2000 TASK-3, a new member of the tandem pore K^+ channel family. J Biol Chem 275:9340–9347
Lahiri S, Rumsey WL, Wilson DF, Iturriaga R 1993 Contribution of in vivo microvascular PO2 in the cat carotid body chemotransduction. J Appl Physiol 75:1035–1043
Lopes CM, Gallagher PG, Buck ME, Butler MH, Goldstein SA 2000 Proton block and voltage gating are potassium-dependent in the cardiac leak channel Kcnk3. J Biol Chem 275:16969–16978
Mills E, Jobsis FF 1970 Simultaneous measurement of cytochrome a3 reduction and chemoreceptor afferent activity in the carotid body. Nature 225:1147–1149
Montoro RJ, Urena J, Fernandez Chacon R, Alvarez de Toledo G, Lopez Barneo J 1996 Oxygen sensing by ion channels and chemotransduction in single glomus cells. J Gen Physiol 107:133–143
Mosqueria M, Iturriaga R 2002 Carotid body chemosensory excitation induced by nitric oxide: involvement of oxidative metabolism. Respir Physiolo Neurobiol 131:175–187
Mulligan E, Lahiri S 1982 Separation of carotid body chemoreceptor responses to O_2 and CO_2 by oligomycin and by antimycin A. Am J Physiol 242:C200–206
Mulligan E, Lahiri S, Storey BT 1981 Carotid body O2 chemoreception and mitochondrial oxidative phosphorylation. J Appl Physiol 51:438–446
Ortega Saenz P, Pardal R, Garcia Fernandez M, Lopez Barneo J 2003 Rotenone selectively occludes sensitivity to hypoxia in rat carotid body glomus cells. J Physiol 548:789–800
Perez Garcia MT, Obeso A, Lopez Lopez JR, Herreros B, Gonzalez C 1992 Characterization of cultured chemoreceptor cells dissociated from adult rabbit carotid body. Am J Physiol 263:C1152–1159
Rajan S, Wischmeyer E, Xin Liu G et al 2000 TASK-3, a novel tandem pore domain acid-sensitive K^+ channel. An extracellular histidine as pH sensor. J Biol Chem 275:16650–16657
Reyes R, Duprat F, Lesage F et al 1998 Cloning and expression of a novel pH-sensitive two pore domain K+ channel from human kidney. J Biol Chem 273:30863–30869
Williams BA, Buckler KJ 2004 Biophysical properties and metabolic regulation of a TASK-like potassium channel in rat carotid body type 1 cells. Am J Physiol Lung Cell Mol Physiol 286:L221–230
Wilson DF, Mokashi A, Chugh D et al 1994 The primary oxygen sensor of the cat carotid body is cytochrome a3 of the mitochondrial respiratory chain. FEBS Lett 351:370–374
Wyatt CN, Buckler KJ 2004 The effect of mitochondrial inhibitors on membrane currents in isolated neonatal rat carotid body type I cells. J Physiol 556:175–191
Zhang M, Zhong H, Vollmer C, Nurse CA 2000 Co-release of ATP and ACh mediates hypoxic signalling at rat carotid body chemoreceptors. J Physiol 525 Pt 1:143–158

DISCUSSION

Duchen: You showed experiments using hypoxia. How do you define hypoxia?

Buckler: Below 5 Torr. It is quite a substantial stimulus, but if we are looking for additivity we need a robust response.

Duchen: If you propose that ATP is the signal, if you are simply slowing mitochondrial respiration to account for your fast responses then ATP would have to fall rapidly. There would need to be an ATP consumer driving ATP down quickly.

Buckler: That is true, and we have data consistent with this, particularly with mitochondrial inhibitors. We have been trying to get magnesium dyes to work. The idea is that ATP is a major intracellular buffer for magnesium so if ATP concentration goes down, Mg^{2+} is released and Mg^{2+} concentration in the cell increases. We should be able to measure this with a Mg^{2+}-sensitive dye. We are struggling with calibrations at the moment, but our best current estimate is that CN or FCCP cause an increase in free Mg^{2+} of about 0.5 mM. The change occurs rapidly, starting within several seconds. This corresponds to apparent changes in intracellular Ca^{2+} that we see in voltage clamp or in Ca^{2+}-free conditions when using Ca^{2+}-selective dyes. Since most Ca^{2+}-sensitive dyes (e.g. indo-1) are also sensitive to Mg^{2+}, it is possible that what we are looking at in Ca^{2+}-free media is actually a rise in Mg^{2+}, due to the ATP depletion. This response (apparent rise in intracellular Ca^{2+} measured using indo-1 in voltage clamped cells) has always correlated with changes in membrane current.

Duchen: My point is that you would need a powerful ATPase of some sort driving ATP down fast in order to generate that response.

Gurney: In relation to the additivity of the effects of hypoxia and mitochondrial inhibitors, after the effect of mitochondrial inhibitors there seemed to be very little current left.

Buckler: That is partly an illusion because there are other currents present. In those slides you were looking at whole cell current. When we directly estimate K^+ conductance from changes in membrane current induced by changing extracellular K^+ we only see a 60% reduction in K^+ conductance with cyanide or FCCP (see e.g. Wyatt & Buckler 2004). We also see a similar reduction when looking at the effects of metabolic inhibitors on single channel activity. So there is enough background K^+ current left with which to detect any effect of hypoxia if there were any.

Evans: Do you see additional effects of pH on carotid body glomus cell function after prior and complete abolition of the hypoxic response?

Buckler: I haven't looked, but it's an interesting question.

Archer: I noticed on the pharmacology slides you have a somewhat non-ohmic looking current. This is atypical of TASK currents which should be linear and ohmic. There was quite marked voltage dependency on several of your pharmacology slides. I mention because one of the things we found with conventional K_v channels is that if they are held open for 200 ms and you repeatedly are going over the entire range, you do get inactivation. However, if you minimize the range over which one depolarizes, K_v currents are active at very negative membrane potentials and contribute to the resting membrane potential of pulmonary artery smooth muscle cells (see Archer et al 2004). It is surprising that you see no K_v current at all at resting potential.

Buckler: It is possible that there is some voltage-activated current towards more positive potentials. Some experiments used voltage ramps from −100 (or −90) up to −30 mV and some to −40 mV. In some recordings we see what looks like the beginnings of activation of Ca^{2+} current at potentials positive to −40 mV. In experiments using voltage ramps from −100 to −40 mV we see no major effects of 4-AP or TEA so we assume that there are no voltage-activated K^+ channels active in this voltage range (Buckler 1997).

Archer: There are lots of channels that require ATP to avoid dephosphorylation and yet don't use ATP as a signalling molecule. If you take Ca^{2+} channels and study them for any length of time, how would you differentiate ATP as a signalling molecule (which would be quite novel) versus the more common phenomenon of dephosphorylation-mediated channel rundown?

Buckler: It is tricky, isn't it? We have an ATP dose–response that would be consistent with a signalling role for ATP. In addition if we can confirm our preliminary data with non-hydrolysable ATP analogues this would also argue for an ATP-sensing protein: something that ATP binds to but is not necessarily involved in phosphorylation.

Gonzalez: In our experiments in rabbit chemoreceptor cells using whole-cell recordings we can see oxygen sensitivity with 5 mM and with no ATP in the pipette, and in experiments with rat chemoreceptor cells with no ATP in the perfusing fluid while recording from inside-out patches. Can you comment on this?

Buckler: That is difficult, because I would have to speculate on what might or might not happen to ATP levels in the whole cell configuration when energy metabolism is inhibited. How quick is diffusion going to be? Are you going to clamp ATP levels by dialysis through the patch electrode or could it change?

Ward: Have you looked at glucose and anoxia? Rick Paul's model of having membrane-associated glycolysis and thus ATP production in the sub-sarcolemmal domain might be relevant here.

Buckler: We have had a preliminary look at glucose with regards to Ca^{2+} signalling and we don't see any sensitivity to glucose removal in acutely isolated cells. It could be that there is an electrical response but that it is just subthreshold for activating Ca^{2+} influx.

López-Barneo: I like part of your data: I like the bit under current clamp, but not when you go to voltage clamp! You never showed clearly that hypoxia produced an increase in membrane resistance. If you want to test the additivity, the right way to do it is to show first the effect of hypoxia, then go to the inhibitors, and then do inhibitor plus hypoxia. You didn't show the effect of hypoxia on its own.

Buckler: I have shown a lot of data on the effects of hypoxia on membrane currents, including a pO2 dose–response.

López-Barneo: I am not comfortable with the experiment in which you add dithionite to help anoxia. Dithionite is known to produce a number of effects on the membrane conductance. If you go to mitochondrial inhibitors, they will produce a marked increase in membrane resistance. You didn't show the effect of hypoxia itself on the currents; you showed the effect of anoxia.

Buckler: If we have healthy cells and get good seals, we always see oxygen sensitivity. The effects of hypoxia on membrane currents are highly reproducible; we have reproduced this now in over 50 cells. We have published many data showing the effects of hypoxia, along with the inhibition of the (hypoxic) oxygen-sensitive current with zinc, barium, quinidine, bupivicaine, acidosis and stimulation with halothane and so on (see Buckler 1997, Buckler et al 2000).

We have also published a dose–response for oxygen: only the zero point was anoxia (i.e. containing $Na_2S_2O_4$; see Buckler 1997). The single-channel data I presented showing changes in channel activity were also made under hypoxic conditions.

López-Barneo: What happened with antimycin A?

Buckler: We didn't try this. This drug is messy to work with as (in our hands) it isn't reversible and does not fully inhibit mitochondrial function instantly. We chose compounds that acted rapidly and were reversible.

Chandel: The membrane starts to depolarise below 40 Torr.

Buckler: Some do respond around 20 Torr, and some need to go to 10 Torr. It depends on the cell.

Chandel: The mitochondria don't get inhibited until 2 Torr, so what is happening at 20 Torr to explain the changes?

Buckler: We don't really know how low pO2 has to be to compromise mitochondrial function in type 1 cells. Mike Duchen showed that in type 1 cell clusters, if you look at mitochondrial potential using rhodamine 123 or NADH autofluorescence, as pO2 is lowered below about 40 Torr you see mitochondrial depolarization and increase in NADH autofluorescence (Duchen & Biscoe 1992a, b). I have no explanation for this.

Chandel: Years ago we took isolated cytochrome oxidase and over time this depressed its V_{max} at 20 Torr. No one has ever characterized the oxygen dependence of the oxidase, to the best of my knowledge.

Duchen: Those measurements that we made were our first foray into mitochondrial measurements, and I didn't have any experience of anything else beforehand. We saw graded changes in NAD, flavoprotein and mitochondrial membrane potential as we dropped the oxygen tension. Since then I have worked in many different cell types and have never seen anything like it again. There is something different about those carotid body cells.

Ward: We see the same thing in pulmonary artery smooth muscle. The difference between the systemic and pulmonary vasculature is consistent with the carotid body. There seems to be something different.

Buckler: We are doing some work with sympathetic neurons and dorsal root ganglion neurons. Although we often see a small rise in Ca^{2+} with anoxia in these cells, we never see anything like the extremely rapid and very robust increase in Ca^{2+} that we see in type 1 cells.

Duchen: One thing about these cells that is also unique is this response to oligomycin. Nothing happens in the short term if you apply oligomycin to most cells, certainly in terms of $[Ca^{2+}]$. With these cells the $[Ca^{2+}]$ goes shooting up. This is remarkable and needs an explanation. You can't account for this response in terms of change in mitochondrial membrane potential, because if anything in these cells the mitochondria hyperpolarize in response to oligomycin.

Prabhakar: To follow that question, the idea that ATP might get decreased during hypoxia is fascinating. If you look at the whole organ response, you start to see an increase in sensory activity if you drop arterial pO2 from 100 to 80 mmHg. The question is whether such a moderate drop in pO2 results in a decrease in ATP levels in the carotid body? Did you ever measure ATP levels in the carotid body in response to varying pO2s?

Buckler: No, we haven't.

Ratcliffe: I was interested in that time course problem. If you postulate that there is a massively active ATPase, then would this produce a lot of ADP? My understanding is that the oxygen sensitivity of mitochrondrial metabolism is greatly dependent on the ATP : ADP ratio, so that an ATPase could both increase the time sensitivity and the oxygen sensitivity.

Buckler: We haven't looked at the ATP : ADP ratio yet.

Ratcliffe: If these cells have a massive ATPase activity then you should be able to see this quite easily.

Murphy: I am trying to think of ways in which you could alter the activity of an ATPase within a cell independently of the oxygen concentration and without altering mitochondrial function too drastically. One possibility might be through use of arsenate. If you dose cells with arsenate, this takes the place of phosphate in ATP synthesis, but the terminal bond linking the ADP and the arsenate is now labile and spontaneously hydrolyses. Consequently, you are effectively setting up a futile cycle akin to activating an intracellular ATPase. You might predict that this would have the same effects as altering the ATP/ADP ratio, but without any major impact on mitochondrial function.

Gurney: Could the mitochondrial depolarization with hypoxia be due to the rise in Ca^{2+}?

Duchen: No. It is not dependent on external $[Ca^{2+}]$. It is a response to the change in oxygen. If you raise Ca^{2+} massively by depolarizing the cells with K^+, the change in fluorescent signal is very small.

Rich: Did you say that you don't have to hydrolyse the ATP, so ADP has the same effect.

Buckler: Yes, at high concentrations it has a similar effect. At about 2 mM ADP.

Rich: So it is not terribly different. I can't see how it can make any difference when you knock out the mitochondrial ATP generation, because the ADP plus ATP level in the cytoplasm is a constant.

Buckler: That isn't my understanding of biochemistry. Under conditions in which ATP resynthesis is inhibited, adenylate kinase converts ADP into ATP and AMP, so that ADP levels are always kept relatively low.

Rich: Let me put it differently: the total amount of adenosine compounds is staying the same.

Buckler: Yes, ATP goes down and AMP (and probably adenosine) go up. ADP might double, but this is from about 50 μM to 100 μM (see e.g. Kupriyanov et al 1996).

Rich: This would be a very insensitive way of designing a regulatory mechanism. My other question is how did you rule out the effects of ionic strength of these compounds, given that they are polyionic materials?

Buckler: We haven't looked. We have been looking at pyrophosphate.

Rich: ATP is quadruply charged so its ionic strength effect is enormous. If you want to simulate the ionic strength effect you need to take this into account. My guess would be that you are looking at an ionic strength effect. Although there is a mediation between mitochondrial function and your channel, it is somewhere else.

Buckler: That wouldn't explain the saturation of the response to ATP. Experiments were conducted using $MgATP^{2-}$ not ATP^{4-}.

Rich: It would: don't forget that the ionic strength is non ideal. Even ionic strength in biochemistry saturates biochemical processes, particularly for multiple charged species such as ATP. In other words, the ionic strength doesn't keep going up as you add more chemical.

Nurse: These are nice data. I am trying to think of differences between your work and that of Chris Peers and José López-Barneo. You have not focused on the BK channels. If one looks at the overall response of the carotid body, note particularly the fact that the receptor cells are organized in clusters and most of José's release experiments are based on an endpoint that goes beyond (i.e. downstream from) the one you are considering. Given that the BK channel can be inhibited by hypoxia as well, I wonder whether we are too constrained by thinking that there is only one sensor. There are at least two channels in the rat type I cell that are oxygen sensitive and these could operate via different mechanisms.

Buckler: There is not much information on possible signalling pathways with BK.

Nurse: Let me ask another question. We have seen spontaneous activity in some of the type I clusters. In hypoxia we see a broadening of the spike, which is consistent with inhibition of BK channels. You showed some recordings where you get depolarization and spike activity with hypoxia. Do you ever see spontaneous spikes occurring under normoxic conditions whose properties you could compare with the spikes seen during hypoxia?

Buckler: I'd have to go back and look. I'm sure there's the odd spontaneous spike.

Peers: You could artificially depolarize with current injection.

Nurse: Within those spikes that you see during hypoxia, could there be a broadening of the action potential?

Buckler: There could well be. With barium (a general K^+ channel inhibitor) we get very broad action potentials.

Aaronson: In your excised patches the activity ran down. Was it these patches that you used to look at the ATP and ADP?

Buckler: Yes, it would be under similar circumstances although it wouldn't necessarily be the same patch.

Aaronson: Is it possible that the reason you didn't see any effect of reducing oxygen was because there wasn't enough activity there anyway?

Buckler: There is a quantifiable level of channel activity so we could easily tell if channel activity were to go down further. The question is, is the channel oxygen-insensitive because it has run down? That's possible, but until we know what is responsible for rundown we can't put everything back and see whether we can restore oxygen sensitivity.

Aaronson: Have you tried putting ATP in, recovering the channel activity and then making the bath hypoxic?

Buckler: No.

López-Barneo: We do patch clamp in which we load cells with and without ATP. When we load cells with zero ATP or MgATP and we inspect the sensitivity to hypoxia two minutes later (once the cell is fully dialysed) we can see an intact response to hypoxia. How would you explain this?

Buckler: If a cell has 5 mM ATP at rest, you attach a patch electrode with 5 mM ATP and you end up with 5 mM ATP in the cell, so it is not 'ATP loaded'.

López-Barneo: We have shown that in this same configuration without ATP, if we record the Ca^{2+} channel currents we see that if the cell is left long enough there is wash out of the Ca^{2+} channel.

Buckler: So you are arguing that run down of Ca^{2+} channel activity is due to a loss of ATP.

López-Barneo: We do this very fast. If we modify the ATP concentration inside the cell and then look at the effect of hypoxia, you would expect that in those cells loaded with ATP, the response to hypoxia would be weaker.

Buckler: I dispute that you are loading the cells with ATP. If you are saying that there is a clear effect of hypoxia in the cells that have had the ATP dialysed out of them, this would be an interesting observation.

López-Barneo: What I am saying is that we are looking at loading the cells with ATP. We make the seal, break the membrane, go to whole cell configuration and allow a solution with a lot of ATP to enter the cells.

Buckler: There will only be net ATP diffusion into the cell if there is a concentration gradient. If you have 6 mm ATP in the pipette and 6 mM in the cell, you will not 'load' the cell with ATP. You are just stopping it from washing out.

Duchen: The difference is between having ATP or dialysing it out of the cell. If you take a pancreatic β cell and make a whole cell recording without ATP you get opening of channels very quickly. In your cells, if you make a whole-cell patch recording, do you lose oxygen sensitivity if you have no ATP in your filling solutions?

Buckler: I have done very few conventional whole-cell recordings. There were one or two where I saw oxygen sensitivity, but it ran down. This was when we were trying to do single channel recordings with very fine-tipped pipettes, some patches spontaneously went into the whole cell configuration. In these cells there was a high access resistance of about 20 mega Ohms. I initially saw oxygen sensitivity but it did go away with time.

Archer: In the pulmonary circulation there are data suggesting that energy doesn't get depleted in the rapid response time that occurs for HPV (Buescher et al 1991, Pillai et al 1986). We found preserved HPV despite a 50% depletion of ATP, achieved by feeding rats beta aminopropionic acid to deplete ATP and phosphocreatine (Archer et al 1989). This change in energetics didn't change oxygen sensing. In this specialized oxygen sensing system the responses are rapid (onset in seconds) and the threshold for the response is high (onset at pO2s that are likely far too high to deplete energy). This would tie the mitochondria to sensing by the flux of electrons down the transport chain, and the leak of electrons which may or may not result in ROS. This is loosely associated with energy, but it is far upstream. Teleologically speaking, it is a great sensor: if the organism waits for ATP depletion to signal hypoxia it is in huge trouble. Any time anyone has tried to measure energetics with brief, moderate hypoxia, energetics are preserved. If you wanted to measure it I guess the tool to use would be freeze clamping and use of a high Tesla magnet. This would give the spectra of frozen carotid bodies. I don't believe that ATP depletion occurs within minutes of moderate hypoxia (pO2 40, 10–12% FiO_2) in the pulmonary circulation.

Duchen: It does happen in the β cell, so there is a precedent.

Sylvester: Our previous studies (Leach et al 1998, 2000) suggested that energy state homeostasis is different in pulmonary arterial smooth muscle versus systemic arterial smooth muscle. For example, in terms of energy state poise during normoxia, pulmonary arterial smooth muscle seems to be at a disadvantage compared to systemic smooth muscle; however, during hypoxia, after an initial fall, energy state recovers in pulmonary arterial smooth muscle, whereas in systemic arterial smooth muscle, it does not. This recovery of energy state is associated with a recovery of tone. This might go along with what Steve Archer was saying.

Nurse: Is the pH sensitivity of TASK-1 or 3 sufficient to explain acidic chemoreception in the physiological range?

Buckler: For TASK-1 it might be, but I am not sure about TASK-3.

Ratcliffe: Is the pH sensitivity the same in your isolated patches as in the whole cell?

Buckler: I have not looked at pH sensitivity in isolated patch.

Peers: pH modulation is a direct effect in the recombinant channel.

Buckler: It is, and we assume it is probably the same in this one.

Kumar: One thing that makes the carotid body unique is that it measures all these other factors as well as hypoxia. Dr Buckler mentioned temperature, pH and CO_2. There is an interaction between these, also. CO_2 potentiates the hypoxia response, and at the same time hypoxia must be potentiating the CO_2 response. If you look at what you think is a dose–response curve to hypoxia because you are changing hypoxia and holding the CO_2 constant, you might actually be altering the CO_2 sensitivity by changing the oxygen concentration. In other words, you might be looking at a CO_2 response and observing how hypoxia is affecting this. The 'AS' in TASK stands for acid sensitive, so this would be quite an obvious way to go looking. This means you would have to work out where the CO_2/O_2 interaction occurs, and this might well come back to the mitochondria. Rather than looking for a hypoxia sensor, perhaps we could look for a pH sensor and see how this is modulated by hypoxia. You can't change one stimulus and just forget about all the others simply by holding them constant.

Archer: Some K_v and BK_{Ca} channels are also acid sensitive; they are just not called acid-sensitive K_v channels. They are inhibited by acidosis and HPV is also modulated by changes in peak CO_2. I agree that this concern would be valid, and not just for the carotid body.

Gurney: There is a difference in the acid sensitivity. K_v channels tend to be inhibited around pH 6, whereas TASK channels are inhibited at pH 7.

Archer: I don't dispute the difference in the intrinsic sensitivity of the channel types. Interestingly, clinically if one wants to reverse pulmonary hypertension one can hyperventilate them, lower pCO2 (create alkalosis) and lower pulmonary vascular resistance and inhibit HPV.

Gonzalez: Somehow, I feel sorry because we are discussing at the same level as Anichkov and Belenkii were in 1963 (Anichkov & Belenkii 1963). They were discussing at that time whether or not ATP was important in hypoxia sensing. We have made very little progress.

Kumar: Professor López-Barneo mentioned earlier about different channels opening when ATP was added in the pipette. We know that type 1 cells have P2X receptors which may be gated by ATP and allow Ca^{2+} entry. Might P2 receptors be involved in any of the things you have mentioned?

López-Barneo: I don't think so, as what we do is to change intracellular ATP and P2X receptors are activated by extracellular ATP. In any instance, the critical point is that in some cells loaded with a solution without ATP (using pipettes filled with internal solution without ATP) sensitivity to hypoxia is maintained.

References

Anichkov SV, Belenkii ML 1963 Pharmacology of the carotid body chemoreceptors. Pergamon, Oxford

Archer SL, Nelson DP, Zimmer S, From AH, Weir EK 1989 Hypoxic pulmonary vasoconstriction is unaltered by creatine depletion induced by dietary beta-guanidino propionic acid. Life Sci 45:1081–1088

Archer SL, Wu XC, Thebaud B et al 2004 Preferential expression and function of voltage-gated, O_2-sensitive K^+ channels in resistance pulmonary arteries explains regional heterogeneity in hypoxic pulmonary vasoconstriction: ionic diversity in smooth muscle cells. Circ Res 95:308–318

Buckler KJ 1997 A novel oxygen-sensitive potassium current in rat carotid body type I cells. J Physiol 498:649–662

Buckler K, Williams B, Honoré E 2000 An oxygen-, acid- and anaesthetic-sensitive TASK-like background potassium channel in rat arterial chemoreceptor cells. J Physiol 525:135–142

Buescher PC, Pearse DB, Pillai RP, Litt MC, Mitchell MC, Sylvester JT 1991 Energy state and vasomotor tone in hypoxic pig lungs. J Appl Physiol 70:1874–1881

Kupriyanov VV, Yang L, Deslauriers R 1996 Cytoplasmic phosphates in Na^+–K^+ balance in KCN-poisoned rat heart: a 87Rb-, 23Na-, and 31P-NMR study. Am J Physiol 270:H1303–1311

Leach RM, Sheehan DW, Chacko VP, Sylvester JT 1998 Effects of hypoxia on energy state and pH in resting pulmonary and femoral arterial smooth muscles. Am J Physiol Lung Cell Mol Physiol 275:L1051–1060

Leach RM, Sheehan DW, Chacko VP, Sylvester JT 2000 Energy state, pH, and vasomotor tone during hypoxia in precontracted pulmonary and femoral arteries. Am J Physiol Lung Cell Mol Physiol 278:L294–304

Pillai RP, Buescher PC, Pearse DB, Sylvester JT, Eichhorn GL 1986 31P NMR spectroscopy of isolated perfused lungs. Magn Reson Med 3:467–472

Wyatt CN, Buckler KJ 2004 The effect of mitochondrial inhibitors on membrane currents in isolated neonatal rat carotid body type I cells. J Physiol 556:175–191

Reactive oxygen species facilitate oxygen sensing

Nanduri R. Prabhakar*, Ying-Jie Peng*, Guoxiang Yuan* and Ganesh K. Kumar†

Departments of Physiology & Biophysics and Biochemistry†, School of Medicine, Case Western Reserve University, 10900 Euclid Avenue, Cleveland, OH 44106, USA*

Abstract. Recent studies suggest that reactive oxygen species (ROS) function as second messengers in a variety of physiological processes. Reflexes arising from carotid body (CB) chemoreceptors play critical roles in the pathophysiology associated with chronic intermittent hypoxia (IH) caused by recurrent apnoeas. In the present study, we examined the potential importance of ROS in O_2 sensing of the CB in a rodent model of IH. Chronic IH elicited selective augmentation of hypoxic sensory response and induced a novel form of functional plasticity manifested as sensory long-term facilitation (LTF). Systemic administration of membrane permeable superoxide dismutase (SOD) mimetic prevented chronic IH-induced changes in the CB activity. H_2O_2 at nanomolar concentration mimicked the effects of chronic IH on CB activity in normoxic animals. ROS levels in the carotid body were elevated in chronic IH exposed animals. Inhibition of complex I of the mitochondrial electron transport chain contribute in part to the increased generation of ROS. Chronic IH facilitated serotonin (5-HT) release by acute hypoxia via ROS dependent mechanisms, and 5-HT receptor antagonist prevented alterations in CB activity induced by chronic IH. These observations suggest that chronic IH facilitates O_2 sensing in the CB by mechanisms involving increased generation of ROS in the chemoreceptor tissue.

2005 Signalling pathways in acute oxygen sensing. Wiley, Chichester (Novartis Foundation Symposium 272) p 95–105

Adequate supply of oxygen is essential for the survival of mammalian cells. Much of the oxygen is normally utilized in the mitochondria for generating ATP, which provides energy for various cellular processes. A small portion of oxygen, however, is reduced to superoxide anion ($O_2^{\bullet -}$), which is then dismutated to hydrogen peroxide (H_2O_2) by superoxide dismutase. H_2O_2 is then converted to H_2O and O_2 by catalase and glutathione peroxidase or to hydroxyl radical ($OH^{\bullet}$) via the Fenton reaction (Halliwell & Gutteridge 1990). Collectively, these metabolites of O_2 are often referred to as reactive oxygen species (ROS). Because ROS oxidize various proteins, it was thought that they are deleterious to cellular functions. However, several recent studies suggest that ROS in fact participate as second messengers in a variety of physiological processes (Wolin 1996, Bauer 2000, Hancock et al 2001). Whether ROS also influence the signalling pathways associated with O_2 sensing remains uncertain. In the carotid body (CB), which is the primary chemoreceptor for detecting changes in arterial pO_2, the stimulation of sensory activity by acute hypoxia is inhibited in the presence of micromolar concentrations of H_2O_2 (Acker et al 1992, Osanai et al 1997). Consequently, it was proposed that ROS function as 'inhibitory' messengers in acute hypoxic sensing at the carotid body. Chronic hypoxia by activating transcription factors such as hypoxia-inducible factors (HIFs) up-regulate certain genes associated with O_2 homeostasis (Semenza 2001) and ROS seem to facilitate transcriptional activation of genes during chronic hypoxia (Merrill & Murphy 1997, Chandel & Schumacker 2000).

Intermittent hypoxia (IH) is more often encountered in people living at sea level than continuous hypoxia. For instance, periodic hypoxia is associated with sleep-disordered breathing manifested as recurrent apnoeas. Patients with chronic IH caused by sleep apnoeas develop hypertension (Young et al 1997, Nieto et al 2000). Reflex augmentation of sympathetic activity arising from the CB has been implicated in the morbidity associated with recurrent apnoeas (Fletcher et al 1992). In this communication, we present evidence that (a) chronic IH augments hypoxic sensing at the CB and IH induces a novel form of functional plasticity manifested as sensory long-term facilitation (LTF), and (b) these effects are mediated by increased generation of ROS in the chemoreceptor tissue.

Augmented hypoxic sensitivity of the carotid body by chronic IH

To assess the effects of chronic IH on carotid body activity, we exposed rats to 15 s 5% O_2 followed by 5 min 21% O_2; 9 episodes/h; 8 h/day, for 10 days. In chronic IH CB, hypoxic sensory response was augmented (Peng & Prabhakar 2004). Similar enhancement of the hypoxic sensory response was also elicited in *ex vivo* carotid bodies harvested from rats conditioned with 10 days of IH, suggesting that the effect of chronic IH was not secondary to cardiovascular changes. The facilitatory effects of IH on the hypoxic sensory response completely reversed after placing the IH-conditioned animals in normoxic environment for 10 days. The effects of chronic were selective to hypoxic sensory response because IH conditioning had no significant effect on the sensory response to hyperoxic hypercapnia (Peng & Prabhakar 2004). These observations suggest that chronic IH selectively up-regulates hypoxic sensing at the CB.

IH induces sensory long-term facilitation of the carotid body

When rats conditioned with chronic IH were subjected to acute IH (15 s 5% O_2 followed by 5 min hyperoxia; 10 episodes) baseline sensory activity progressively increased with each episode of hypoxia. The increased activity persisted for hours even after the termination of acute IH challenges despite the maintenance of arterial O_2 levels close to baseline control conditions. The persistent increase in baseline sensory activity elicited by chronic IH resembled the LTF of breathing (Turner & Mitchell 1997), and hence is referred to as 'sensory LTF' (Peng et al 2003). Induction of sensory LTF is not secondary to cardiovascular alterations or circulating vasoactive hormones because it can be elicited in *ex vivo* CB, wherein the influences from cardiovascular alteration are effectively absent. Furthermore, sensory LTF was a time-dependent, reversible phenomenon, and was not associated with gross morphological changes of the CB (Peng et al 2003). These observations demonstrate that chronic IH, in addition to enhancing the hypoxic sensing, also induces a novel form of functional plasticity in the CB manifested as sensory LTF.

In contrast, exposing rats to 4 h of continuous hypoxia (CH), which is equivalent to the total duration of hypoxia accumulated during 10 days of IH, or exposure to 4 h of hypoxia/day for 10 days, neither enhanced the hypoxic sensory response nor elicited sensory LTF (Peng et al 2003, Peng & Prabhakar 2004). These observations suggest that although hypoxia occurs in both IH and CH, only chronic IH with repetitive exposures to low O_2 is needed to induce functional plasticity in the CB.

Evidence for ROS in chronic IH-induced functional plasticity of the carotid body

What makes IH more unique than CH? IH is characterized by periodic re-oxygenations, which are absent in CH. It is likely that ROS are generated in the re-oxygenation phases, which in turn may induce functional changes in the CB. If ROS are involved then: (a) ROS scavengers should prevent IH-induced changes in the CB; (b) IH should increase ROS levels in the glomus tissue; and (c) exogenous ROS should mimic the effects of IH on carotid body sensory activity in control animals.

Systemic administration of a membrane permeable superoxide dismutase (SOD) mimetic (manganese [III] tetrakis [1-methyl-4-pyridyl] porphyrin pentachloride; MnTMPyP), a potent $O_2^{\bullet-}$ scavenger, prevented enhanced hypoxic sensory response and sensory LTF in chronic IH animals (Peng et al 2003, Peng & Prabhakar 2004). Aconitase enzyme activity, an index of $O_2^{\bullet-}$ generation (Gardner et al 1994), was markedly down-regulated in chronic IH compared to control CB (Peng et al 2003), suggesting increased generation of ROS. H_2O_2 (100–500 nM) significantly potentiated the hypoxic sensory response and induced sensory LTF in response to acute IH in *ex vivo* CB harvested from normoxic rats.

Mitochondrial electron transport chain is source of ROS generation during chronic IH

Studies thus far suggest that ROS are involved in chronic IH-induced changes in the CB. Cellular sources of ROS include the mitochondrial electron transport chain (ETC) at complex I and/or III (Ambrosio et al 1993), as well as cytosolic and membrane-bound oxidases (Halliwell & Gutteridge 1990). To begin to understand which of these sources are affected by chronic IH, we examined the effects of chronic IH on the mitochondrial ETC. The activity of complex I was down-regulated in carotid bodies from chronic IH rats whereas activity of complex III was unaltered (Peng et al 2003). Comparable, cumulative duration of CH, however, had no effect either on complex I or III activities. These observations suggest that inhibition of complex I of the ETC contributes in part to increased ROS generation during chronic IH.

How might ROS affect O_2 sensing at the carotid body? Evidence for 'priming' of transmitter release

Previous studies on adrenal chromaffin cells showed that disruption of the mitochondrial ETC enhances transmitter release in response to depolarizing stimulus (Cuchillo-Ibanez et al 2004). It is well established that release of neurotransmitters from glomus cells is obligatory for afferent nerve activation by hypoxia. We therefore tested whether enhanced transmitter release by acute hypoxia would account for chronic-IH-induced augmented hypoxic sensitivity and sensory LTF. Our recent data revealed that hypoxia-evoked serotonin (5-HT) and dopamine (DA) releases were enhanced by five- and threefold, respectively, in chronic IH compared to control carotid bodies. The effects of chronic IH on 5-HT and DA release are mediated by ROS as evidenced by blockade of their facilitated release by SOD mimetic. These observations led us to postulate that chronic IH via increased generation of ROS 'primes' the secretory machinery facilitating the release of neurotransmitters.

To understand the functional consequence of enhanced DA and 5-HT release, we examined the effects of DA and 5-HT receptor antagonists on CB sensory activity in chronic IH conditioned rats. 5-HT_2 receptor blocker (ketanserin, 1 μM) completely prevented sensory LTF, with modest effect on the hypoxic sensory response. By contrast, DA receptor antagonists neither affected the sensory LTF nor the enhanced hypoxic sensory response. These observations suggest that chronic IH facilitates 5-HT release by hypoxia via ROS-dependent mechanisms, and 5-HT plays a major role in the induction of sensory LTF in chronic IH animals.

In summary, the results presented above demonstrate that chronic IH such as occurs with recurrent apnoeas facilitates O_2 sensing in the CB by mechanisms involving increased generation of ROS in the chemoreceptor tissue.

Acknowledgement

This work is supported by grants from National Institutes of Health, Heart, Lung, and Blood Institute HL-25830 and HL-66448.

References

Acker H, Bolling B, Delpiano MA, Dufau E, Gorlach A, Holtermann G 1992 The meaning of H2O2 generation in carotid body cells for PO2 chemoreception. J Auton Nerv Syst 41: 41–51

Ambrosio G, Zweier JL, Duilio C et al 1993 Evidence that mitochondrial respiration is a source of potentially toxic oxygen free radicals in intact rabbit hearts subjected to ischemia and reflow. J Biol Chem 268:18532–18541

Bauer G 2000 Reactive oxygen and nitrogen species: efficient, selective, and interactive signals during intercellular induction of apoptosis. Anticancer Res 20:4115–4139

Chandel NS, Schumacker PT 2000 Cellular oxygen sensing by mitochondria: old questions, new insight. J Appl Physiol 88:1880–1889

Cuchillo-Ibanez I, Lejen T, Albillos A et al 2004 Mitochondrial calcium sequestration and protein kinase C cooperate in the regulation of cortical F-actin disassembly and secretion in bovine chromaffin cells. J Physiol 560:63–76

Fletcher EC, Lesske J, Behm R, Miller CC 3rd, Stauss H, Unger T 1992 Carotid chemoreceptors, systemic blood pressure, and chronic episodic hypoxia mimicking sleep apnea. J Appl Physiol 72:1978–1984

Gardner PR, Nguyen DD, White CW 1994 Aconitase is a sensitive and critical target of oxygen poisoning in cultured mammalian cells and in rat lungs. Proc Natl Acad Sci USA 91:12248–12252

Halliwell B, Gutteridge JM 1990 Role of free radicals and catalytic metal ions in human disease: an overview. Methods Enzymol 186:1–85

Hancock JT, Desikan R, Neill SJ 2001 Role of reactive oxygen species in cell signalling pathways. Biochem Soc Trans 29:345–350

Merrill JE, Murphy SP 1997 Regulation of gene expression in the nervous system by reactive oxygen and nitrogen species. Metab Brain Dis 12:97–112

Nieto FJ, Young TB, Lind BK et al 2000 Association of sleep-disordered breathing, sleep apnea, and hypertension in a large community-based study. Sleep heart health study. JAMA 283:1829–1836

Osanai S, Mokashi A, Rozanov C, Buerk DG, Lahiri S 1997 Potential role of H2O2 in chemoreception in the cat carotid body. J Auton Nerv Syst 63:39–45

Peng YJ, Prabhakar NR 2004 Effect of two paradigms of chronic intermittent hypoxia on carotid body sensory activity. J Appl Physiol 96:1236–1242 (discussion 1196)

Peng YJ, Overholt JL, Kline D, Kumar GK, Prabhakar NR 2003 Induction of sensory long-term facilitation in the carotid body by intermittent hypoxia: implications for recurrent apneas. Proc Natl Acad Sci USA 100:10073–10078

Semenza GL 2001 Hypoxia-inducible factor 1: control of oxygen homeostasis in health and disease. Pediatr Res 49:614–617

Turner DL, Mitchell GS 1997 Long-term facilitation of ventilation following repeated hypoxic episodes in awake goats. J Physiol 499:543–550

Wolin MS 1996 Reactive oxygen species and vascular signal transduction mechanisms. Microcirculation 3:1–17

Young T, Peppard P, Palta M et al 1997 Population-based study of sleep-disordered breathing as a risk factor for hypertension. Arch Intern Med 157:1746–1752

DISCUSSION

Duchen: It is surprising to me that everything should be attributable to the carotid body (CB). Adrenals, kidneys and other areas are going to be seeing intermittent hypoxia (IH) also.

Prabhakar: Dr Cory Smith in our group is looking into the effects of IH on adrenal medullary secretion.

Peers: I was interested in the stabilization of hypoxia-inducible factor (HIF) for so long after the IH. What is the mechanism behind this?

Prabhakar: I can't really say. There are several possibilities including increased protein stabilization, perhaps involving the prolyl hydoxylases. Also, there might be increase in protein synthesis. Currently, we are looking into these possibilities.

Chandel: How do you think complex I is being inhibited?

Prabhakar: That is a good question. In other systems there is some evidence that ROS themselves can target complex I. We have some evidence that IH increases glutathione peroxidase activity, without altering glutathione reductase activity. The altered redox state of the mitochondria might impact on complex I activity. It is just one possibility.

Chandel: So it is conceivable that the complex I inhibition still might be a downstream effect of reactive oxygen species (ROS).

Prabhakar: Yes, it is quite likely.

López-Barneo: How precisely can you measure mitochondrial complex I in the CB?

Prabhakar: In the experiments involving rat CB we couldn't fractionate mitochondria, because the protein levels were so low. However, in studies with PC12 cells, we were able to fractionate mitochondria, and measure complex I and III activities as well as monitor ROS.

López-Barneo: You showed that after the CB was exposed to IH it contained 5-HT, and the amount of 5-HT being released was increased. Can you envisage a system that will reproduce an enhancement of release by some vesicles and not by others? How do you think that 5-HT release from the CB is influencing sensitivity to hypoxia of the glomus cells?

Prabhakar: Basal efflux of serotonin was evident in our HPLC profiles. However, 5 min of hypoxia had no effect on basal release. We also measured the content of 5-HT in the tissues. I believe that the basal release could be putting a brake on further release by interacting with 5-HT_1 autoreceptors.

Weir: Do you measure the antioxidant defences, such as superoxide dismutase (SOD) catalase and glutathione peroxidase? An alternative idea would be that the repeated ROS challenges were increasing cellular antioxidant defences. If that is true and if one believes that ROS go down in hypoxia, you are making an environment that stimulates more prolonged hypoxia.

Prabhakar: We just started looking at this issue in a cell culture model of IH. We measured activities of five pro- and antioxidant enzymes, and found increases in glutathione peroxidase activity, and glutathione reductase activity was unaltered. There were no changes in catalase activity.

Murphy: Which glutathione peroxidase?

Prabhakar: GPX1.

Murphy: There are mitochondrial and cytosolic isoforms of GPX1.

Prabhakar: We are looking in the mitochondrial fraction.

Weir: So this went up with the repeated ROS challenges?

Prabhakar: Glutathione peroxidase activity went up following IH. We did not test the effects of repeated ROS challenges.

Weir: Another explanation would be that all the phenomena you are seeing are a result of decrease in ambient ROS, secondary to the increase in glutathione peroxidase, rather than being the result of an intermittent increase in ROS.

Prabhakar: I don't know. We have not done experiments specifically testing this possibility.

Schumacker: This is an interesting idea. But when you applied exogenous H_2O_2, you found that this mimicked the effects of chronic intermittent hypoxia. So if Ken Weir's idea is correct, how would this explain the data with the exogenous oxidants?

Prabhakar: I don't know.

Schumacker: In some of your earlier data you were able to blunt the enhanced hypoxic sensitivity of the CB by giving a cell-permeant SOD mimetic. This converts superoxide into hydrogen peroxide. This would suggest that superoxide was required or was participating in this enhanced sensitivity. Yet you were able to mimic the effects later on by preconditioning with low concentrations of H_2O_2, which amplified the effects of the IH. This would suggest that H_2O_2 is involved, rather than superoxide.

Prabhakar: We did two types of experiments. In one we tested the effects of superoxide anions by giving xanthine–xanthine oxidase, and in the other nanomolar concentrations of H_2O_2 on sensory activity of the *ex vivo* carotid bodies. The effects of xanthine–xanthine oxidase were not that robust, whereas H_2O_2 significantly enhanced hypoxic sensitivity and induced sensory long-term facilitation in response to acute IH. I do not know whether SOD mimetic produces H_2O_2.

Murphy: I have a comment on the SOD mimetics. In some cases they do not act as real SOD mimetics in that they don't reduce superoxide to H_2O_2. Instead they can react with the superoxide to reduce the manganese but without producing H_2O_2. The production of H_2O_2 is seldom measured in an *in vivo* situation. When we have done this with other mimetics we don't find a lot of hydrogen peroxide production. However, I can't say for sure if this is the case with your mimetic.

Prabhakar: We used MnTyPP.

Rich: I had a question which perplexed me about your hydrogen peroxide. You said you injected in 100 nM.

Prabhakar: We superfused the *ex vivo* CB with 100 nM of H_2O_2. These experiments were not done in intact animals.

Nurse: You describe 'enhanced' 5-HT release after IH. A simple explanation might be that this isn't really an enhanced release of 5-HT, but rather a down-regulation of the 5-HT transporter.

Prabhakar: That is a possibility.

Acker: H_2O_2 by itself is sluggish in reaction. It has to be another ROS species which generates the IH response. It is unclear which species is reacting on your channels.

Prabhakar: That is correct. We do not know the nature of the reactive species affecting ion channels in our system.

Acker: Was there any change in the aconitase activity under long-term hypoxia?

Prabhakar: We didn't find any significant reduction.

Sylvester: You said that IH decreased the membrane potential of glomus cells. Do you mean it depolarized them?

Prabhakar: Yes, there was a tendency for depolarization. The resting membrane potential of the glomus cells from IH exposed CB was positive by about 15 mV more than control cells.

Sylvester: Does 5-HT alter ROS production?

Prabhakar: We are looking into this.

Kemp: Crucial to your mechanism is the protein kinase C (PKC)-dependent step. In your *ex vivo* system, does pharmacological manipulation of PKC cause the same effect?

Prabhakar: PKC inhibitor prevented the induction of sensory long term facilitation (LTF).

Kemp: Secondly, PKC is a very promiscuous phosphorylator. You looked specifically at MARCKS as a phosphorylation target. Have you any idea which other proteins are being phosphorylated, and what is the evidence that it must be a MARCKS-dependent process?

Prabhakar: You are right: there could be several targets for the PKC. The reason I focused on MARCKS was because of its purported role in synaptic vesicle docking to the membrane. My hypothesis was to see if IH facilitates transmitter release, which prompted me to focus on the MARCKS protein. I am not ruling out other PKC targets.

Gonzalez: In the classical reperfusion damage produced by ROS, many people have implicated different origins for ROS. It is not only mitochondrial. Can you exclude the participation of NADH oxidase or other sources?

Prabhakar: As an initial step we started focusing on the mitochondria. I am not ruling out the role of oxidases. In fact we are currently looking into this.

Aaronson: The H_2O_2 by itself didn't do anything. It was only when you combined it with a stimulus. I have two questions, one specific and one more general. Specifically, have you pulsed them with the hydrogen peroxide to try to replicate IH? Second, how do you separate ROS being facilitatory from those being causative?

Prabhakar: The scenario I tried to present is that ROS by themsleves are not effective in eliciting sensory LTF. However, by interacting with hypoxia, ROS lead to sensory LTF.

Aaronson: Perhaps certain levels of ROS are needed for normal cell function to occur. When you put in an antioxidant you are affecting cell function. How do you separate that from it being a specific causative mechanism? When you use an antioxidant you are lowering the basal level of ROS, and it may affect all sorts of signal transduction systems, changing the system so that your specific stimulus no longer works.

Prabhakar: In SOD mimetic-treated rats, basal hypoxic sensory response is still preserved. The *enhanced* hypoxic sensitivity and sensory LTF were absent following treatment with SOD mimetic. We believe that the effects of the SOD mimetic are specific.

Aaronson: Was there any effect of antioxidants in control animals?

Prabhakar: We have not systematically examined the effects of antioxidants in control animals.

Duchen: You almost seem to be accepting that we can expect an intermittent period of hypoxia to generate free radicals. How do you think this actually works?

Prabhakar: The culprit is probably not the IH, but the reoxygenation: in cell culture experiments we changed the duration of hypoxia in each episode and looked at the *c-fos* gene expression as our readout. The magnitude of the c-*fos* response was the same for 15 s, 30 s or 1 min of hypoxia. On the other hand, increasing the duration of the reoxygenation phase from 30 s, to 1 min, to 4 min progressively increased the magnitude of c-*fos* expression. My belief is that free radicals are being generated during the reoxygenation phase.

Duchen: In all the discussions so far we have accepted that it is reasonable to accept an increase in free radical production with short episodes of hypoxia followed by normoxia. What mechanism is generating free radicals?

Prabhakar: I can't speculate. All we know is that it is a time-dependent process in intact animals, because exposure to 1 day of IH has no effect, whereas 3 and 10 day exposures caused progressive enhancement of hypoxic sensitivity of the CB.

Murphy: We can speculate that we have a fully reduced ubiquinone pool and NADH pool which are then exposed to an influx of oxygen.

Duchen: If that is the case, wouldn't you expect that longer episodes of hypoxia would give a bigger response?

Murphy: You would reduce these pools fairly quickly during anoxia.

Duchen: It depends how low the oxygen tension is.

Chandel: It would have to be really low: close to anoxic.

Prabhakar: In intact animals, the effects of IH on CB activity are not dependent on the severity of hypoxia. For instance, the magnitude of changes in the CB sensory activity were the same whether the animals were conditioned with IH either 5% or 10% of inspired O_2 during each episode.

Acker: So the steepness of the influx is critical.

Archer: It is concerning that when one uses biochemical surrogates for hypoxia, such as dithionite, radical production increases rapidly (the opposite of what we see with authentic hypoxic ventilation of the lung) (Archer et al 1995, 1989).

We haven't measured it in the CB, but in isolated pulmonary artery or isolated lung, brief periods of hypoxia (pO2 40 mmHg for 10–30 min) lower ROS production and we don't usually see an overshoot in ROS levels with restoration of normoxia. This situation is quite different from cardiac ischaemia reperfusion where reperfusion does cause an overshoot increase in ROS above normoxic baseline (as measured using lucigenin-enhanced chemiluminescence) (Henry et al 1990).

When we talk about what makes radicals go up and down we should do comparative physiology. Interestingly, when we use enhanced chemiluminescence to measure ROS in the heart, whether with lucigenin or luminol, we find that there is an identical pattern to that seen with electron spin resonance (spin-trapping). Specifically, with normoxic perfusion we see ROS at a certain level, with cessation of perfusion (ischaemia) there is a dramatic fall in ROS, and with reperfusion there is a dramatic overshoot. This is inhibited by SOD, to a large extent. In general, over 15–20 years of measuring, I haven't seen this very often in isolated lungs or pulmonary arteries. This is with moderate hypoxia. I am surprised you get this re-oxygenation overshoot in ROS within a couple of minutes, unless the antioxidant defence system is somehow deficient in this model.

Duchen: With your heart measurements, you are talking about anoxia or ischaemia. Here we are talking modest hypoxia. Are these animals affected by the hypoxia? Is it enough to alter their state of consciousness?

Prabhakar: They are awake animals.

Archer: In your work on the heart (as presented today), this was with profound drops in pO2, not brief modest drops.

Schumacker: I have done several experiments with IH. We superfuse cells using a system where we can oscillate the source of perfusate from normoxic and hypoxic reservoirs, while we measure ROS production. We find that even with fairly severe hypoxia—down to 5 mmHg—there is an increase in ROS production, and this increase occurs during the hypoxia, rather than during the reoxygenation. The only time that we see reoxygenation-induced ROS production is in the heart when it has been ischaemic for some time. A reoxygenation ROS burst occurs, but this is associated with massive cell death and probably ROS production from multiple sources. I think that the ROS signal that Nanduri Prabhakar is seeing is coming from the mitochondria, but during the hypoxia rather than during reoxygenation.

Prabhakar: In fact, there is a recent paper reporting increased generation of reactive oxygen species from neutrophils in patients with obstructive sleep apnea (Dyugovskaya et al 2002). When these patients were treated with continuous positive airway pressure (CPAP) the ROS production decreased.

References

Archer SL, Nelson DP, Weir EK 1989 Simultaneous measurement of O2 radicals and pulmonary vascular reactivity in rat lung. J Appl Physiol. 67:1903–1911

Archer SL, Hampl V, Nelson DP, Sidney E, Peterson DA, Weir EK 1995 Dithionite increases radical formation and decreases vasoconstriction in the lung. Evidence that dithionite does not mimic alveolar hypoxia. Circ Res 77:174–181

Dyugovskaya L, Lavie P, Lavie L 2002 Increased adhesion molecules expression and production of reactive oxygen species in leukocytes of sleep apnea patients. Am J Respir Crit Care Med 165:934–939

Henry TD, Archer SL, Nelson D, Weir EK, From AH 1990 Enhanced chemiluminescence as a measure of oxygen-derived free radical generation during ischemia and reperfusion. Circ Res 67:1453–1461

Oxygen sensing in neuroepithelial and adrenal chromaffin cells

C. A. Nurse*, J. Buttigieg*, R. Thompson*[1], M. Zhang* and E. Cutz†

*Department of Biology, McMaster University, Hamilton, Ontario L8S 4K1, and †Division of Pathology, The Hospital For Sick Children, Toronto, Ontario M5G 1X8, Canada

Abstract. Oxygen sensing and initiation of appropriate physiological responses to hypoxia are crucial for survival. The molecular identity of the sensor has generally sparked considerable interest and controversy in O_2-sensitive cells. In mammals, pulmonary neuroepithelial bodies (NEBs) and adrenal chromaffin cells (AMCs) are O_2 sensitive, particularly during the transition from intrauterine to air-breathing life. In NEBs, there is good evidence that the O_2 sensor is a plasma membrane-bound NADPH oxidase which during hypoxia, signals K^+ channel inhibition, membrane depolarization and neurosecretion via changes in reactive oxygen species (ROS) (e.g. H_2O_2). Accordingly, hypoxic sensitivity is lost in NEBs from transgenic mice deficient in the gp91^{phox} subunit of NADPH oxidase; it is, however, retained in neonatal AMCs from these transgenic mice. A search for the O_2 sensor in neonatal rat AMCs suggests a role for the mitochondrial electron transport chain. For example, the complex I blocker, rotenone (1 μM), mimics hypoxia in causing K^+ channel inhibition and ATP secretion, and occludes hypoxic sensitivity. The evidence is consistent with hypoxia and rotenone acting via a *decrease* in ROS. In contrast, the complex IV blocker cyanide (2 mM) did not mimic the effects of hypoxia. We propose that changes in ROS serve as a common link between the O_2 sensor and secretion in perinatal NEBs and chromaffin cells. However, the subcellular localization of the O_2 sensor appears to be different between these two cell types.

2005 Signalling pathways in acute oxygen sensing. Wiley, Chichester (Novartis Foundation Symposium 272) p 106–118

The ability to sense oxygen and make appropriate physiological responses to low pO2 (hypoxia) is crucial for the survival of aerobic organisms. Thus, the mechanisms underlying cellular responses to decreased O_2 availability are of both biological and clinical importance. While it is recognized that probably all cells have some ability to detect pO2, in recent years considerable attention has been directed to the more specialized receptor cells, as is evident from the papers included in this symposium volume. These O_2-sensitive cells include the prototypic carotid body

[1]Current address: Department of Psychiatry, The Brain Research Centre, University of British Columbia, 2211 Wesbrook Mall, Vancouver, British Columbia, Canada.

chemoreceptors, vascular smooth muscle cells, pulmonary neuroepithelial bodies (NEBs), and adrenal chromaffin cells (Lopez-Barneo et al 2001). This paper will contrast O_2-sensing mechanisms in the latter two cell types. The NEBs are distributed as scattered clusters of intrapulmonary neuroepithelial cells, whose apical surfaces are exposed to the airway lumen (Cutz & Jackson 1999). These epithelial corpuscles are preferentially situated near airway branch points and receive a sensory innervation from the vagus nerve. Lauweryns et al (1977) have demonstrated that NEBs respond to hypoxia with increased exocytosis of their serotonergic dense cored granules, suggesting they act as sensors of airway hypoxia. The prominence of NEBs in the perinatal period has suggested that their principal role is to act as airway pO2 sensors during the transition from intrauterine to extrauterine or air-breathing life (Cutz & Jackson 1999). Likewise, an important role of adrenomedullary chromaffin cells (AMCs) as direct pO2 sensors during the transition of the neonate from intrauterine to air-breathing life has been established. These cells release catecholamines (CA) in response to the natural stressors, e.g. hypoxia or asphyxia, associated with delivery and the birthing process. This CA release is critical to survival of the neonate since it promotes regulation of cardiac function, and prepares the lung for air-breathing by stimulating surfactant secretion and by transforming the lung epithelium from a state of net fluid secretion to one of net fluid absorption (Seidler & Slotkin 1985, Slotkin & Seidler 1988). Interestingly, though neonatal AMCs act as 'direct' PO2 sensors in rodents and humans, this property is gradually lost postnatally along a time course that roughly parallels the sympathetic innervation of these cells by the splanchnic nerve (Slotkin & Seidler 1988). Thus, a comparison of the components of the hypoxia signalling pathway in neonatal (postnatal day P1–P2) versus juvenile (P15–21) chromaffin cells in rodents may give clues about the O_2-sensing mechanisms in these cells. In the subsequent sections we review evidence suggesting that though NEBs and neonatal AMCs may share common steps in the signalling pathway, they use different pO2 sensors.

O_2 sensing by neuroepithelial bodies

Using whole-cell patch clamp techniques, Youngson et al (1993) demonstrated that NEB cells from fetal rabbit lungs were electrically excitable and expressed a voltage-activated outward K^+ current that was reversibly inhibited (~25%) by hypoxia (pO_2 = 25–30 mmHg). Under current clamp conditions, hypoxia caused an increase in firing frequency in these cells, presumably leading to increased entry of extracellular Ca^{2+} through voltage-dependent Ca^{2+} channels and enhanced neurosecretion. These responses were not simply the result of the artificial culture conditions since similar hypoxic inhibition of outward K^+ current was seen in NEB cells studied in intact tissue slices of both fetal rabbit (Fu et al 1999) and mouse (Fu et al 2000) lung. Moreover, in fetal rabbit lung slices hypoxia induced a dose-dependent, tetrodotoxin-sensitive increase in 5-HT secretion as detected by carbon fibre amperometry (Fu et al 2002). This secretion was largely inhibited by the L-type Ca^{2+} channel blocker, nifedipine. There is compelling evidence that the pO2 sensor in these NEB cells is the NADPH oxidase complex, which includes two membrane proteins ($gp91^{phox}$ and $p22^{phox}$) that form a b-type cytochrome and two cytosolic protein subunits ($p47^{phox}$ and $p67^{phox}$). First, several components of the NADPH oxidase complex were demonstrated in NEBs of fetal rabbit lung by immunocytochemistry (Youngson et al 1997). Second, activity of the NADPH oxidase complex, which catalyzes the one-electron reduction of molecular O_2 to superoxide (O_2^-), was demonstrated using dihydrorhodamine fluorescence as a probe for H_2O_2 generation (see Cutz & Jackson 1999). Third, application of the NADPH oxidase inhibitor, diphenylene iodonium (DPI; 1 μM) suppressed outward K^+ current in NEB cells similar to hypoxia, and the presence of DPI occluded the hypoxic response (Youngson et al 1993, Fu et al 1999). Fourth and most compelling, NEB cells in tissue slices from transgenic mice, deficient in the $gp91^{phox}$ subunit of the oxidase complex, failed to show hypoxic inhibition of outward K^+ current, in contrast to wild-type cells (Fu et al 2000). Similarly, DPI had no effect on K^+ current in NEB cells from oxidase deficient (OD) mice. Taken together, these data lend support to a model where the gating properties of the O_2-sensitive K^+ channels are regulated by by-products of NADPH oxidase activity, e.g. H_2O_2 (Cutz & Jackson 1999). Indeed, expression of mRNA encoding the H_2O_2-sensitive, voltage-gated K^+ channel subunit $K_V3.3a$ has been demonstrated in NEB cells of fetal rabbit lung, together with mRNA for the membrane components, i.e. $gp91^{phox}$ and $p22^{phox}$ (Wang et al 1996). Thus, under hypoxic conditions a decrease in generation of reactive oxygen intermediates presumably leads to closing of the H_2O_2-sensitive, voltage-gated K^+ channels, increased depolarization or spike frequency, and enhanced neurosecretion which activates the afferent sensory pathway (Cutz & Jackson 1999).

O_2 sensing by neonatal adrenomedullary chromaffin cells

In the perinatal period, AMCs secrete catecholamines in response to hypoxic stress via a direct pO_2 sensing mechanism (Slotkin & Seidler 1988, Mojet et al 1997, Thompson et al 1997). Similar to neonatal NEB cells, hypoxia inhibits outward K^+ current and depolarizes neonatal rat AMCs, after short-term (1–2 days) culture (Thompson et al 1997, Thompson & Nurse 1998). A variety of K^+ currents appear to be modulated by hypoxia in neonatal rat AMCs. Among these are an iberiotoxin-sensitive Ca^{2+}-dependent (BK) K^+ current and a delayed rectifier-type K^+ current which are inhibited by hypoxia, and a glibenclamide-sensitive, presumptive ATP-dependent K^+ current (K_{ATP}) which is augmented by hypoxia (Thompson & Nurse

1998, see also Keating et al 2001). Following isolation, these cells also release CA in a Ca^{2+}-dependent manner during hypoxia, though there are discrepancies concerning the minimum pO2 required to evoke release and the responsive age of the animals (Mojet et al 1997, Mochizuki-Oda et al 1997, Thompson et al 1997, Lee et al 2000).

In contrast to NEB cells (see above), the NADPH oxidase complex does not appear to function as the key pO2 sensor in neonatal AMCs. This conclusion is based on data from mice deficient in the $gp91^{phox}$ subunit of the NADPH oxidase complex. In neonatal AMCs derived from these oxidase-deficient (OD) mice, hypoxic sensitivity was retained and appeared comparable to that seen in wild-type cells (Thompson et al 2002). For example, in the latter study hypoxia caused a reversible inhibition of outward K^+ current by ~27% in wild-type and ~29% in OD neonatal chromaffin cells. Additionally, hypoxia depolarized both cell types, caused a broadening of the action potential, and evoked a four–sixfold stimulation of CA secretion as determined by HPLC (Thompson et al 2002). This raises the question of the identity and location of the pO2 sensor in AMCs. Mojet et al (1997) proposed that the sensor was located in the mitochondrial electron transport chain (ETC) since various blockers of the ETC, including rotenone (complex I blocker) and cyanide (complex IV blocker) mimicked the effects of severe hypoxia. For example, both rotenone and cyanide caused a rise in intracellular Ca^{2+} and stimulated CA secretion, and the effects of severe hypoxia and cyanide on these parameters were non-additive, suggesting a common mechanism (Mojet et al 1997). In studies on neonatal AMCs in short-term culture, we have found that only rotenone mimicked and occluded the effects of hypoxia, raising the possibility that complex I may act as the pO2 sensor (Thompson 2000, Nurse et al 2003). For example, in Fig. 1 (upper traces), both hypoxia (pO2 ~ 5 mmHg) and rotenone (0.3 μM) cause a reversible inhibition of outward K^+ current in neonatal AMCs and, moreover, the combined effect of hypoxia and rotenone (h + r) was non-additive. In contrast, the effects of cyanide were variable. Surprisingly, in most cells cyanide (2 mM) had no significant effect on outward K^+ current (Fig. 1; lower right), though in a few cases it caused a potentiation of the current (Fig. 1; lower left). Similarly, in several cells cyanide had negligible effect on the resting membrane potential, though in a few cases it caused membrane hyperpolarization which is opposite to the effect of hypoxia (Thompson 2000, Nurse et al 2003). The cyanide-induced hyperpolarization may be the result of a dominant effect in activating a K_{ATP} current since it was reversed by glibenclamide. The reasons for the variable effects of cyanide are unclear, but it may have several other non-specific effects (at least in this system) in addition to blocking complex IV.

If hypoxia and rotenone act through a common pO2 sensor protein (e.g. complex I), what is the intervening signal that couples the sensor to K^+ channel inhibition? Complex I is one of several sites in the mitochondrial ETC where ROS are thought

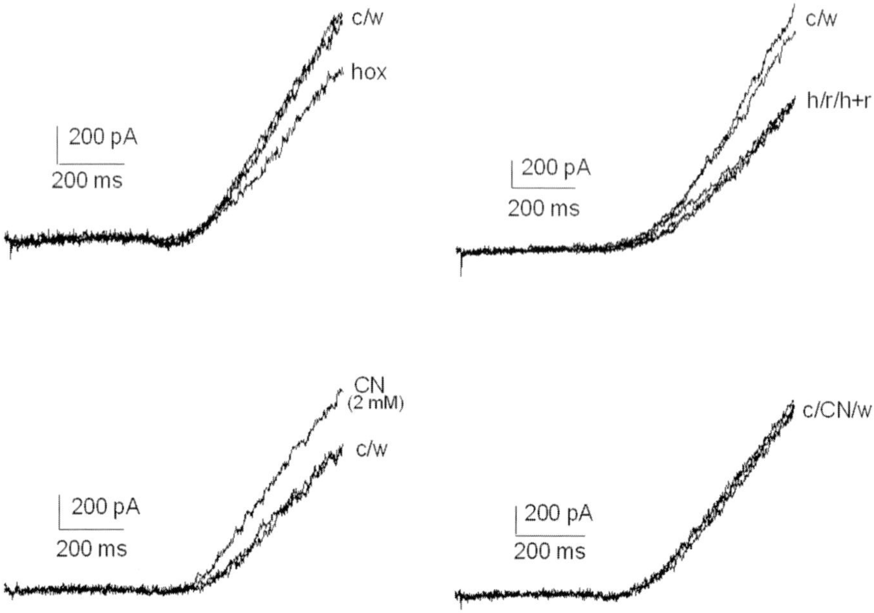

FIG. 1. Effects of hypoxia and mitochondrial inhibitors on K^+ currents in isolated neonatal rat adrenomedullary chromaffin cells. Traces in each plot represent the current-voltage (I–V) relationship elicited during voltage ramps from −100 mV to +50 mV over a period of 1 s; holding potential was −60 mV. In the upper left panel, hypoxia (hox; pO_2 = 5 mmHg) reversibly inhibits outward K^+ current; control response (c) and recovery after drug washout (w) are also shown. In the upper right panel, the complex I inhibitor rotenone (0.3 μM; r) mimics hypoxia (h) in causing inhibition of outward K^+ current, and the presence of rotenone occluded the effect of hypoxia (h + r). Though in most cells the complex IV blocker cyanide (CN) had negligible effects on outward current (lower right panel), it enhanced the current in a few cases (lower left panel). Recordings were obtained at 35 °C from enriched cultures of chromaffin cells after 1–2 days *in vitro.*

to be generated as a result of electron leak, though there is controversy as to whether hypoxia causes an increase or a decrease in ROS (Waypa et al 2001, Archer & Michelakis 2002). We have measured ROS in neonatal AMCs using luminol chemiluminescence and found that both hypoxia and rotenone consistently produced a decrease in light signal, consistent with a decrease in production of reactive oxygen intermediates (Fig. 2A). In a few experiments, 2 mM cyanide appeared to increase ROS generation over the first few minutes of exposure (Fig. 2A,B). Other mitochondrial ETC blockers including myxothiozol (complex III blocker) mimicked hypoxia in causing a decrease in ROS but the effects were additive, suggesting separate pathways were involved (not shown). We also tested whether the secretory functions of neonatal AMCs were present under conditions where ROS

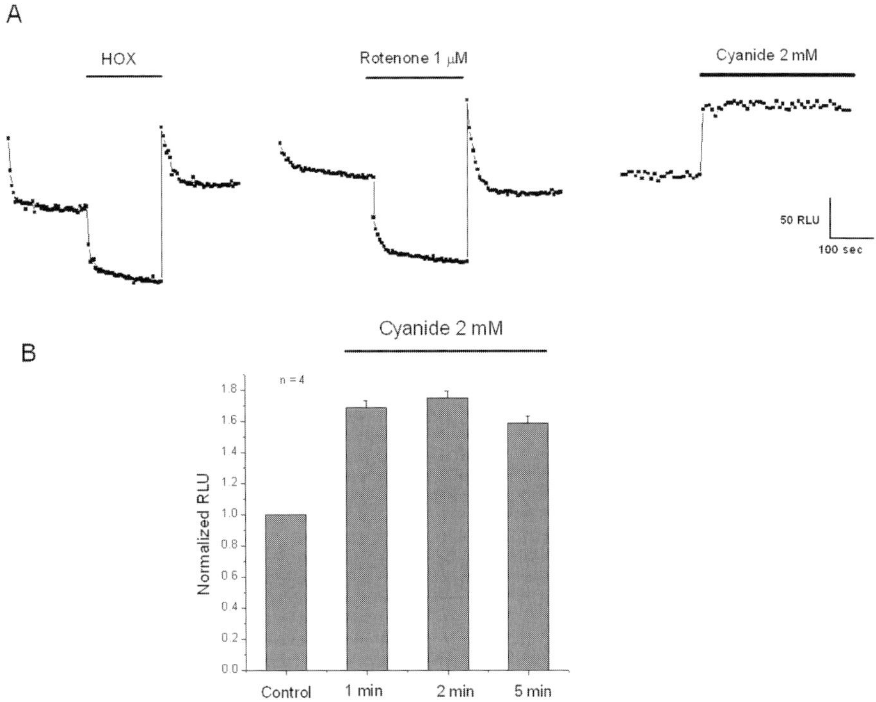

FIG. 2. Effects of hypoxia and mitochondrial blockers on levels of reactive oxygen species (ROS) in neonatal rat adrenomedullary chromaffin cells. ROS measurements were obtained using the chemiluminescent probe luminol and data were recorded as relative light units (RLU). In A, hypoxia (hox; PO_2 15–20 mmHg) and rotenone (1 μM) decrease ROS, whereas cyanide (2 mM) caused an increase in ROS. Time course of the relative increase in ROS after the application of CN is shown in B. Chemiluminescence data were obtained at 35 °C from enriched cultures of chromaffin cells after 1–2 days *in vitro*.

measurements were obtained. Since ATP is normally co-released with catecholamines from chromaffin granules, we used the sensitive luciferin–luciferase chemiluminescence assay to detect extracellular ATP from similar cultures to those used for ROS measurements. As illustrated in Fig. 3, hypoxia caused a significant increase in extracellular ATP (normalized to normoxic control) and similar effects were seen with rotenone (1 μM; not shown). On the other hand, cyanide (2 mM) failed to stimulate secretion under these conditions (Fig. 3), in contrast to the data reported by Mojet et al (1997). Taken together, our results support the hypothesis that hypoxia sensing by neonatal AMCs occurs at a site in the proximal ETC (probably complex I), and a decrease in ROS acts as the intermediary signal linking this stimulus to K^+ channel inhibition, membrane depolari-

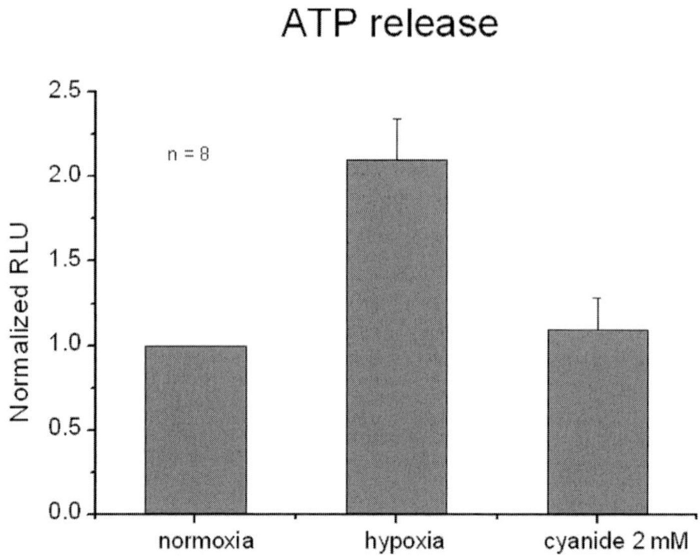

FIG. 3. Effects of hypoxia and cyanide on secretory activity of isolated neonatal rat adrenomedullary chromaffin cells. Secretory activity was measured as the change in extracellular ATP levels, determined by the sensitive luciferin–luciferase chemiluminescence assay, under similar conditions to those used to determine ROS levels in Fig. 2. Note hypoxia stimulated ATP secretion whereas cyanide had negligible effect.

zation and exocytosis. In support of this idea, in recent studies the complex II substrate succinate (5 mM) was found to reverse the effects of hypoxia and rotenone on K^+ channel inhibition (our unpublished observations), presumably by bypassing complex I and supplying electrons to other ROS generating sites (e.g. complex III; see Leach et al 2001).

Conclusions

Both pulmonary NEBs and AMCs act as hypoxia sensors in the perinatal period, but appear to use different molecules to detect pO2. Whereas the NADPH oxidase complex appears to fulfil this role in NEB cells, this is not the case for AMCs. A search for the sensor in the latter cells supports a role for the mitochondrial ETC. Controversies still surround the actual site of the O_2 sensor in the ETC and the role of ROS intermediates. Our recent studies favour the proximal ETC as the sensor and a decrease in ROS as the link between hypoxia sensing and K^+ channel inhibition, leading to CA secretion.

Acknowledgements

This work was supported by operating grants to CAN from the Heart and Stroke Foundation of Ontario and the Natural Sciences and Research Engineering Council of Canada, and to EC from the Canadian Institutes of Health Research.

References

Archer S, Michelakis E 2002 The mechanism(s) of hypoxic pulmonary vasoconstriction: potassium channels, redox O(2) sensors, and controversies. News Physiol Sci 17:131–137

Cutz E, Jackson A 1999 Neuroepithelial bodies as airway sensors. Respir Physiol 115:201–214

Fu XW, Nurse CA, Wang YT, Cutz E 1999 Selective modulation of membrane currents by hypoxia in intact airway chemoreceptors from neonatal rabbit. J Physiol 514:139–150

Fu XW, Wang D, Nurse CA, Dinauer MC, Cutz E 2000 NADPH oxidase is an O_2 sensor in airway chemoreceptors: evidence from K^+ current modulation in wild-type and oxidase-deficient mice. Proc Natl Acad Sci USA 97:4374–4379

Fu XW, Nurse CA, Wong V, Cutz E 2002 Hypoxia-induced secretion of serotonin from intact pulmonary neuroepithelial bodies in neonatal rat. J Physiol 539: 503–510

Keating DJ, Rychkov GY, Roberts ML 2001 Oxygen sensitivity in the sheep adrenal medulla: role of SK channels. Am J Physiol Cell Physiol 281:C1434–C1431

Lauweryns JM, Cokelaere M, Beleersnyder M, Liebens M 1977 Intrapulmonary neuroepithelial bodies in newborn rabbits. Influence of hypoxia, hypercapnia, nicotine, reserpine, L-dopa and 5-HT. Cell Tiss Res 182:425–440

Leach RM, Hill HM, Snetkov VA, Robertson TP, Ward JPT 2001 Divergent roles of glycolysis and the mitochondrial electron transport chain in hypoxic pulmonary vasaconstriction of the rat: identity of the hypoxic sensor. J Physiol 536:211–224

Lee J, Lim W, Eun S, Kim S, Kim J 2000 Inhibition of apamin-sensitive K^+ current by hypoxia in adult rat adrenal chromaffin cells. Pflugers Arch 439:700–704

Lopez-Barneo J, Pardal R and Ortega-Saenz P 2001 Cellular mechanism of oxygen sensing. Ann Rev Physiol 63:259–287

Mochizuki-Oda N, Takeuchi Y, Matsumura K, Oosawa Y, Watanabe Y 1997 Hypoxia-induced catecholamine release and intracellular Ca^{2+} increase via suppression of K^+ channels in cultured rat adrenal chromaffin cells. J Neurochem 69:377–387

Mojet M, Mills E, Duchen MR 1997 Hypoxia-induced catecholamine secretion in isolated newborn rat adrenal chromaffin cells is mimicked by inhibition of mitochondrial respiration. J Physiol 504:175–189

Nurse CA, Fearon IM, Jackson A, Thompson RJ 2003 Oxygen sensing by neonatal adrenal chromaffin cells: a role for mitochondria? *In:* Lahiri S, Semenza G, Prabhakar N (eds) Oxygen sensing: responses and adaptations to hypoxia. Marcel Dekker Inc, New York, p 603–618

Seidler FJ, Slotkin T 1985 Adrenomedullary function in the neonatal rat: responses to acute hypoxia. J Physiol 385:1–16

Slotkin TA, Seidler FJ 1988 Adrenomedullary catecholamine release in the fetus and newborn: secretory mechanisms and their role in stress and survival. J Devel Physiol 10:1–16

Thompson RJ 2000 Mechanisms of O_2-chemosensing in developing chromaffin cells. Ph.D. Dissertation, McMaster University, Hamilton, Ontario, Canada

Thompson RJ, Nurse CA 1998 Anoxia differentially modulates multiple K^+ currents and depolarizes neonatal rat adrenal chromaffin cells. J Physiol 512:421–434

Thompson RJ, Jackson A, Nurse CA 1997 Developmental loss of hypoxic chemosensitivity in rat adrenomedullary chromaffin cells. J Physiol 498:503–510

Thompson RJ, Farragher SM, Cutz E, Nurse CA 2002 Developmental regulation of O_2 sensing in neonatal adrenal chromaffin cells from wild-type and NADPH-oxidase-deficient mice. Pflugers Arch 444:539–548

Wang D, Youngson C, Wong V et al 1996 NADPH-oxidase and a hydrogen peroxide—sensitive K^+ channel may function as an oxygen sensor complex in airway chemoreceptors and small cell lung carcinoma cell lines. Proc Natl Acad Sci USA 93:13182–13187

Waypa G, Chandel N, Schumacker P 2001 Model for hypoxic pulmonary vasoconstriction involving mitochondrial oxygen sensing. Circ Res. 88:1259–1266

Youngson C, Nurse C, Yeger H, Cutz E 1993 Oxygen sensing in airway chemoreceptors. Nature 365:153–155

Youngson C, Nurse C, Yeger H et al 1997 Immunocytochemical localization on O_2-sensing protein (NADPH oxidase) in chemoreceptor cells. Microsc Res Tech 37:101–106

DISCUSSION

Kummer: I have a comment on the rotenone effects. In some cell types, such as adrenomedullary cells or PC12 cells, rotenone has an effect that mimics hypoxia, and in other cells it doesn't (e.g. Höhler et al 1999, Waypa et al 2002). Something must be different in these neural cells. These effects have mostly been ascribed to complex I. In a series of experiments we treated PC12 cells with thiamphenicol, which inhibits mitochondrial translation. Thus, we depleted the mitochondria of these cells from proteins which are encoded by the mitochondrial genome. Efficiency of this treatment was controlled using antibodies against cytochrome c oxidase. The stimulatory effect of rotenone was entirely there in these thiamphenicol-treated cells, even though we expected it to be gone because it should act on a mitochondrial-encoded protein. This special effect on the PC12 cells is still present at the same level in thiamphenicol-treated cells (W. Kummer, B. Höhler, A. Sell, K. Hoffmann, A. Goldenberg and R. Paddenberg, unpublished results). I am therefore in favour of the data we heard yesterday that rotenone might act, at least in some cell types, on something outside the mitochondria.

Murphy: How long after thiamphenicol treatment did you add the rotenone?

Kummer: 5 d.

Murphy: All the proteins encoded by mitochondrial DNA would be gone by then.

Kummer: Not all. There are at least 40 proteins in complex I and most of them are encoded by the nuclear genome. But those which are encoded by the mitochondrial genome are gone.

López-Barneo: We have similar data. Looking at the proteins encoded by the mitochondrial genome, after 3 d, they are almost all gone.

Nurse: There is always a problem with the drugs. We hope they are acting via complex I, but this is an assumption

López-Barneo: We have been recently looking at the oxygen sensitivity in adrenal medulla slices. It is true that with postnatal maturation the number of cells responding to hypoxia is reduced very much, but still we find that about 10% of the cells are sensitive to hypoxia. The disappearance of the oxygen sensitivity induced by age is not complete.

Nurse: There could be reasons for that given that as yet there are no good cases of hypoxic inhibition in juvenile cells. In adults one doesn't know. The question arises whether all the chromaffin cells are innervated? I can imagine a situation where the innervation is widespread in the first few weeks of postnatal life, and then there is a pruning back. Maybe you need to have a nerve contact with a chromaffin cell in order to see the disappearance of oxygen sensitivity. I don't think anyone has done the experiment of looking at the innervation pattern in the adult adrenal gland.

López-Barneo: You have proposed that a decrease in reactive oxygen species (ROS) is what mediates the change in membrane K^+ conductance in hypoxia. If the channels are the same how do you envisage that this changes with development?

Nurse: Perhaps there are some changes in the channels. There may be differences in terms of their modulation.

López-Barneo: Have you tried other blockers of complex I in your model?

Nurse: Not yet.

Buckler: Why is succinate reversing the effects of hypoxia?

Nurse: My assumption is that succinate can supply electrons via complex II, possibly to complex III. Complex III is a site where ROS are generated. So ROS levels may increase as a result of adding succinate.

Buckler: So the assumption is that normally ROS production is limited by the availability of substrate. There simply isn't enough substrate feeding electrons into the electron transport chain (ETC) to maintain high ubisemiquinone levels so as to generate ROS. If this is the case, if you add succinate under normoxia, do you see an increase in ROS production?

Nurse: The succinate experiments for the most part have been done when complex I is blocked. We haven't looked at succinate on its own in sufficient detail to comment on changes in ROS levels.

Buckler: Presumably when you see a reduction in ROS with hypoxia it is because of reduction in availability of oxygen with which to form ROS at complex I and III.

López-Barneo: In the carotid body, methyl succinate does not prevent occlusion of the sensitivity to hypoxia by rotenone.

Acker: It is important now to show which Nox isoforms can be identified in the adrenal medulla, so you can rule out the oxygen sensing contribution by another specific Nox isoform. We have been talking a lot about ROS formation in the mitochondrial chain. New data show that almost no ROS are produced by mitochondria at resting state when mitochondrial membrane potential is about −120 mV. Only when mitochondrial membrane potential increases to −140 mV ROS production increases (Lee et al 2001). The ROS produced by the normal functioning mitochondria don't appear in the cytosol due to highly effective mitochondrial scavenging systems. It is only under stress conditions that they are found in the cytosol. This raises an interesting question about how rotenone really works at the molecular level in mitochondrial complex I or other structures.

Rich: That's not correct. Rotenone is a natural poison that was used originally by South American Indians to paralyse and catch fish. It has a very specific chemical structure and specifically goes into the ubiquinone binding site of complex I without affecting ubiquinone binding sites of other redox proteins. Complex I contains flavin mononucleotide (FMN), not flavin adenine dinucleotide (FAD), and this is associated with the NADH oxidizing part of the enzyme that is not directly affected by rotenone. When you say that cytochrome oxidase is phosphorylated, this isn't correct: there is some weak evidence that it can be controlled by ATP and ADP binding, but this is controversial. Mitochondria do produce ROS in non-stressed conditions, but only when there is a high membrane potential, such that the ubiquinone pool is partly reduced and the NADH dehydrogenase components are also fairly reduced. You can also probably get some ROS production from succinate dehydrogenase and bc complex under high state 4 conditions. Mitochondria *in vivo*, of course, are neither in state 3 or state 4, but are in a state in between, somewhat substrate limited but with a significant membrane potential. When rotenone is added, ROS production is increased specifically by complex I because the ubiquinone reduction site is blocked. All of the components within complex I are therefore highly reduced and some or all can autooxidize to make ROS. In general, the lower the redox potential of the component, the faster the expected rate of autooxidation to make ROS, though there are other factors to consider for different types of chemical species. So you add rotenone, and this is now an artificial situation: all the iron sulfur centres and the FMN go reduced and, because their reduction level is greater, all will produce ROS at a much faster rate than they would have done without rotenone present simply because you have blocked electron transfer out of the complex.

Acker: In normal operating mitochondria about 1% of the oxygen consumption is due to mitochondrial ROS formation.

Rich: I don't know what you mean by 'normal' mitochondria.

Acker: In state 3.

Rich: In state 3 mitochondria, where there is a low phosphate potential and therefore a relatively lower membrane potential, the amount of ROS production is small. This goes up dramatically in state 4, with a higher membrane potential. The status of mitochondria *in vivo* is somewhat in between the two: there are all sorts of substrate limitations.

Murphy: The figure of 1% usually comes from Britton Chance's papers in the early 1970s (reviewed in Chance et al 1979). This figure is repeatedly quoted, but I don't think anyone has a good idea of what is going on in the living organism. In intact isolated mitochondria, it is easy to measure hydrogen peroxide efflux by several techniques. The techniques that are usually used produce a small amount of hydrogen peroxide efflux in very artificial conditions. There is some evidence, however, that reverse electron transport through complex I, which will produce far more hydrogen peroxide than direct inhibition of complex I (this is where there are a lot of electrons coming in from succinate and a high membrane potential, so the electrons go backwards through complex I) will produce hydrogen peroxide at about 10 times the levels of complete inhibition of complex I. This produces a very large efflux of hydrogen peroxide from mitochondria, but it isn't clear whether this occurs *in vivo*.

Kemp: Hardcore biochemistry notwithstanding, there is evidence that at the concentrations of rotenone which most of us use, it is a fantastic K^+ channel blocker. This occurs with or without functional mitochondria. A few years ago we published data on rhoQ-style cells where functional mitochondria are knocked out genetically, but that rotenone at 100 nM causes a 90–95% inhibition of all K^+ channels. The crunch here is not whether rotenone inhibits anything; it is whether this can be reversed with succinate. If you ever see a rotenone effect and it hasn't been followed by succinate rescue, then you are not necessarily looking at mitochondrial influence.

Nurse: This was one of the nice things about the chromaffin cells at a juvenile stage, rotenone had no effect on the K^+ currents at the same concentration that caused inhibition in neonatal cells.

Kemp: Can you discriminate between the hypoxic effect on electron transport from electron transport effects on channels? What is wrong? Is it the coupling between the hypoxic stimulus and ETC, or ETC and the channel?

Nurse: I think both. There are two things here given the observation that hypoxia doesn't change ROS levels in the juvenile cells that do not sense hypoxia. Decreased ROS in those same cells, as seen with rotenone or NAC, appears to have no significant effect on the K^+ channels or ATP secretion. Since hypoxia decreases ROS in neonatal cells it looks to me that there are at least two changes that are associated with the shift from neonatal to the juvenile stage.

Gonzalez: We have been measuring ROS in carotid body chemoreceptor cells. We have found that hypoxia increases ROS production. If we use knockouts for p47 subunit of NADPH oxidase there is no modification in the production during hypoxia. This means that mitochondrial leak of ROS doesn't change with hypoxia in the chemoreceptor cells.

Nurse: How long was hypoxia applied for?

Gonzalez: We applied hypoxia for 3–5 min.

Nurse: Are you talking about a pure population of receptor cells?

Gonzalez: We have done this with isolated identified chemoreceptor cells. Do you know what isoforms of oxidase you have here?

Nurse: No, we don't.

Gonzalez: How efficient is the system to transport succinate inside the cells? When you put succinate in the bathing solution you lost the entire effect implying that succinate has entered inside the cells. I am not aware of any efficient system for transporting succinate across the plasma membrane.

Nurse: I don't know the system either, but I presume it is getting into the cells in our experiments.

Duchen: My impression is that succinate doesn't cross the membrane well. You can see effects but they are very slow. Most people would use methyl succinate as a membrane permeant form.

Chandel: You showed some effects of myxothiazol mimicking rotenone.

Nurse: We used a concentration of 2 μM. When we have myxothiazol and hypoxia together, the release appeared additive and not occlusive as was the case with rotenone.

Chandel: If we are going to use pharmacological agents, we always like to use myxothiazol in one set of experiments and then follow it up with antimycin. This is because myxothiazol can allow decrease in ROS production or a Q cycle, while antimycin within the Q cycle will generate the ROS.

Nurse: We did some experiments with antimycin, but we found it problematic. The reversibility was a problem when we tried to do further experiments.

Chandel: Did it give opposite results?

Nurse: No, in the same direction, it's just that we couldn't reverse the effects easily on wash out. The concentrations were relatively large, as well, in the 5 μM range.

References

Chance B, Sies H, Boveris A 1979 Hydroperoxide metabolism in mammalian organs. Physiol Rev 59:527–605

Höhler B, Lange B, Holzapfel B et al 1999 Hypoxic upregulation of tyrosine hydroxylase gene transcription is paralleled, but not induced, by increased generation of reactive oxygen species in PC12 cells. FEBS Lett 457:53–56

Lee I, Bender E, Arnold S, Kadenbach B 2001 New control of mitochondrial membrane potential and ROS formation—a hypothesis. Biol Chem 382:1629–1636

Waypa GB, Marks JD, Mack MM, Boriboun C, Mungai PT, Schumacker PT 2002 Mitochondrial reactive oxygen species trigger calcium increases during hypoxia in pulmonary arterial myocytes. Circ Res 91:719–726

Hypoxic regulation of Ca^{2+} signalling in astrocytes and endothelial cells

C. Peers*, P. Kang*, J. P. Boyle*, K. E. Porter*, H. A. Pearson†, I. F. Smith‡ and P. J. Kemp§

Schools of Medicine and Biological Sciences†, University of Leeds, Leeds LS2 9JT, UK, ‡Department of Neurobiology and Behaviour, University of California, Irvine, CA 92697-4545, USA, and §School of Biosciences, University of Cardiff, Cardiff CF10 3US, UK*

Abstract. Acute hypoxia is well known to modulate plasmalemmal ion channels in specific tissue types, thereby modulating $[Ca^{2+}]_i$. Alternative mechanisms by which acute hypoxia could modulate $[Ca^{2+}]_i$ are less well explored, particularly in non-excitable cells. Here, we describe experiments employing microfluorimetric recordings from Fura-2-loaded rat cortical astrocytes and human saphenous vein endothelial cells designed to explore any effects of hypoxia (pO_2 20–30 mmHg) on $[Ca^{2+}]_i$. In both cell types, hypoxia evoked small rises of $[Ca^{2+}]_i$ in the majority of cells during perfusion with a Ca^{2+}-free solution, indicating hypoxia can release Ca^{2+} from an intracellular pool. Capacitative Ca^{2+} entry was observed when Ca^{2+} was subsequently restored to the extracellular solution. These effects were abolished by pre-treatment of cells with thapsigargin or prior application of inositol 1,4,5-trisphosphate (IP_3)-generating agonists. Antioxidants fully prevented this effect of hypoxia in both cell types. Mitochondrial uncoupling significantly enhanced the effects of hypoxia in astrocytes, yet markedly suppressed the effects of hypoxia in endothelial cells. Our findings indicate that hypoxia can modulate $[Ca^{2+}]_i$ in non-excitable cells; most importantly, it can evoke Ca^{2+} release from intracellular stores via a mechanism which involves reactive oxygen species. The involvement of mitochondria in this effect appears to be tissue specific.

2005 Signalling pathways in acute oxygen sensing. Wiley, Chichester (Novartis Foundation Symposium 272) p 119–130

Our understanding of the ways in which rapid changes in local O_2 levels can modulate cellular activity has expanded rapidly in recent years. Since the original report of hypoxic inhibition of K^+ channels in carotid body glomus cells (Lopez-Barneo et al 1988), the realization that hypoxia can evoke rapid cellular responses via modulation of ion channels has expanded into numerous tissue types, including both pulmonary (Weir & Archer 1998) and systemic (Franco-Obregon et al 1995) vascular smooth muscle, neuroepithelial bodies of the lung (Youngson et al 1993), cardiac myocytes (Hool 2001), central neurons (Jiang & Haddad 1994) and immature adrenal chromaffin cells (Rychkov et al 1998). Hypoxic modulation of specific channel types can also be reproduced in recombinant expression systems (Fearon et al 2000, Lewis et al 2001, 2002, Williams et al 2004), providing an opportunity for us to decipher the molecular mechanism(s) of O_2 sensing by ion channels (see Kemp et al 2005, this volume).

The consequences for cell function of hypoxic ion channel modulation are, where studied, reasonably well established. In most cases, an appropriate response to hypoxia (such as systemic vasodilation, pulmonary vasoconstriction or carotid body glomus cell transmitter release—see Lopez-Barneo et al 2001 for review) involves modulation of $[Ca^{2+}]_i$ and this occurs primarily via modulation of Ca^{2+} influx (it should be noted that in the case of hypoxic pulmonary vasoconstriction this is a contentious issue, and compelling evidence to support an important role for intracellular Ca^{2+} stores has been published; see e.g. Evans & Dipp 2002). Ca^{2+} influx can be regulated either indirectly, through control of membrane potential via modulation of K^+ channel activity (Buckler & Vaughan-Jones 1994, Wyatt et al 1995, Osipenko et al 1997, Weir & Archer 1998), or via a more direct effect on Ca^{2+} channel activity (Franco-Obregon et al 1995, Hool 2001).

To date, very little attention has been paid to the effects of hypoxia on electrically non-excitable cells, which express voltage-gated Ca^{2+} channels only at very low levels, if at all. Such cells, exemplified by astrocytes and endothelial cells, are worthy of study with respect to hypoxic responses, since they are in intimate contact with—and so influence the physiological activity of—neighbouring O_2-sensitive, excitable cells (respectively, neurons and vascular smooth muscle). Furthermore, many of their vital roles depend on the precise control of $[Ca^{2+}]_i$. For these reasons, we have explored the ability of acute hypoxia to modulate $[Ca^{2+}]_i$ in both astrocytes and endothelial cells.

For these studies, primary cultures of rat cortical astrocytes were established as previously described (Smith et al 2003) and used for a maximum of two passages. Endothelial cells were obtained from donated samples of saphenous vein of patients undergoing elective coronary artery bypass surgery according to previously described methods (Budd et al 1991) and, as with the astrocytes, used only for a maximum of two passages. $[Ca^{2+}]_i$ was monitored in Fura-2-loaded cells following incubation with the acetoxymethylester form of the dye. Cells were excited alternately at 340 and 380 nm and emitted light collected at 510 nm while cells were perfused with a HEPES-buffered physiological saline. For Ca^{2+}-free solutions, Ca^{2+} was omitted and replaced with 1 mM EGTA. All protocols and procedures have been detailed elsewhere (e.g. Smith et al 2003).

In the absence of extracellular Ca^{2+}, application of agonists (ATP for endothelial cells, bradykinin [BK] for astrocytes), which are known to generate inositol 1,4,5-trisphosphate (IP_3) via receptor-dependent phospholipase C activation, caused transient rises of $[Ca^{2+}]_i$ (Fig. 1A,C). Following washout of agonist, and in the continued absence of extracellular Ca^{2+}, exposure of either cell type to hypoxia (pO2

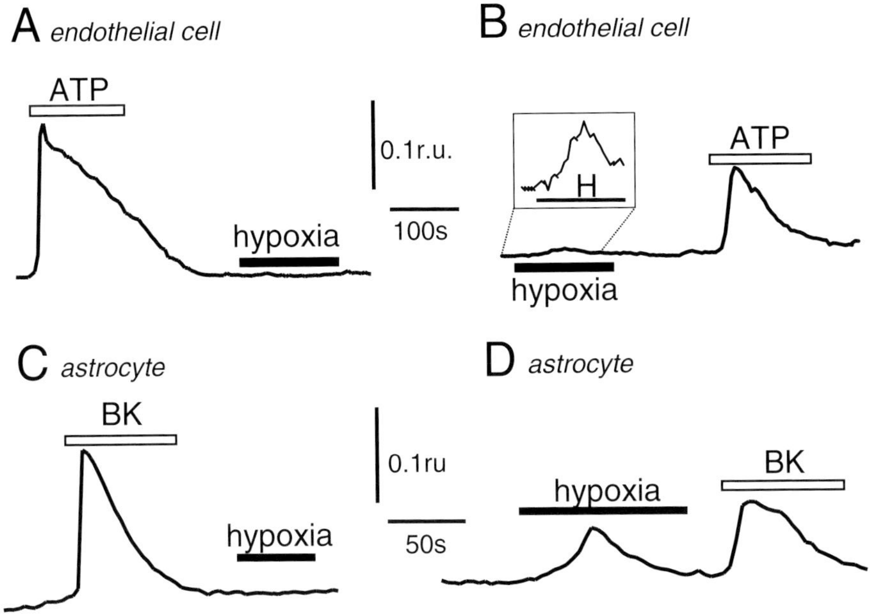

FIG. 1. Acute hypoxia mobilizes Ca^{2+} from an intracellular pool. Example recordings of $[Ca^{2+}]_i$ in endothelial cells (A,B) and astrocytes (C,D). In each case, cells were exposed to an agonist (either ATP; 10 μM, or bradykinin [BK]; 100 nM) for the periods indicated by the open horizontal bars, and to hypoxia (pO2 20–30 mmHg) for the periods indicated by the solid horizontal bars (in B, this has been enlarged in the boxed section for clarity). Throughout these experiments, Ca^{2+} was omitted from the extracellular solution and replaced with 1 mM EGTA.

20–30 mmHg) was without effect on $[Ca^{2+}]_i$ (Fig. 1A,C). However, when this protocol was reversed, and cells were exposed firstly to hypoxia, small transient rises of $[Ca^{2+}]_i$ were apparent in the majority of recordings (>80%) from both cell types (Fig. 1B,D). Following restoration of normoxia, subsequent application of agonist also evoked transient rises of $[Ca^{2+}]_i$ (Fig. 1B,D), but these were markedly attenuated when compared to responses evoked before prior exposure to hypoxia. These data suggested that hypoxia could mobilize Ca^{2+} from an IP_3-sensitive pool in both endothelial cells and astrocytes. Further support for this idea came from the observation that in both cell types neither hypoxia nor the relevant agonist could evoke changes in $[Ca^{2+}]_i$ following store depletion by pre-treatment of cells for 20 min with the endoplasmic reticulum Ca^{2+}-ATPase inhibitor, thapsigargin (1 μM; data not shown).

A growing body of evidence suggests that some effects of hypoxia arise from a seemingly paradoxical increase in reactive oxygen species (ROS) generation (Chandel et al 1998, Chandel & Schumacker 2000, Duranteau et al 1998, Pearlstein et al 2002, Waypa et al 2001, Leach et al 2001). To examine whether the ability of hypoxia to mobilize Ca^{2+} from an intracellular pool required ROS generation, we tested the ability of two structurally and mechanistically distinct antioxidants to interfere with such hypoxic signalling. These were trolox (6-hydroxy-2,5,7,8-tetramethylchroman-2-carboxylic acid, a cell-permeable, water-soluble derivative of vitamin E—see McClain et al 1995) and TEMPO (4-hydroxy-2,2,6,6-tetramethylpiperidine-1-oxyl), applied together with catalase. TEMPO catalyses the conversion of superoxide to H_2O_2, which is then removed by catalase (see Abramov et al 2004). Cells were pretreated with 500 μM trolox for 30 min, or 500 μM TEMPO plus 250 U/ml catalase for 45 min, and the compounds were present during the recordings (when Ca^{2+} was omitted from the perfusate). As exemplified in Fig. 2, both agents fully prevented rises in $[Ca^{2+}]_i$ evoked by hypoxia in endothelial cells (Fig. 2A) and astrocytes (Fig. 2B). These data indicate that hypoxic mobilization of Ca^{2+} from an intracellular pool(s) requires ROS formation.

There are numerous intracellular sites where ROS can be generated and, of these, mitochondria have received much recent attention. A reduction in O_2 levels has been proposed to lead to a rise of ROS derived from specific site(s) within the electron transport chain (Chandel et al 1998, 2000, Chandel & Schumacker 2000). To investigate mitochondria as a possible source of ROS, we examined responses to hypoxia (again during perfusion with Ca^{2+}-free solutions) during mitochondrial inhibition using the uncoupler carbonyl cyanide p-trifluoromethoxyphenylhydrazone (FCCP; 10 μM) which was applied together with 2.5 μg/ml oligomycin to prevent ATP consumption by the F_1F_0-ATP synthase. In endothelial cells, FCCP and oligomycin caused a transient rise of $[Ca^{2+}]_i$ (Fig. 3A) and, while hypoxic rises of $[Ca^{2+}]_i$ could still be detected during mitochondrial inhibition, they were markedly reduced as compared with responses evoked in cells with fully functional mitochondria (Fig. 3A,B). Responses observed in astrocytes were strikingly different (Fig. 3C,D). FCCP and oligomycin caused small rises of $[Ca^{2+}]_i$ (see also Smith et al 2003) but, in its presence, responses to hypoxia were markedly increased as compared with those observed in the absence of mitochondrial inhibitors (Fig. 3C,D).

The fact that hypoxia was capable of at least partially depleting intracellular stores in both cell types raised the possibility that it may be capable of initiating capacitative Ca^{2+} entry (CCE), a Ca^{2+} influx pathway important for numerous physiological functions (Putney 2001, Putney et al 2001). To investigate this, we first exposed both endothelial cells and astrocytes to Ca^{2+}-free perfusate (under normoxic conditions), then re-admitted Ca^{2+} to the perfusate. As exemplified in Fig. 4 (A,C), this procedure caused no marked change in $[Ca^{2+}]_i$ in either cell type, indicating that this period of exposure to Ca^{2+}-free perfusate was not sufficient to deplete intracellular stores. However, when the cells were exposed to hypoxia during the period of perfusion with Ca^{2+}-free solution, subsequent re-addition of Ca^{2+} to the perfusate always caused marked rises of $[Ca^{2+}]_i$ in each cell type and, further-

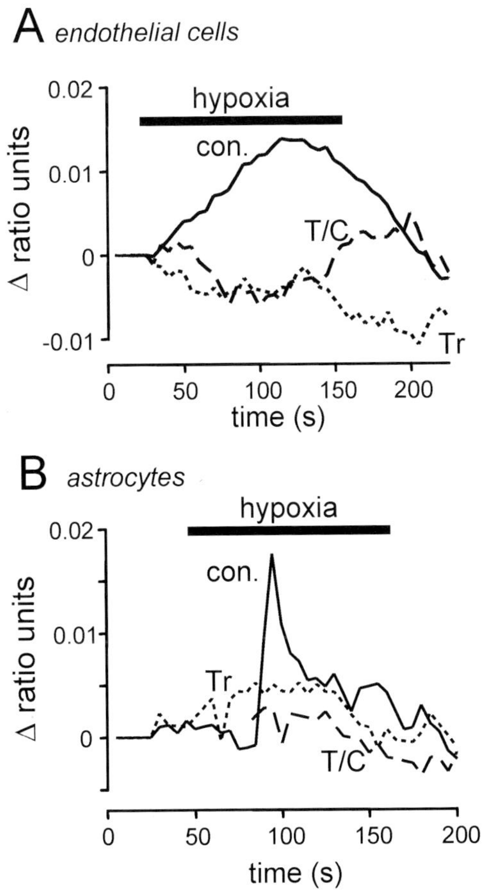

FIG. 2. Antioxidants prevent rises of $[Ca^{2+}]_i$ evoked by hypoxia. (A) Example recordings of $[Ca^{2+}]_i$ from three individual endothelial cells during exposure to hypoxia (applied for the period indicated by the horizontal bar) in the absence of antioxidants (con.; solid line) or in the presence of either 500 μM trolox (Tr; dotted line), or 500 μM TEMPO plus 250 U/ml catalase (T/C; dashed line). (B) as (A), except recordings were made from astrocytes. In both panels, initial baselines were aligned for illustrative purposes. Ca^{2+} was omitted from the perfusate throughout these experiments.

more, these rises could be blocked fully by co-application of Gd^{3+} (1mM; Fig. 4B,D). These data indicate that acute hypoxia, by stimulating Ca^{2+} release from an intracellular pool, can activate CCE.

A number of important issues concerning cellular responses to hypoxia arise from the present work. Most importantly, we have demonstrated in two cell types that hypoxia can mobilize Ca^{2+} from an intracellular pool that is also susceptible to

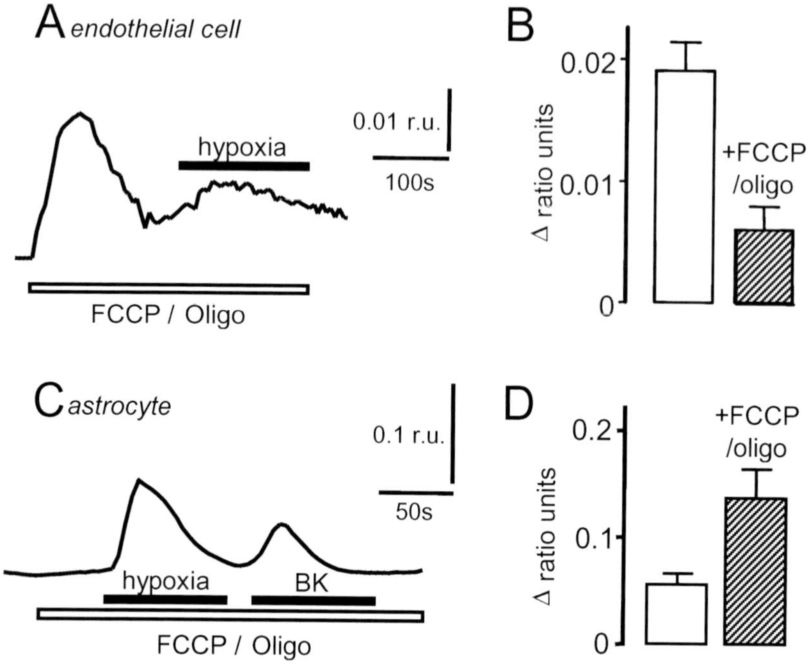

FIG. 3. Modulation of hypoxic responses by mitochondrial inhibition. (A) *Left*, example recording of $[Ca^{2+}]_i$ in an endothelial cell which was exposed to 10μM FCCP together with 2.5μg/ml oligomycin for the period indicated by the open bar. For the period indicated by the solid bar, the cell was also exposed to hypoxic solution (pO2 20–30 mmHg). Bar graph shows mean (±SEM) peak rise of $[Ca^{2+}]_i$ evoked by hypoxia alone (open bar) or in the presence of FCCP and oligomycin (hatched bar). (B) As (A), except that data were acquired in astrocytes. Note the opposite effect on hypoxic changes of $[Ca^{2+}]_i$ caused by mitochondrial inhibition. Ca^{2+} was omitted from the perfusate throughout these experiments.

depletion via agonists which induce IP_3 generation; presumably, this pool is the endoplasmic reticulum. Whilst the rises of cytosolic $[Ca^{2+}]$ were modest when compared to those evoked by agonists, they were sufficient to trigger CCE, an important Ca^{2+} influx pathway for various cellular functions (Putney 2001, Putney et al 2001). Clearly, these responses to hypoxia were dependent on cellular production of ROS, since they were abolished in both cell types by antioxidants. However, cell-specific responses to hypoxia were revealed during mitochondrial uncoupling, indicating that potential roles for mitochondria in hypoxic Ca^{2+} signalling differed between the two cell types. In endothelial cells hypoxic responses were suppressed by FCCP and oligomycin, a finding consistent with the idea that mitochondria are the primary source of ROS in these cells during hypoxia. By contrast, hypoxic

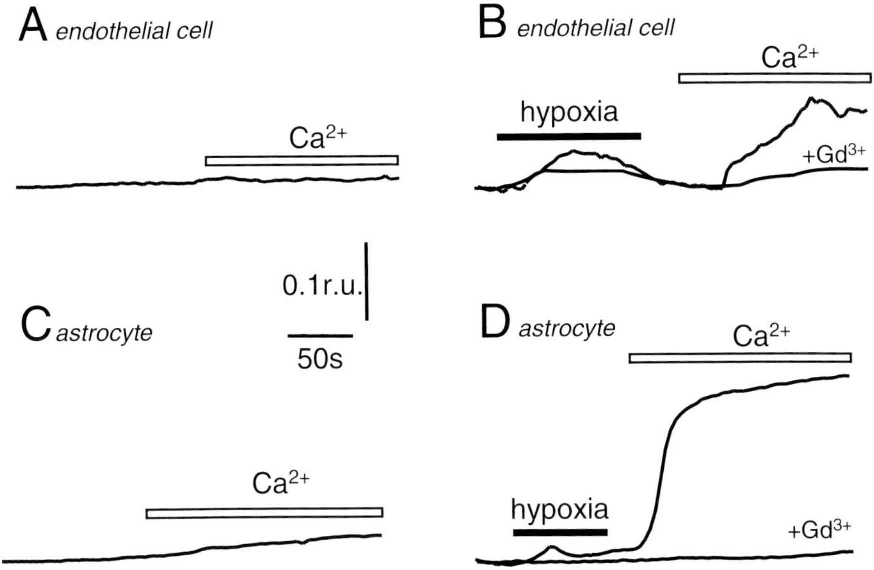

FIG. 4. Acute hypoxia stimulates capacitative Ca^{2+} entry. (A, C) Example recordings of $[Ca^{2+}]_i$ in an endothelial cell (A) and an astrocyte (C) during perfusion with a Ca^{2+}-free perfusate in normoxia. For the period represented by the open horizontal bar in each case, Ca^{2+} (2.5 mM) was readmitted to the perfusate. (B, D) as (A, C), except that before re-admission of Ca^{2+} to the perfusate, cells were exposed to hypoxia (pO2 20–30 mmHg) for the period indicated by the horizontal bar. Note the marked rise of $[Ca^{2+}]_i$ when Ca^{2+} is added to the perfusate. Also shown are example traces where 1 mM Gd^{3+} was co-applied with Ca^{2+} to prevent capacitative Ca^{2+} entry. Scale bars apply to all traces.

responses in astrocytes were strikingly potentiated by FCCP and oligomycin. This finding suggests that ROS are generated in astrocytes during hypoxia from a non-mitochondrial source and also that mitochondria serve an additional function, since their inhibition caused potentiation of the hypoxic response. In this regard, our previous studies have shown that mitochondria can be an important sink for Ca^{2+} in these cells (Smith et al 2003), raising the possibility that mitochondria are effective as buffers for hypoxia-evoked rises of cytosolic $[Ca^{2+}]$. Thus, their inhibition uncovers a much greater effect of hypoxia to mobilize Ca^{2+}.

Both endothelial cells and astrocytes have a strict requirement for closely controlled regulation of $[Ca^{2+}]_i$ in order to perform many of their specific functions (Adams & Hill 2004, Verkhratsky et al 1998). Clearly, episodes of hypoxia are likely to interfere with such functions through disruption of Ca^{2+} signalling. However, it should be borne in mind that both cell types exist under physiological conditions in environments of relative hypoxia, and so the effects of hypoxia described here may reflect physiological rather than potentially pathophysiological effects of hypoxia.

Acknowledgements

This work was supported by The British Heart Foundation, The Wellcome Trust, The Medical Research Council, and Pfizer Central Research (through CASE studentships to PK and IFS).

References

Abramov AY, Canevari L, Duchen MR 2004 Beta-amyloid peptides induce mitochondrial dysfunction and oxidative stress in astrocytes and death of neurons through activation of NADPH oxidase. J Neurosci 24:565–575

Adams DJ, Hill MA 2004 Potassium channels and membrane potential in the modulation of intracellular calcium in vascular endothelial cells. J Cardiovasc Electrophysiol 15:598–610

Buckler KJ, Vaughan-Jones RD 1994 Effects of hypoxia on membrane potential and intracellular calcium in rat neonatal carotid body type I cells. J Physiol 476:423–428

Budd JS, Allen KE, Bell PR 1991 Effects of two methods of endothelial cell seeding on cell retention during blood flow. Br J Surg 78:878–882

Chandel NS, Schumacker PT 2000 Cellular oxygen sensing by mitochondria: old questions, new insight. J Appl Physiol 88:1880–1889

Chandel NS, Maltepe E, Goldwasser E, Mathieu CE, Simon MC, Schumacker PT 1998 Mitochondrial reactive oxygen species trigger hypoxia-induced transcription. Proc Natl Acad Sci USA 95:11715–11720

Chandel NS, McClintock DS, Feliciano CE et al 2000 Reactive oxygen species generated at mitochondrial complex III stabilize hypoxia-inducible factor-1alpha during hypoxia: a mechanism of O2 sensing. J Biol Chem 275:25130–25138.

Duranteau J, Chandel NS, Kulisz A, Shao Z, Schumacker PT 1998 Intracellular signaling by reactive oxygen species during hypoxia in cardiomyocytes. J Biol Chem 273:11619–11624

Evans AM, Dipp M 2002 Hypoxic pulmonary vasoconstriction: cyclic adenosine diphosphate-ribose, smooth muscle Ca^{2+} stores and the endothelium. Respir Physiol Neurobiol 132:3–15

Fearon IM, Varadi G, Koch S, Isaacsohn I, Ball SG, Peers C 2000 Splice variants reveal the region involved in oxygen sensing by recombinant human L-type Ca^{2+} channels. Circ Res 87:537–539

Franco-Obregon A, Urena J, Lopez-Barneo J 1995 Oxygen-sensitive calcium channels in vascular smooth muscle and their possible role in hypoxic arterial relaxation. Proc Natl Acad Sci USA 92:4715–4719

Hool LC 2001 Hypoxia alters the sensitivity of the L-type Ca^{2+} channel to alpha- adrenergic receptor stimulation in the presence of beta-adrenergic receptor stimulation. Circ Res 88:1036–1043

Jiang C, Haddad GG 1994 A direct mechanism for sensing low oxygen levels by central neurons. Proc Natl Acad Sci USA 91:7198–7201

Kemp PJ, Williams SEJ, Mason HS et al 2005 Functional proteomics of BK potassium channels: defining the acute oxygen sensor. Wiley, Chichester (Novartis Found Symp 272), p 141–150

Leach RM, Hill HM, Snetkov VA, Robertson TP, Ward JP 2001 Divergent roles of glycolysis and the mitochondrial electron transport chain in hypoxic pulmonary vasoconstriction of the rat: identity of the hypoxic sensor. J Physiol 536:211–224

Lewis A, Hartness ME, Chapman CG et al 2001 Recombinant hTASK1 is an O_2-sensitive K^+ channel. Biochem Biophys Res Comm 285:1290–1294

Lewis A, Peers C, Ashford MLJ, Kemp PJ 2002 Hypoxia inhibits human recombinant maxi K^+ channels by a mechanism which is membrane delimited and Ca^{2+} sensitive. J Physiol 540:771–780

Lopez-Barneo J, Lopez-Lopez JR, Urena J, Gonzalez C 1988 Chemotransduction in the carotid body: K^+ current modulated by PO_2 in type I chemoreceptor cells. Science 241:580–582
Lopez-Barneo J, Pardal R, Ortega-Saenz P 2001 Cellular mechanism of oxygen sensing. Annu Rev Physiol 63:259–287
McClain DE, Kalinich JF, Ramakrishnan N 1995 Trolox inhibits apoptosis in irradiated MOLT-4 lymphocytes. FASEB J 9:1345–1354
Osipenko ON, Evans AM, Gurney AM 1997 Regulation of the resting potential of rabbit pulmonary artery myocytes by a low threshold, O_2-sensing potassium current. Br J Pharmacol 120:1461–1470
Pearlstein DP, Ali MH, Mungai PT, Hynes KL, Gewertz BL, Schumacker PT 2002 Role of mitochondrial oxidant generation in endothelial cell responses to hypoxia. Arterioscler Thromb Vasc Biol 22:566–573
Putney JW 2001 The pharmacology of capacitative calcium entry. Molecular Interventions 1:84–94
Putney JW, Broad LM, Braun FJ, Lievremont JP, Bird GS 2001 Mechanisms of capacitative calcium entry. J Cell Sci 114:2223–2229
Rychkov GY, Adams MB, McMillen IC, Roberts ML 1998 Oxygen-sensing mechanisms are present in the chromaffin cells of the sheep adrenal medulla before birth. J Physiol 509:887–893
Smith IF, Boyle JP, Plant LD, Pearson HA, Peers C 2003 Hypoxic remodeling of Ca2+ stores in type I cortical astrocytes. J Biol Chem 278:4875–4881
Verkhratsky A, Orkand RK, Kettenmann H 1998 Glial calcium: homeostasis and signaling function. Physiol Rev 78:99–141
Waypa GB, Chandel NS, Schumacker PT 2001 Model for hypoxic pulmonary vasoconstriction involving mitochondrial oxygen sensing. Circ Res 88:1259–1266
Weir EK, Archer SL 1998 The mechanism of acute hypoxic pulmonary vasoconstriction: the tale of two channels. FASEB J 9:183–189
Williams SE, Wootton P, Mason HS et al 2004 Hemoxygenase-2 is an oxygen sensor for a calcium-sensitive potassium channel. Science 306:2093–2097
Wyatt CN, Wright C, Bee D, Peers C 1995 O_2-sensitive K^+ currents in carotid-body chemoreceptor cells from normoxic and chronically hypoxic rats and their roles in hypoxic chemotransduction. Proc Natl Acad Sci USA 92:295–299
Youngson C, Nurse C, Yeger H, Cutz E 1993 Oxygen sensing in airway chemoreceptors. Nature 365:153–155

DISCUSSION

Archer: We have also looked at this, and we found the same things in terms of Ca^{2+} (i.e. hypoxia rapidly increases cytosolic Ca^{2+} levels in pulmonary artery endothelial cells) (Hampl et al 1995). We didn't look at the signalling mechanism but the finding was associated with increased nitric oxide (NO) synthesis. Have you figured out the consequences of this Ca^{2+} signal in your system? Did you explore any kind of functional readout beyond the Ca^{2+} signal to look at the physiological role of this redox-invoked Ca^{2+} elevation?

Peers: The exposure to hypoxia that we are giving these cells is much closer to their normoxic environment *in vivo*. It could be that oxygen tension plays a tonic regulatory role in how easily the Ca^{2+} can be mobilized into or out of intracellular stores. It could be a background determinant of Ca^{2+} fluxes throughout the cell.

Archer: Do you postulate that this pathway is relevant to regulation of thrombosis or vascular tone? If so, is this effect mediated through NO?

Peers: I don't know.

Ward: Have you looked at other types of endothelial cells? HUVECs are a special circulation in some senses in terms of oxygen.

Peers: Lots of circulations are specialized. We are trying to do this in arterial cells because we get mammary artery grafts routinely, but they occur less frequently and it is tougher to culture the endothelial cells from them.

Schumacker: You saw an interesting difference between the astrocytes and the endothelial cells with respect to the Ca^{2+} buffering by mitochondria. All the classical studies of mitochondrial biology suggest that they take up Ca^{2+} very avidly. But the relative number of mitochondria in one cell type might be very different from another cell type. For example, a cell type with very few mitochondria might have a lesser ability to buffer changes of cytosolic Ca^{2+} released from the endoplasmic or sarcoplasmic reticulum. What do you know about differences in mitochondrial density in these two cell types?

Peers: We haven't studied this in depth. We have stained both cell types with MitoTracker. The highest concentration is close to the nucleus, and further out towards the periphery they become less dense. There is not much difference between the cell types. However, *in vivo* these cells take on very different shapes, particularly the astrocytes: they are not nice flat shapes that stick to coverslips.

Chandel: I have a comment about the effects that you see with FCCP and the relationship of these to the effect seen with rotenone. They are almost identical. Rotenone blocks complex I and blocks a hypoxic increase in free radical production. Rotenone will also depolarise the mitochondrial membrane potential. FCCP will depolarize the mitochondrial membrane potential, but it hasn't had a major effect in modulating reactive oxygen species (ROS) production under hypoxia. Both give membrane potential depolarization, but only one gives the reduction in ROS.

Peers: Agreed, but I can only show you what we have.

Acker: In your scheme you ruled out that the NADPH oxidase generates fewer ROS under hypoxia. At the same time, mitochondria generate more ROS under hypoxia. What is the explanation?

Peers: I believe our data suggest that mitochondrial ROS increase in hypoxia in accordance with the 'Chicago theory', but that ROS derived from NADPH oxidase decrease under hypoxia simply because of reduction in the substrate, O_2.

Acker: Is there any other explanation? I doubt this ROS increase under hypoxia. I think it is mostly due to severe technical problems with the dyes.

Duchen: Our experience in neurons has been that we can see changes in glutamate responses in the presence of antioxidants which have nothing to do with buffering free radicals, but rather the modulation of redox-sensitive sites on proteins. You can't automatically assume that experimental effects of antioxidants mean that free radicals are being generated (Vergun et al 2001).

Murphy: ROS production in the absence of oxygen seems unlikely, but it is quite easy to speculate that ROS production in the presence of oxygen leads to something such as a change in the reduction potential of particular protein thiols or of the reduction potential of the glutathione pool, for example. Such a change could be accessed during hypoxia without a direct intervention from hydrogen peroxide or other ROS.

Acker: You are saying that the redox state changes due to a similar effect.

Murphy: I am speculating that the ROS signalling and hypoxia converge on some common intermediates, and that possibilities for consideration are the reduction potentials of the glutathione pool, or of thioredoxin.

Sylvester: As you mentioned, hypoxia can cause Ca^{2+} release in pulmonary arterial smooth muscle. We have recently done some experiments in which we measured Ca^{2+} release before and after exposure to cyclopiazonic acid (CPA). CPA abolished hypoxia-induced release, suggesting complete depletion of the Ca^{2+} stores that were sensitive to hypoxia. Furthermore, hypoxia increased CCE measured in the presence of CPA (Wang et al 2005). If we assume that CPA completely depleted stores sensitive to hypoxia, this would imply that hypoxia facilitated CCE through some effect on signal transduction or on store-operated Ca^{2+} channels. Do you have any data relevant to this?

Peers: I don't. Hypoxia could hyperpolarize the cell and increase the CCE very simply this way.

Duchen: There were a few things in your paper which surprised me a bit. As I understand it, the received wisdom is that free radicals tend to sensitize IP_3 Ca^{2+} release and increase the probability of Ca^{2+} release through the IP_3 receptor. You were suggesting that somehow your tonic activity was suppressing IP_3 production. Is this based on the effects of the antioxidants on the agonist responses?

Peers: Yes.

Archer: Getting back to the role of this in real life, when you take a pulmonary artery (or indeed any artery) and expose it to a vasoconstrictor (e.g. hypoxia for the pulmonary circulation), there is a homeostatic production of NO. To some extent it also makes EDHF (endothelium-derived hyperpolarizing factor). Measurement of this pathway is difficult, but it is easy to measure NO online (Gragasin et al 2004). Have you tried?

Peers: No, but we intend to do this, bearing in mind the data of Moncada and colleagues, who suggest NO causes redistributing of oxygen within the cell.

Duchen: Along these lines, you could argue that if they are producing NO and free radicals, this would generate peroxynitrite. When you are putting on your antioxidants, you are leaving the NO there.

Aaronson: The implication of your work is that the mitochondrial ROS must be localized differently than the ROS from NADPH oxidase. One is going up and the other is down, and you don't know what is going to happen to the global ROS. Is your idea that somehow the ryanodine receptor is closer to the mitochondria, and the IP_3 receptor would be closer to the NADPH oxidase? Might there be two stores involved?

Peers: Quite possibly. The location of the source will likely be a key factor. This is something that needs exploring.

Duchen: Referring back to some of the earlier discussion, you are giving these cells quite modest levels of hypoxia, but you are seeing changes in Ca^{2+} signalling which you think are involved in free radical production. How is this being sensed and transduced into a signal? The pO2 changes aren't enough to do anything dramatic to mitochondrial function, and yet changes are seen.

Weir: What pO2 were you seeing when you bubbled with nitrogen?

Peers: 20–30 mmHg, CO_2 is nothing, and the pH doesn't change from 7.4.

References

Gragasin FS, Michelakis ED, Hogan A et al 2004 The neurovascular mechanism of clitoral erection: nitric oxide and cGMP-stimulated activation of BKCa channels. FASEB J 18:1382–1391

Hampl V, Cornfield D, Cowan N, Archer SL 1995 Hypoxia potentiates nitric oxide synthesis and transiently increases cytosolic calcium levels in pulmonary artery endothelial cells. Eur Resp J 8:515–522

Vergun O, Sobolevsky AI, Yelshansky MV, Keelan J, Khodorov BI, Duchen MR 2001 Exploration of the role of reactive oxygen species in glutamate neurotoxicity in rat hippocampal neurones in culture. J Physiol 531:147–163

Wang J, Shimoda LA, Weigand LA, Wang W, Sun D, Sylvester JT 2005 Acute hypoxia increases intracellular [Ca^{2+}] in pulmonary arterial smooth muscle by enhancing capacitative Ca^{2+} entry. Am J Physiol Lung Cell Mol Physiol, in press

General discussion II

Kumar: Professor Nurse's work showed that P0/P1 baby rats had a hypoxic sensitivity, whereas juveniles did not. The implication was that postnatal innervation removed the intrinsic hypoxic sensitivity. Physiologically, a near-term fetal animal will have an arterial pO2 of approximately 25 mmHg. Within one postnatal day that pO2 will rise to about 75 mmHg. This is a threefold increase in oxygen on day 1. In Professor Nurse's experiments he was using normoxia at 20%, which is a high level, and using 5% hypoxia, which returns the animal back to its 'fetal' state. The implication could be that in his control there was a production of reactive oxygen species (ROS) because of the relatively high oxygen, and then he was returning the animal back towards a 'control' state where it had been for many days before. The fall in ROS may reflect a return to a relatively normoxic level rather than a hypoxic one. The adult animals were seeing hypoxia. Professor Prabhakar's model of sleep apnoea presented earlier is actually very different in that his animals are breathing very hard during the hypoxic period and are presumably very aware of this, whereas most people with sleep apnoea are not aware that they aren't breathing. We should be aware of the differences between the models. The differences we are seeing with the ROS story might, in part, be related to the models being used.

Archer: I have a comment regarding the fact that the ROS production is dynamic over a range that is normal, much like the carotid body is sensing pO2 at very high levels. For that matter, the pulmonary smooth muscle cells start sensing hypoxia at pO2s of 70 mmHg. There is some early work by Freeman & Crapo (1981) in which they measured isolated lung mitochondrial production of radicals, showing that it was linear and proportional to pO2 over a broad range. 95% oxygen had much higher radical production than normoxia and it carried on down into hypoxia. This is why it is an excellent signal. If it were not dynamic over this range it wouldn't be a candidate oxygen sensor at all. All these hypoxia responses of the specialized tissues in the body's specialized oxygen sensing system detect hypoxia beginning at relatively high pO2 levels.

Duchen: What is the process that is sensing this?

Archer: I think it is related directly to the flux of electrons down the electron transport chain (ETC). There is a leak of about 5% of ROS as the pO2 is processed by the ETC, due to non-coupled electron flux. There is a leak off the electron transport chain of superoxide dismutase (SOD) that gets dismutated by manganese SOD to H_2O_2. I believe this occurs in proportion to electron flux, and that's why it is a good sensor. The diffusible H_2O_2 is then free to modulate the activity of redox-sensitive K^+ channels.

Schumacker: The study reported by Bruce Freeman and James Crapo in the early 1980s (Freeman & Crapo 1981) in isolated perfused lungs did show that ROS production increased during hyperoxia. Many people who haven't read the paper don't realize that these lungs weren't normal: they were treated with antimycin A, a mitochondrial inhibitor that increases ROS production. Without adding antimycin A to stimulate ROS production from the mitochondria they didn't see any increase in ROS with hyperoxia. The notion that ROS production by cells is a linear function of pO2 has to be treated with a degree of caution. It may not be true in all cells.

Archer: I wasn't aware of that; I read the paper but I missed this. In our experiments we find the same thing, as Gale Gurtner and Paky did at Hopkins. They also found that radicals increased linearly as pO2 increased (Paky et al 1987, 1993). Whether ROS goes up or down with hypoxia, it is essential that the change occurs rapidly (in seconds) over a physiologic range of pO2 (40–80 mmHg) if this is going to be considered as a signal.

Ward: Mark Evans, in your first cADP ribose (cADPR) paper you published with Tony Galliano (Wilson et al 2002), you were proposing that it was the change in NAD/NADH ratio that was affecting cADPR production. Do you still support this hypothesis?

Evans: We have no evidence to discount a facilitatory effect of β-NADH on cyclic ADP-ribose (cADPR) accumulation from β-NAD$^+$. At the time we carried out the initial experiments that suggested such an effect, I had a rather naïve hypothesis that β-NADH might be a better substrate for cADPR synthesis than β-NAD$^+$ and that an increase in β-NADH levels on exposure to hypoxia might thereby increase cADPR accumulation. β-NADH turned out to be a very poor substrate for cADPR synthesis. We took a smooth muscle homogenate, and looked at the enzyme activity for cADPR synthesis (ADP-ribosyl cyclase). When we added β-NADH to the homogenate, 25 mM yielded only a third as much cADPR within 1 hour as did the addition of 2.5 mM β-NAD$^+$. The referees of the journal were insistent that we came up with some sort of mechanism. I suggested putting β-NADH and β-NAD$^+$ together. Over β-NADH/β-NAD$^+$ ratios similar to those that one might see inside the cell, we found that β-NADH facilitated the production of cADPR from β-NAD$^+$. It was a marked synergistic effect. There were a few papers published recently identifying β-NADH-dependent regulation of other enzymes. Thus, there is a possibility that some enzymes might be regulated by β-NADH via its interaction with allosteric sites. More than this, I can't really say. However, when considering this possible regulatory mechanism we also have to bear in mind that if β-NAD$^+$ is the substrate, a marked increase in β-NADH levels may compromise substrate (β-NAD$^+$) supply. Thus, β-NADH may facilitate cADPR accumulation within a window determined by β-NAD$^+$ availability and the kinetics of ADP-ribosyl cyclase/cADPR hydrolase. It should be noted, however, that thus far all such experiments were carried out on smooth muscle homogenates in a cuvette. We weren't looking at the cell and we have not really got any way of judging whether this mechanism of facilitation will occur in the cell.

Ward: As an alternative to the hypothesis that ROS act as signalling intermediaries in hypoxia, do you think that an increase in ETC substrate concentrations, due to reduced consumption during hypoxia, is a likely candidate?

Evans: It is a possible player. There are bound to be numerous signalling pathways that may influence the response of oxygen-sensing cells to hypoxia, and this may be one of them.

Archer: They have to be linked. Whether or not you believe that radicals go up or down, you have to believe that electrons come into the chain from the substrate (NADH and FADH). Since both theories involve changes in the function of the chain, there has to be a change in redox state that is coupled with the altered ROS production (whether it is increased or decreased). As you affect flow of electrons into the chain, if there is a functional inhibition in the chain proximally there will be an upstream accumulation of the electron donors that are normally supplying electrons. This is true in bacteria also. In bacteria the supply of O_2 is modulated by aerotaxis, locomotion of the organism toward the optimal environmental pO2. The haem sensor that mediates aerotaxis transduces its signal via a phospho-relay, involving soluble chemotaxis proteins, which when phosphorylated, interact with the flagellar motor. In *Escherichia coli*, the Aer protein is the O_2 sensor and it has a flavin adenine dinucleotide (FAD) binding site at its amino terminus. The FAD is oxidized or reduced in response to redox changes in the bacterial ETC. The resulting conformational change alters Aer and thus changes the phosphorylation of the chemotaxis proteins.

Duchen: The key debate about whether or not mitochondria can be involved in oxygen sensing has been the oxygen sensitivity of the ETC. The argument has always been that mitochondria are so sensitive to it—their respiratory chain has such a high affinity for oxygen—that the pO2 levels won't affect mitochondrial respiration until they get vanishingly low. If these sorts of effects are going to be attributed to mitochondria, then you have to argue that mitochondria are somehow responsive to higher oxygen levels than you would predict from the effects of isolated mitochondria.

López-Barneo: Some people in the field of sleep apnoea believe that the number of arousals in the night is related to the morbidity of this condition. This relates to the question of the importance of the duration of the intermittent hypoxia (IH), versus the frequency of the episodes. Arousal implies activation of the sympathetic system and this is not related at all to ROS. Whether or not the animals are suffering arousal is important.

Prabhakar: Carotid bodies might be contributing to arousals in sleep apnoea patients. Sometime back, we did chronic denervation of the carotid bodies, and

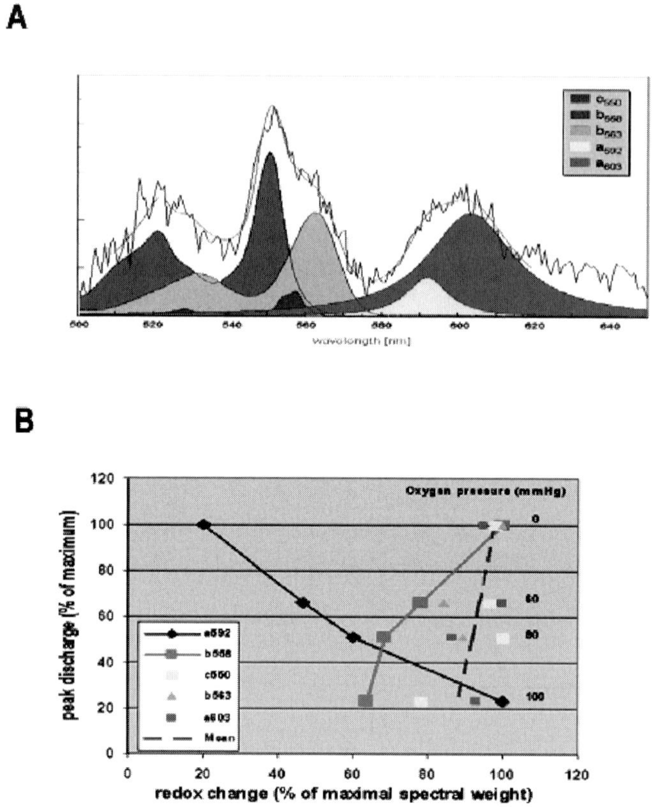

FIG. 1. (Acker) (A) N_2 versus normoxia steady state light absorption difference spectrum of one single rat carotid body. (B) The relative redox change of the different carotid body cytochromes under different pO2 conditions. See text for further explanation.

these animals were rather drowsy compared to sham operated control animals with intact carotid bodies. I must, however, point out that apnoeas are more complex than we think. Hypoxia is only one of the several variables that change during apnoea.

Acker: I would like to draw your attention to the paper by Streller et al (2002) where we identified by absorption spectrophotometry two cytochromes as putative oxygen sensor proteins. Figure 1 (*Acker*) shows in *A* a N_2 versus normoxia steady state light absorption difference spectrum of one single rat carotid body. Various spectra of single cytochromes have been used for deconvolution of the measured spectrum like mitochondrial cytochrome c, b563 of complex III and cytochrome a3 of complex IV. Furthermore b558 of the NADPH oxidase as well as a unique a592 probably a blue-shifted cytochrome a of complex IV were necessary to fit the measured spectrum. This deconvolution procedure enabled us to calculate the relative redox change of the different carotid body cytochromes under different pO2 conditions as shown in *B*. The redox changes are related to the peak chemoreceptor discharge at different pO2 conditions. Cytochrome c, b563 and a3 show a redox change in a linear relation to the actual pO2. Cytochrome a592 and b558 however reveal a non-linear relationship to the actual pO2 as it is known from the relationship between chemoreceptor discharge and pO2. We speculate that the carotid body NADPH oxidase as oxygen sensor decreases ROS production under hypoxia triggering the inhibition of potassium channels to increase the Ca^{2+} influx into chemoreceptor cells. Cytochrome a592 might posses a low midpoint potential as described in bacteria (Kannt et al 1999) shifting complex IV to more negative values and hence decreasing mitochondrial membrane potential (MMP) at normal tissue pO2 values. Small decreases of the ambient pO2 would decrease MMP even more and impair the mitochondrial calcium buffering capacity leading in concert with b558 to an unusual pO2 sensitivity of carotid body transmitter release and chemoreceptor discharge. Cytochrome a592 would represent the unique low pO2 affinity and cytochrome a3 the high pO2 affinity component of the carotid body complex IV as already suggested by Mills & Jöbsis (1972).

Chandel: In your total spectrum there was a nice cytochrome C peak that is easily distinguishable. You should be able to use the redox state of cytochrome C and look at the oxygen dependence of that redox state. In other words, you could go from 100 Torr to 0. In most cells the redox states start to increase around 10 Torr. This model predicts that at 40 Torr you would start to see this.

Acker: What we determined is the contribution of a single cytochrome to the spectrum, according to its redox state. The haem A is contributing a lot to the spectrum at low pO2 values, and at low pO2 values there is no further reduction. It is a non-linear reduction, whereas the other cytochromes follow this more linearly. We shouldn't forget that we have measured the pO2 gradient on the whole organ. Some cells here have a lower pO2 than the other cells.

Chandel: At what pO2 does cytochrome C start to be reduced? In most systems this doesn't happen until below 10 Torr.

Acker: It is linear to the actual ambient pO2.

Chandel: This would argue that the cytochrome oxidase is not so different. It is different in some sense, but not in this ability to do electron transport.

Acker: No. I have shown that we have two pathways of the electron transport in carotid body mitochondrial complex IV. One is the high pO2 affinity employing cytochrome a3. The other one is the low affinity pathway employing the unique blue shifted cytochrome a592. The electrons coming from cytochrome c are passing mostly along the high affinity chain. The low affinity part is easily reduced at high pO2 values.

López-Barneo: Why doesn't the high affinity pathway contribute to the membrane potential, whereas the low affinity pathway does?

Acker: The electrical potential between mitochondrial complex I and complex IV establishes the proton motive force and hence the MMP. The unique a592 would according to our model make complex IV more negative even at high pO2 levels and would therefore lower the mitochondrial membrane potential and impair Ca^{2+} buffering.

López-Barneo: Your model suggests that part of the Ca^{2+} required for excitation–secretion coupling comes from mitochondria.

Acker: Not necessarily. It can be, but it could also be that the buffer capacity is impaired so that the Ca^{2+} influx is potentiated.

Schumacker: I am trying to understand how you did these experiments. You have intact carotid bodies that you are superfusing with different oxygen tensions, and at the same time you are measuring the spectral absorption of these cytochromes as you change the extracellular pO2s. What is going on? As it is an intact organ, aren't you extrapolating from the extracellular buffer pO2 what the intracellular pO2 is in the centre of the organ? It could be much lower. How do you know what the intracellular pO2 is in the cells in which you are seeing an effect?

Acker: We published years ago the pO2 gradient in the superfused carotid body. The gradient decreases with depth of puncture into the tissue but it doesn't go to zero due to the linear pO2 dependence of the carotid body oxygen consumption (Delpiano & Acker 1980). An exact correlation between extracellular pO2 of the superfusion medium and cytochrome reduction behaviour is very difficult. However the linear relationship of cytochrome c, b563 and a3 as well as the non-linear relationship of cytochrome a592 and b558 argues for the O_2-sensing properties of the later cytochromes.

Duchen: It amuses me that the arguments are exactly the same as they were in 1970 with the Mills & Jobsis (1970) data.

Rich: I have spent a fair bit of time looking at such spectra in difficult materials. It was impressive signal to noise that you got, given the nature of the material; nevertheless the signal to noise is low. When you do this type of deconvolution, which involves global fitting over all the wavelengths, there is a problem. First, if you are analysing the cytochrome b and c region you can't possibly analyse for the NADPH oxidase as well. As you probably know, there are two B haems in the mitochondrial cytochrome b and they have entirely different spectra. You are never going to get an accurate analysis by deconvoluting with a single B spectrum. Secondly, succinate dehydrogenase has a b-560 which is almost identical to the two haems in the NADPH oxidase. On the cytochrome oxidase part, I agree that the cytochrome oxidase profile is broader than it should be for a conventional reduced minus oxidized spectrum. Your interpretation that there is an additional haem A component, blue shifted, could be correct. But only 'could be' correct. The reason for this is that when you try to do this type of thing in whole cells, you are not taking oxidized minus reduced different spectra: you are taking steady state minus reduced different spectra. The steady state of cytochrome oxidase contains many of the intermediates, all of which have different spectra to reduce and to oxidise. For example, the peroxy state, which will be a 25% occupancy in your 'oxidized' state, looks a bit like reduced haem A. Your sample is haemoglobin free, but is there myoglobin in any of the cells?

Acker: No.

Rich: So that isn't a problem. There probably is an additional component there which you assume is a novel form of haem A. But it is more likely to be a form of haem A_3, possibly with a ligand bound to it. To give you an example, if you were to add CO to that material you would get a similar spectrum, with a shoulder at about 592 nm, with the CO binding to reduced haem A_3. Haem A_3 is relatively invisible in a reduced minus oxidized difference spectrum until such a ligand is bound. I don't know where you got your figures from but the actual number for contributions to the 606 nm peak in unligated oxidase is approximately 80% haem A contribution and 20% haem A_3. When you add a ligand such as CO to reduced haem A_3 it gets an extinction coefficient that is closer to that of haem A but is blue-shifted, in the case of CO ligation to 592 nm.

Acker: For the cytochrome b region we could easily include more spectra, but we don't have isolated spectra of the other cytochrome bs.

Rich: In that case the extra peak is almost certainly the haem A_3 cyanide-ligated complex. This is also close to 592 nm. When you reduce haem A_3 in unligated oxidase, it has a weak contribution to the reduced minus oxidized difference spectrum. However, when you do a reduced minus oxidised difference spectrum in the presence of cyanide you get pretty much what your data are showing. The reduced form of haem A_3 with cyanide bound is visible at around 590 nm, rather like the shoulder seen in your data.

Acker: The spectra which I have shown are N_2 versus normoxia steady state difference spectra. The blue-shifted cytochrome a592 is unique for the carotid body and could not be found in other tissues by us.

Rich: If you look in the literature for a spectrum of reduced plus cyanide minus oxidized plus cyanide you would get that spectrum.

Duchen: The comment about possible binding of CO or NO may be an interesting topic to return to later.

Acker: CO has two different effects on carotid body discharge: under normoxia you get a light sensitive stimulation and under hypoxia you get a light sensitive inhibition.

Rich: The cyanide and the CO compounds of the ferrous form of haem A_3 have similar spectra, and both are very similar to the extra component.

Acker: What about the argument that this is only found in carotid body?

Rich: There is a subtlety in the cyanide case. When CO is bound to haem A_3 its redox potential goes up and it is easy to form. When cyanide is bound to haem proteins the redox potential goes down and, as a result, there can be difficulties in providing enough reducing power to reduce the haem A_3, so you would have to be careful. You could get variable levels of reduced haem A_3 in the presence of cyanide. To see it consistently between cell types might depend on how much reducing power they have. If you added dithionite this should help by consistently providing sufficient reductant.

Acker: Under dithionite the reduction of all other cytochromes is so strong that the contribution of cytochrome a592 to the total N_2 versus normoxia steady state light absorption spectrum is small.

Rich: I would suggest it is the inability of some cells to produce enough reducing potential to reduce the haem A_3 when cyanide is bound to it.

Prabhakar: What is the current notion on the tissue pO_2 of the carotid body?

Acker: The mean pO_2 seems to be as in other organs at about 30 Torr. There is a pO2 distribution ranging from 0 to 90 Torr. Therefore I think the carotid body mirrors the O_2 supply situation in other organs. Our explanation for this is a large shunt blood flow meaning that only 10% of the total flow goes through the specific tissue. The underlying vascular structure would favour a predominant plasma flow through the specific carotid body tissue.

Duchen: We have talked about the oxygen sensitivity of the mitochondrial respiratory chain, and people have also talked about NADH oxidase as the oxygen sensor. What is the oxygen sensitivity of this process?

Acker: The K_m is about 10 Torr.

Gonzalez: According to Cross & Jones (1991) it is in the range of 10–15 Torr.

Evans: I have a question on IH. Looking at the experiments you have presented and considering the proposal that hypoxia triggers carotid body excitation by inducing an increase in ROS, there is a clear conflict. If IH is raising ROS and bringing about the facilitation of the observed carotid body response, and hypoxia doesn't bring about such facilitation, do your data not argue against a role for ROS in mediating carotid body excitation by hypoxia?

Prabhakar: Let me be explicit about this: I haven't done the experiments to test the role of ROS in normal carotid bodies. I don't know what role they might be playing under normal conditions. In the paradigm of IH, which I presented today, ROS are facilitatory to the hypoxic sensing both in the carotid body and in cell culture system.

Evans: But, given your data, would you expect ROS to facilitate the response of the carotid body in both circumstances?

Prabhakar: Possibly.

Evans: It is generally accepted that if you push tissues to hyperoxic conditions you get ROS generation (Freeman et al 1982, Jamieson et al 1986). There are papers showing that hyperoxia turns off the carotid body, i.e. afferent fibre discharge goes down (Alcayaga et al 1997, Wang & Fitzgerald 2002). Have you looked at hyperoxia followed by a return to normoxia to see whether hyperoxia on its own generates the facilitation you observe?

Prabhakar: We have not done these experiments. Some years ago Dr Lahiri and his group tested the effects of hyperbaric oxygen, and found after 72 h of hyperbaric hyperoxia marked increases in basal sensory activity. They attributed this increase to oxidative stress.

Kemp: Weren't they hypoxaemic at that point? They must have been close to death: 72 h of 100% oxygen is enough to kill most people, unless they are in the ICU.

Prabhakar: They are not dead.

Sylvester: We have patients in the ICU who go for weeks on high oxygen.

Evans: Did anyone look at the impact on afferent fibre discharge in the acute phase of the response?

Gonzalez: We have done this in many experiments, going back and forth from breathing room air and 100% oxygen. When you go from air to 100% oxygen there is a decrease in the afferent fibre discharge, and in the opposite direction there is a regular increase in activity. I do not recall seeing any overshoot.

Kumar: If you keep the carotid body in hypoxia for 2–3 minutes and then go to hyperoxia you always get an undershoot in afferent discharge, yet there is never an overshoot going to hypoxia.

Evans: Is there any facilitation of the hypoxic response after hyperoxia for a short period?

Prabhakar: I do not know.

Gonzalez: This hasn't been systematically studied.

Prabhakar: There are studies in humans and rats showing enhanced ventilatory response to hypoxia after 30 min of hyperoxia. I am not sure anyone has systematically examined the role of the carotid bodies in hypoxic ventilatory response following brief hyperoxic exposures.

References

Alcayaga J, Iturriaga R, Zapata P 1997 Time structure, temporal correlation and coherence of chemosensory impulses propagated through both carotid nerves in cats. Biol Res 30:125–133

Cross AR, Jones OT 1991 Enzymic mechanisms of superoxide production. Biochim Biophys Acta 1057:281–298

Freeman BA, Crapo JD 1981 Hyperoxia increases oxygen radical production in rat lungs and lung mitochondria. J Biol Chem 256:10986–10992

Freeman BA, Topolosky MK, Crapo JD 1982 Hyperoxia increases oxygen radical production in rat lung homogenates. Arch Biochem Biophys 216:477–484

Jamieson D, Chance B, Cadenas E, Boveris A 1986 The relation of free radical production to hyperoxia. Annu Rev Physiol 48:703–719

Kannt A, Pfitzner U, Ruitenberg M et al 1999 Mutation of Arg-54 strongly influences heme composition and rate and directionality of electron transfer in Paracoccus denitrificans cytochrome c oxidase. J Biol Chem 274:37974–37981

Mills E, Jobsis FF 1970 Simultaneous measurement of cytochrome a3 reduction and chemoreceptor afferent activity in the carotid body. Nature 225:1147–1149

Mills E, Jobsis FF 1972 Mitochondrial respiratory chain of carotid body and chemoreceptor response to changes in oxygen tension. J Neurophysiol 35:405–428

Paky A, Farrukh I, Michael J et al 1987 Basal superoxide production in the isolated perfused rabbit lung. Fed Proc 46:995

Paky A, Michael J, Burke-Wolin T et al 1993 Endogenous production of superoxide by rabbit lungs: effects of hypoxia or metabolic inhibitors. J Appl Physiol 74:2868–2874

Streller T, Huckstorf C, Pfeiffer C, Acker H 2001 Unusual cytochrome a592 with low PO2 affinity correlates as putative oxygen sensor with rat carotid body chemoreceptor discharge. FASEB J 16:1277–1279

Wang HY, Fitzgerald RS 1997 Muscarinic modulation of hypoxia-induced release of catecholamines from the cat carotid body. Brain Res 927:122–137

Wilson HL, Dipp M, Thomas JM, Lad C, Galione A, Evans AM 2002 ADP-ribosyl cyclase and cyclic ADP-ribose hydrolase act as a redox sensor: a primary role for cADPR in hypoxic pulmonary vasoconstriction. J Biol Chem 276:11180–11188

Functional proteomics of BK potassium channels: defining the acute oxygen sensor

Paul J. Kemp, Sandile E. J. Williams, Helen S. Mason, Phillippa Wootton*, David E. Iles†, Daniela Riccardi and Chris Peers*

*Cardiff School of Biosciences, Museum Avenue, Cardiff University, Cardiff, CF10 3US, *School of Medicine, Worsley Buiding, University of Leeds, LS2 9JT, and †School of Biology, Manton Building, University of Leeds, LS2 9JT, UK*

Abstract. Recombinant and native large conductance, Ca^{2+}-activated K^+ (BK) channels often demonstrate O_2 sensitivity in cell-free membrane patches suggesting that a significant component of the O_2-sensing machinery must be closely associated with the channel protein complex. Until recently, however, the identity of the O_2 sensor itself had remained elusive. Employing functional proteomics we have defined the molecular nature of such an O_2 sensor of BK channels. Using immunoprecipitation, 1D and 2D gel electrophoresis, and mass spectroscopy we identified the constitutive form of haem oxygenase, haem oxygenase 2 (HO-2), as a BK α-subunit protein partner. Functional measurement of hypoxic modulation of BK channel activity during manipulation of HO-2 enzyme substrates and reaction products, followed by protein knock-down of HO-2 using small interfering RNA, indicated that this enzyme is directly involved in hypoxic inhibition of BK channels. Furthermore, good correlation was observed between data obtained from recombinant BK channels and those from acutely isolated rat carotid body glomus cells, suggesting strongly that HO-2 also acts as an O_2 sensor in native arterial chemoreceptors.

2005 Signalling pathways in acute oxygen sensing. Wiley, Chichester (Novartis Foundation Symposium 272) p 141–156

A fundamental mechanism of adaptation in the mammalian kingdom, and the major subject of this Novartis Symposium entitled 'Signalling pathways in acute oxygen sensing', is detecting and reacting to acute perturbation in the partial pressure of oxygen. As oxygen availability becomes compromised a number of cellular systems respond, often co-operatively, in order to maximize oxygen uptake by the lungs and to optimize delivery to the metabolically most active tissues. Thus, during periods of reduced oxygen tension, ventilation rate and depth are increased to maximize air flow across the respiratory units, local lung perfusion rates become rapidly matched to local alveolar ventilation and systemic arteriolar dilatation ensures that tissue and cerebral blood flow become swiftly optimized. Each of these homeostatic processes depends crucially on regulation by oxygen of particular ion channels (see Lopez-Barneo et al 2001, Peers & Kemp 2001, Patel & Honore 2001 for recent reviews), the most widely expressed of which is almost certainly the large conductance, Ca^{2+}-activated K^+ channel; cloned initially from *Drosophila* and named *slo* (Elkins et al 1986) and subsequently from human tissues (Tseng-Crank et al 1994, Dworetzky et al 1994) where it is known variously as maxiK, BK_{Ca} and BK.

At the cellular level, BK channels are strongly implicated as effectors in the acute oxygen signalling cascade in a number of systems, including: (a) carotid body chemoreceptors (Peers 1990, Riesco-Fagundo et al 2001), where low arterial oxygen is detected by BK channels and the resulting depolarizing signal is ultimately transduced into increased ventilation; (b) fetal and postnatal pulmonary arteriolar myocytes, where BK channels may contribute to both persistent prenatal (Cornfield et al 1996) and acute postnatal hypoxic pulmonary vasoconstriction (Peng et al 1999, Cornfield et al 1996) in order to match ventilation to perfusion; (c) neonatal adrenomedullary chromaffin cells (Thompson & Nurse 1998), where hypoxic inhibition of BK channels is involved in the surge of catecholamines crucial for preparing the newborn's lung for air-breathing by activating alveolar fluid reabsorption and surfactant secretion; and (d) central neurons (Liu et al 1999, Jiang & Haddad 1994a,b), where hypoxic depression of BK channel activity may contribute to the excitotoxicity which results from increased neuronal excitability. It is generally accepted that such channels are acutely and reversibly inhibited as oxygen supply to these tissues becomes limited (Jiang & Haddad 1994a,b, Riesco-Fagundo et al 2001, Wyatt et al 1995, Lewis et al 2002) and that the resulting cell depolarization and voltage-gated Ca^{2+} influx induces hypoxia-dependent transmitter release (Lopez-Barneo et al 2001). However, the molecular nature of the oxygen sensor in the arterial chemoreceptor or elsewhere remains unknown. Acute regulation by oxygen of both native (Riesco-Fagundo et al 2001, Jiang & Haddad 1994a, Wyatt & Peers 1995) and recombinant (Lewis et al 2002) BK_{Ca} channels is variably retained. Thus, in experimental protocols where the soluble fraction of the cytosolic *milieu* is significantly disrupted/diluted (i.e. excised and classical whole-cell configurations of the patch clamp technique) some reports show hypoxic inhibition (Riesco-Fagundo et al 2001, Jiang & Haddad 1994a) whilst others do not (Liu et al 1999, Wyatt & Peers 1995). Clearly, the O_2 sensor appears to be either cell-specific or developmentally regulated, or both. However, we have shown that human recombinant BK channels retain their oxygen-sensitivity in inside-out patches (Lewis et al 2002) prompting us, as well as others, to suggest that a significant component of the oxygen-sensing machinery must be closely associated with the channel protein complex (Jiang & Haddad 1994a, Lopez-Barneo et al 2001, Lewis et al 2002). To investigate this supposition, we have employed the integrated approach of functional proteomics (identification of protein partners by gel electrophoresis and MALDI/TOF mass spectroscopy followed by post transcriptional gene suppression and functional assay using electrophysiology) to identify and characterize proteins associated with the BK α-subunit which might act as acute oxygen sensors (Williams et al 2004a).

Immunoprecipitation from a HEK293 cell line stably expressing BK channels (Ahring et al 1997, Hartness et al 2003, Lewis et al 2002) with a BKα antibody allowed proteins which potentially associate with recombinant human BK to be isolated and separated by 2D (Fig. 1A, right panel) and 1D (Fig. 1B, right lane) gel electrophoresis. For comparison, parallel immunoprecipitations were performed on untransfected, wild type HEK293 cells (Fig. 1A, left panel; Fig. 1B, left lane). Of the unique proteins that immunoprecipitated from the stable BK$\alpha\beta$ cell line, peptide mass mapping using mass spectroscopy of trypsin digests consistently identified γ glutamyl transpeptidase (GGT) and haem oxygenase-2 (HO-2) as candidate protein partners. Although GGT associates with BKα, we have recently demonstrated that it is not involved in hypoxic inhibition of BK channels (Williams et al 2004b). Western blot of BK$\alpha\beta$ cell lysates demonstrated constitutive expression of HO-2 in both cell lines (Fig. 1C). Only in the BK expressing cells was a tight biochemical interaction between BKα subunit and HO-2 confirmed by co-immunoprecipitation of BKα with an HO-2 antibody, and *vice versa* (Fig. 1D). That this interaction is specific was demonstrated by the inability of BKα to immunoprecipitate either endothelial nitric oxide synthase (eNOS) or the α-subunit of the Na^+/K^+-ATPase (Fig. 1D) despite both proteins being abundantly expressed (Fig. 1C). Co-localization of BKα with HO-2 was confirmed immunocytochemically by confocal microscopy (Fig. 1E).

In the presence of O_2 and NADPH, haem oxygenases catalyse the breakdown of haem to biliverdin, iron and CO (Prabhakar 1999). BK$\alpha\beta$ channel activity was robustly and reversibly activated by 30 μM of the chemical CO donor, $[Ru(CO)_3Cl_2]_2$; 30 μM of the breakdown product of this compound, $RuCl_2(DMSO)_4$, which does not release CO, did not affect channel activity indicating that CO strongly activates BK channels in inside-out patches (Fig. 1 F). Normalised NPo was increased 15-fold by the CO donor (Fig. 1F and 1G). 10 μM biliverdin evoked a fourfold activation of BK$\alpha\beta$ channel activity (Fig. 1H) but in patches treated sequentially with biliverdin and the CO donor, the activation was additive with the CO donor causing a further increase to 28-fold above control (Fig. 1H). Wild-type HEK293 cells did not display BK currents (data not shown but see Lewis et al 2002) and no activation was observed upon addition of the CO donor (data not shown).

Consistent with earlier reports (Lewis et al 2002, Williams et al 2004a), acute hypoxia (20–30 mmHg) resulted in a modest depression in NPo of inside–out patches excised from BK$\alpha\beta$ cells (Fig. 2A–D). In the presence of O_2, addition of

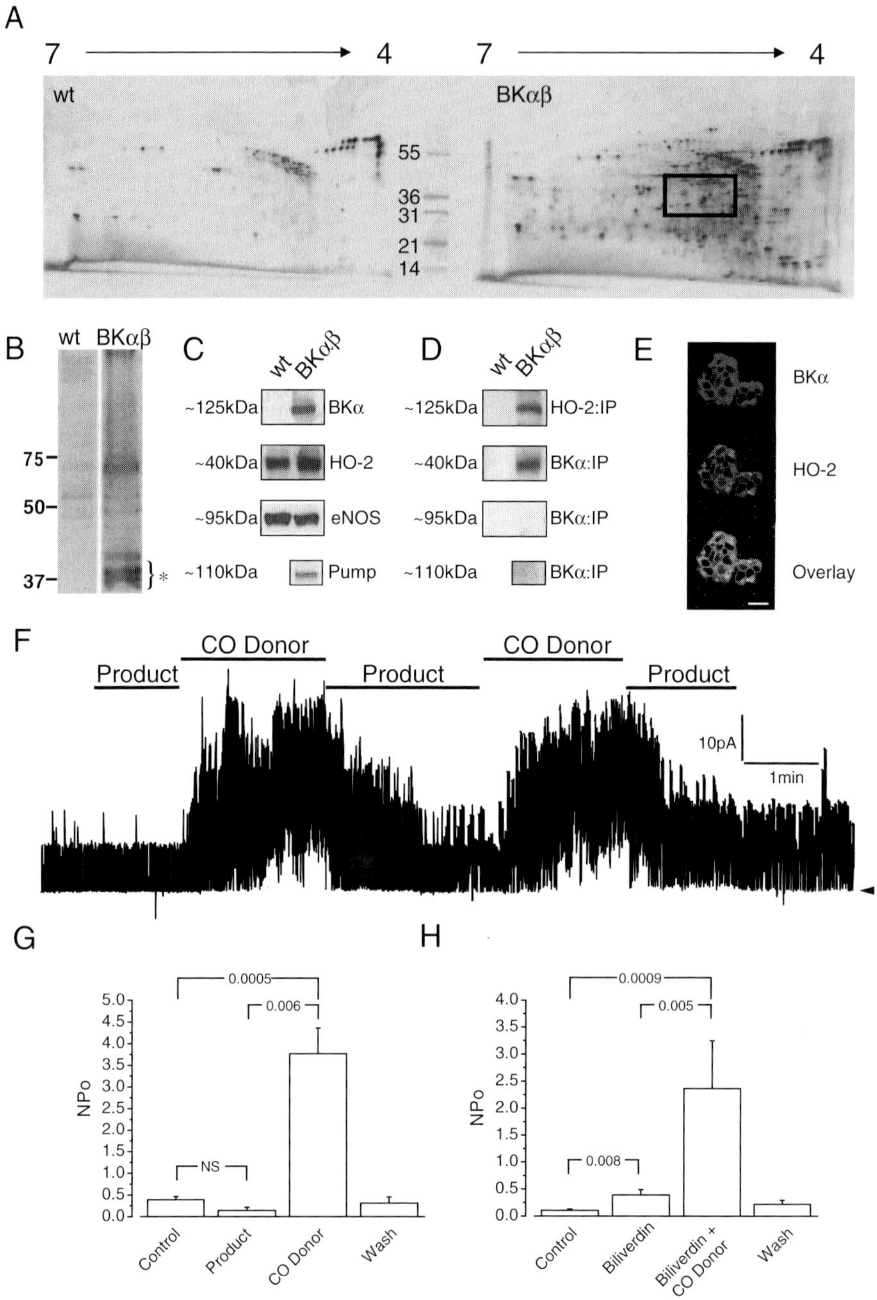
A
7
4
7
4
wt
BKαβ
55
36
31
21
14
B
wt
BKαβ
75
50
37
*
C
wt
BKαβ
~125kDa
BKα
~40kDa
HO-2
~95kDa
eNOS
~110kDa
Pump
D
wt
BKαβ
~125kDa
HO-2:IP
~40kDa
BKα:IP
~95kDa
BKα:IP
~110kDa
BKα:IP
E
BKα
HO-2
Overlay
F
Product
CO Donor
Product
CO Donor
Product
10pA
1min
G
NPo
0.0005
0.006
NS
5.0
4.5
4.0
3.5
3.0
2.5
2.0
1.5
1.0
0.5
0.0
Control
Product
CO Donor
Wash
H
NPo
0.0009
0.005
0.008
4.0
3.5
3.0
2.5
2.0
1.5
1.0
0.5
0.0
Control
Biliverdin
Biliverdin + CO Donor
Wash

the HO-2 co-substrates, haem (1 nM) and NADPH (1 μM), evoked a large increase in patch NPo (Fig. 2E–2H). Importantly, in the continued presence of the HO-2 co-substrates, hypoxia evoked a decrease in channel activity of over 70% suggesting that the enzymatic activity of HO-2 confers a significant enhancement on the O_2-sensing ability of the HO-2/BK protein complex (Fig. 2I–L). Thus, O_2 sensing by recombinant human BK$\alpha\beta$ channels consists of two components of which the HO-2-dependent component is quantitatively more important.

Selective knockdown of HO-2 protein at 48 h was achieved by transfecting cells with siRNA species. Successful transfection of Cy3-labelled siRNA, designed against either a scrambled human GAPDH coding sequence (Fig. 3A*i*) or the human HO-2 coding sequence (Fig. 3A*iv*) was followed using fluorescence microscopy prior to patch clamp. No knockdown of HO-2 immunoreactivity was observed using the scrambled siRNA (Fig. 3A*ii*). In complete contrast, almost total loss of HO-2 immunoreactivity was achieved with the specific HO-2 siRNA (Fig. 3A*v*). That positive transfection of siRNA species results in selective knockdown of HO-2, is reinforced by the overlayed fluorescent images of Fig. 3A*iii* and 3A*vi*. Furthermore, efficient knockdown of protein only in cells treated with the HO-2 siRNA was confirmed by Western blotting (Fig. 3B). The HO-2-dependent hypoxic suppression seen in untreated cells (Fig. 2I–L) was maintained following 48 h incubation with the scrambled, control siRNA (Fig. 3C–F). Following post-transcriptional gene suppression of HO-2 for 48 h with HO-2 siRNA, mean patch NPo was dramatically depressed and NADPH/haem-dependent hypoxic suppression was

FIG. 1. Haem oxygenase 2 (HO-2) as a functional BK$\alpha\beta$ channel-associated protein. (A) 2-D gel electrophoresis of proteins immunoprecipitated with a BKα antibody from wild-type (wt) and BK$\alpha\beta$ HEK293 cells. Boxed area indicates location of protein spots selected for MALDI/TOF analysis. (B) SDS-PAGE of immunoprecipitates from wt and BK$\alpha\beta$ cells. Bands removed for MALDI/TOF analysis are indicated by the asterisk. Linear pH gradients and/or molecular weight markers (in kDa) are shown. (C) Western blot analyses from lysates of wt and BK$\alpha\beta$ cells show HO-2, endothelial nitric oxide (eNOS) and α-subunit of Na^+/K^+-ATPase (pump) are constitutively expressed. (D) Western blot identification of BKα and HO-2 following immunoprecipitation (IP) with the antibodies shown to the right (top two blots). Neither eNOS nor the pump immunoprecipiated with BKα (lower two blots). Pump Western blot and IP were not performed on wt cells. (E) Confocal images of BK$\alpha\beta$ cells showing co-localization (yellow) of BKα (red) and HO-2 (green) with specific antibodies (colours not visible on this greyscale rendition of figure). Scale bar is 20 μm and applies to all. (F) Exemplar current recording from an inside-out membrane patch excised from a BK$\alpha\beta$ cell. Periods of application of 30 μM CO donor and 30 μM of its control (Product) shown above trace. (G) Mean NPo plot showing effect of 30 μM CO donor and Product on BK$\alpha\beta$ channel activity. (H) Mean NPo plot showing effect of 10 μM biliverdin and additive effects of 30 μM CO donor and 10 μM biliverdin. Patch potential (–Vp) = +20 mV, $[Ca^{2+}]_i$ = 335 nM, in this and all subsequent figures. Adapted from Williams et al (2004a).

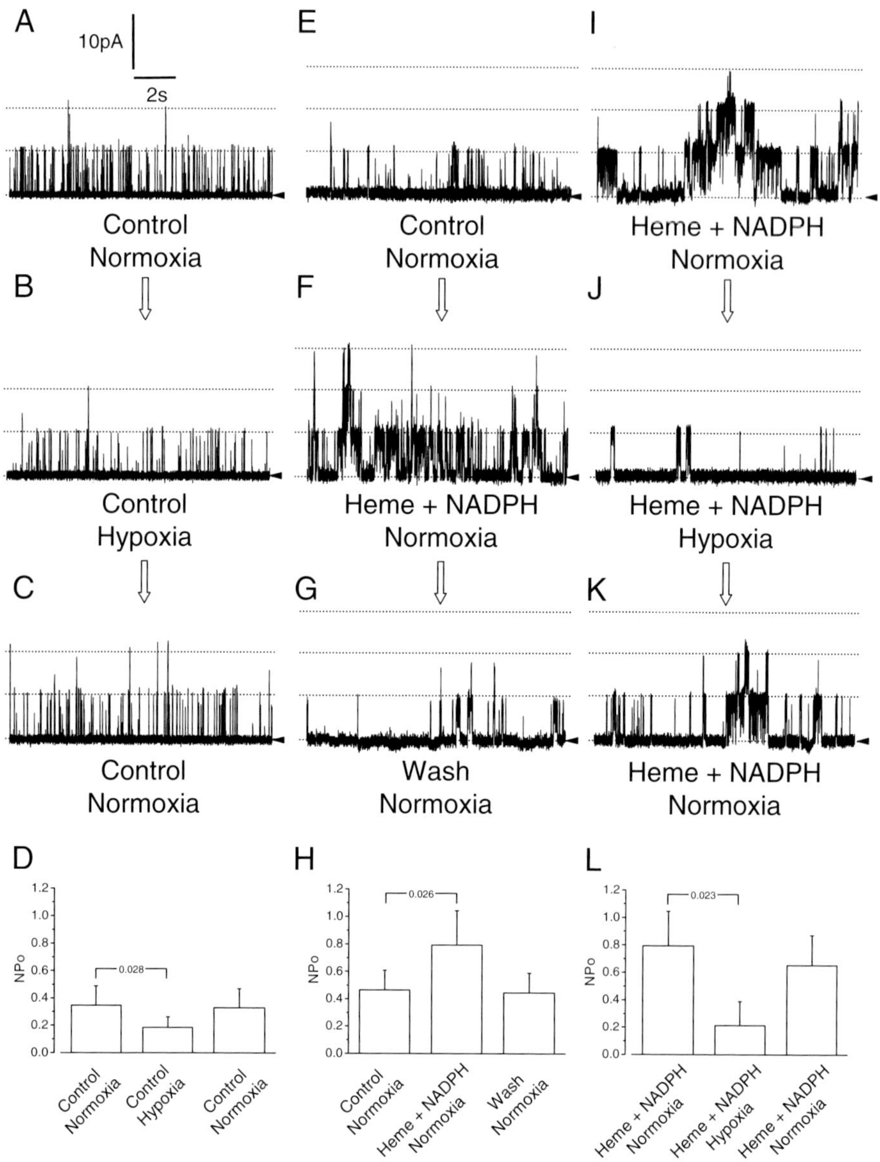

FIG. 2. Haem oxygenase substrates augment BK$\alpha\beta$ channel activity and hypoxic inhibition. Exemplar traces and mean NPo plots indicating modest hypoxic channel inhibition in untreated patches (A–D), increased baseline channel activity by 1 nM haem plus 1 μM NADPH (E–H) and augmentation of the hypoxic inhibition in the continued presence of haem plus NADPH (I–L). All traces are from inside–out membrane patches from BK$\alpha\beta$ cells. Here and in subsequent figures, normoxic pO2 ≈ 150 mmHg, hypoxic pO2 ≈ 15–25 mmHg. Adapted from Williams et al (2004a).

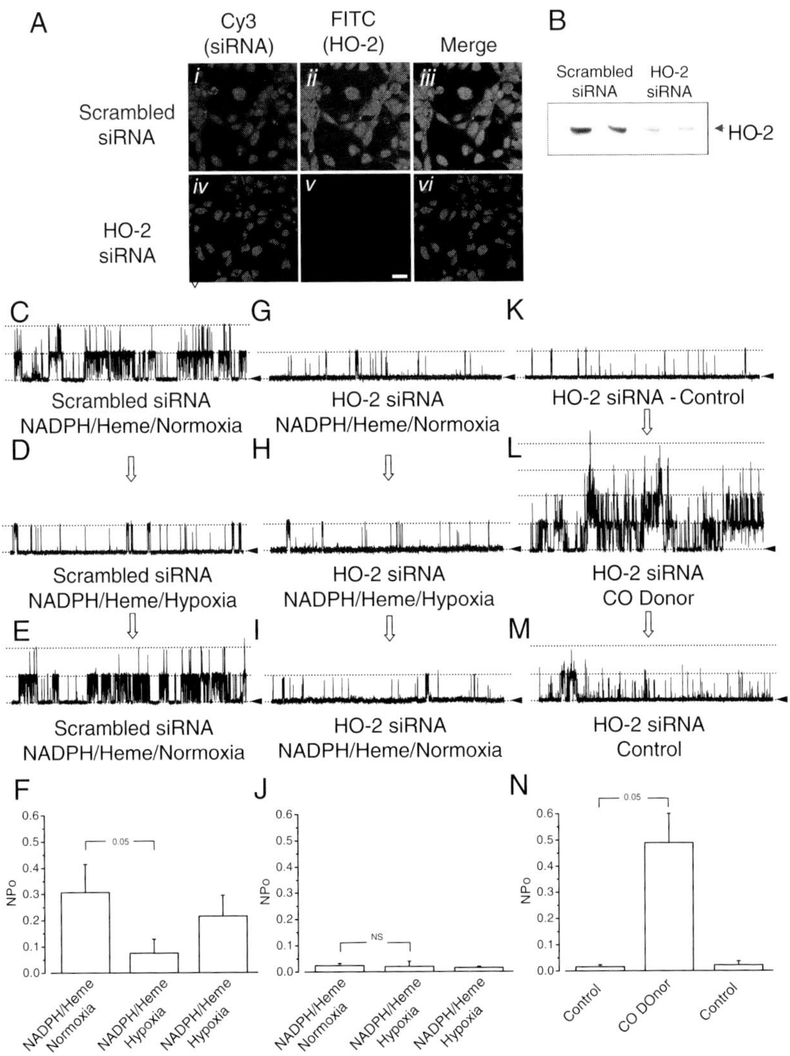

FIG. 3. Modulation of haem- and NADPH-dependent hypoxic inhibition of BK$\alpha\beta$ channels following knockdown of HO-2 expression by siRNA. Cy3-labelled transfected cells are shown in A*i* and A*iv*. HO-2 immunostaining using an FITC-labelled secondary antibody shows the persistent expression of HO-2 following scrambled siRNA treatment (A*ii*) and knockdown of HO-2 expression following HO-2 siRNA treatment (A*v*). Panels A*iii* and A*vi* show the merged images. Scale bar in A*v* = 20 μm and applies to all. (C) Western blot of BK$\alpha\beta$ cells transfected with scrambled siRNA (left panel) and HO-2 siRNA (right panel) shows ≈90 % knockdown of HO-2 expression by HO-2 siRNA-treatment. Exemplar traces and mean NPo plots of NADPH- and haem-dependent hypoxic BK$\alpha\beta$ channel inhibition in scrambled siRNA treated patches (C–F), almost complete loss of channel activity in HO-2 treated patches (G–J) and rescue of channel activity by the CO donor in HO-2 treated patches (K–N). siRNA-positive cells were selected by Cy3 fluorescence prior to patch clamp. Adapted from Williams et al (2004a).

completely absent (Fig 3G–J). Crucially, the CO donor was able to rescue this loss-of-function (Fig. 3K–N).

The physiological relevance of this novel enzyme-linked O_2 sensing by BK channels is illustrated in Fig. 4 which shows the result of activating haem oxygenase in inside–out patches excised from the membrane of rat carotid body glomus cells. Consistent with previous data obtained in native carotid body (Riesco-Fagundo et al 2001), the BK channel was only modestly inhibited by hypoxia (Fig. 4A–C and G). Similar to the recombinant system, supplying the channel complex with haem oxygenase co-substrates (Fig. 4G) or addition of the CO donor (data not shown) increased patch NPo. More importantly, NADPH/haem-dependent hypoxic inhibition was greatly augmented (to a level which has hitherto never been achieved by any investigator employing any cellular system), suggesting that the HO-2-dependent O_2 system is fully operable in native carotid body glomus cells (Fig. 4D).

Amongst the numerous proteins which directly associate with the α-subunit of BK, HO-2 is notable in that it is concentrated in neuronal and chemosensing tissues, including carotid body glomus cells, where it is constitutively expressed (Prabhakar 1999, Maines 1997, Verma et al 1993, Prabhakar et al 1995). Such constitutive expression also holds true for the recombinant system in which we have studied human BK channels, HEK 293 cells (Ahring et al 1997). HO-1 immunoreactivity was not detected in HEK293 cells and has previously been discounted in rat carotid glomus cells (Prabhakar et al 1995). Importantly, immunoprecipitation of proteins from BK$\alpha\beta$ cells provides direct evidence that HO-2 is associated with the BK α-subunit. That the system is still functionally intact in excised patches suggests strongly that the protein–protein interaction is membrane-delimited. Whether this interaction is direct or whether it occurs via intermediate proteins is uncertain; either way, it is clear that such a co-localization of BKα with HO-2 is necessary for both basal and O_2-dependent activity. Such an interaction activity is necessary for basal BK activity since HO-2 knockout results in a dramatic loss of channel activity which is fully rescued by the HO-2 product, CO. Activation by CO gas has been reported in glomus cells, supporting our suggestion that HO-2 activity is crucial to native BK channel regulation (Riesco-Fagundo et al 2001). Together with the data presented herein, the presence of HO-2 in the BK channel complex provides a molecular explanation for the observation that HO inhibition results in carotid body excitation (Prabhakar et al 1995). In the proposed model, O_2-sensing is conferred upon the BK channel by co-localization with HO-2. In normoxia, tonic HO-2 activity generates CO and biliverdin, both of which maintain the open state probability of the channel at a relatively high level. Our data show that the presence of CO and biliverdin together evoke BK channel activation which is more than additive. This behaviour may represent a means by which the normoxic signal is amplified. At this juncture, one can only speculate on the mechanism of CO action. In the

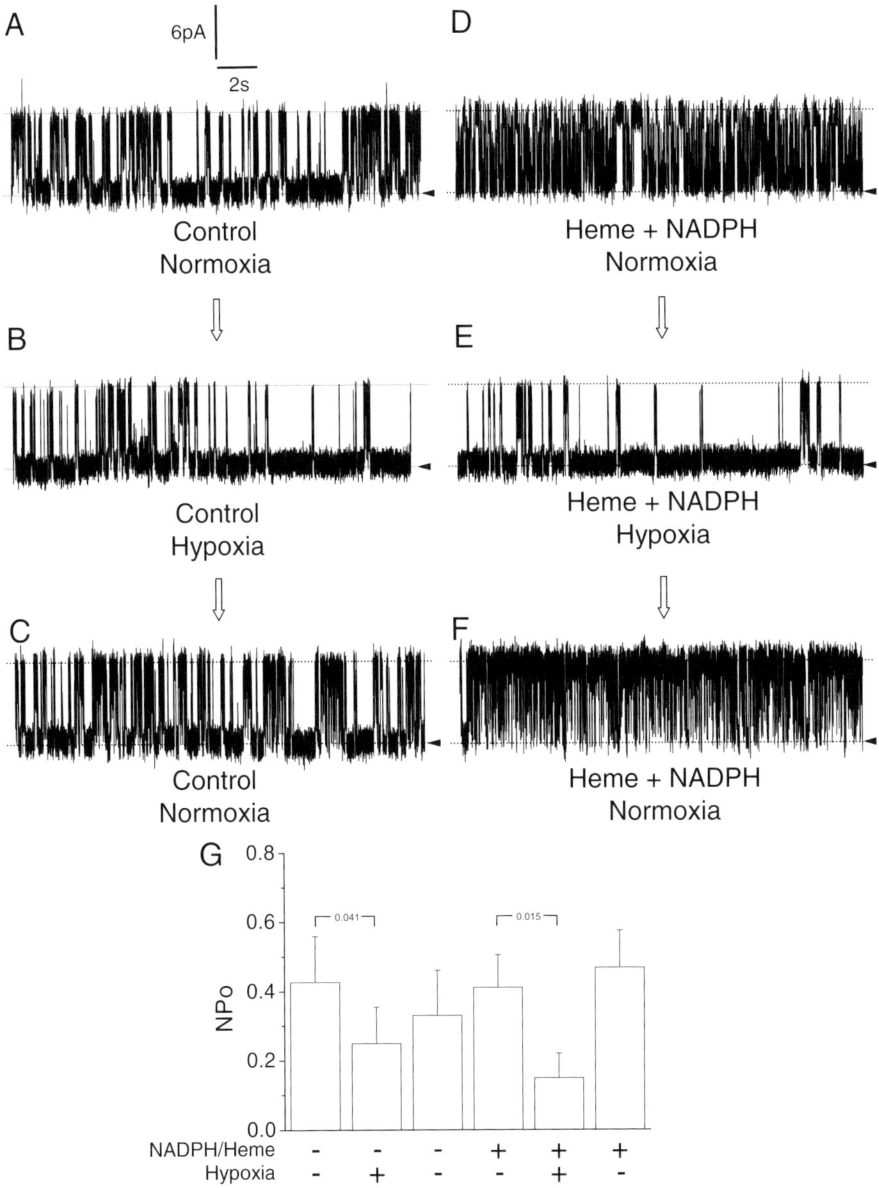

FIG. 4. Augmentation of carotid body glomus cell BK channel activity by haem oxygenase substrates. Exemplar traces indicating the modest hypoxic channel inhibition observed in untreated patches (A–C), increased baseline channel activity by 1 nM haem plus 1 μM NADPH (C–D) and augmentation of the hypoxic inhibition in the continued presence of heme/NADPH (D–F). Corresponding mean NPo values shown in (G). All traces are from inside–out membrane patches excised from carotid body glomus cells. Adapted from Williams et al (2004a).

absence of other second messenger systems (such as gas-activation of guanylate cyclase) an appealing candidate, based on earlier data in native vascular tissue (Wang & Wu 1997), is conformation regulation through direct interaction of CO with a histidine residue, potentially in the haem-binding domain of BKα (Wood & Vogeli 1997, Tang et al 2003). Whatever the molecular nature of the CO effect, cellular CO levels are reduced during hypoxic challenge as HO-2 substrate (O_2) becomes scarce, and rapidly fall below the critical threshold for the maintenance of BK channel activity at the tonically high level. In other words, HO-2 functions as a sensor of acute reduction in environmental O_2 by regulating BK channel activity primarily through the production of CO.

Acknowledgements

This work was funded by The British Heart Foundation, The Wellcome Trust and a Commonwealth Scholarship to SEJW.

References

Ahring PK, Strobaek D, Christophersen P, Olesen S-P, Johansen TE 1997 Stable expression of the human large-conductance Ca^{2+}-activated K^+ channel α- and β-subunits in HEK 293 cells. FEBS Lett 415:67–70

Cornfield DN, Reeve HL, Tolarova S, Weir EK, Archer S 1996 Oxygen causes fetal pulmonary vasodilation through activation of a calcium-dependent potassium channel. Proc Natl Acad Sci USA 93:8089–8094

Dworetzky SI, Trojnacki JT, Gribkoff VK 1994 Cloning and expression of a human large-conductance calcium-activated potassium channel. Brain Res Mol Brain Res 27:189–193

Elkins T, Ganetzky B, Wu CF 1986 A Drosophila mutation that eliminates a calcium-dependent potassium current. Proc Natl Acad Sci USA 83:8415–8419

Hartness ME, Brazier SP, Peers C, Bateson AN, Ashford ML, Kemp PJ 2003 Post-transcriptional control of human maxiK potassium channel activity and acute oxygen sensitivity by chronic hypoxia. J Biol Chem 278:51422–51432

Jiang C, Haddad GG 1994a A direct mechanism for sensing low oxygen levels by central neurons. Proc Natl Acad Sci USA 91:7198–7201

Jiang C, Haddad GG 1994b Oxygen deprivation inhibits a K^+ channel independently of cytosolic factors in rat central neurons. J Physiol 481:15–26

Lewis A, Peers C, Ashford MLJ, Kemp PJ 2002 Hypoxia inhibits human recombinant maxi K^+ channels by a mechanism which is membrane delimited and Ca^{2+}-sensitive. J Physiol 540:771–780

Liu H, Moczydlowski E, Haddad GG 1999 O_2 deprivation inhibits Ca^{2+}-activated K^+ channels via cytosolic factors in mice neocortical neurons. J Clin Invest 104:577–588

Lopez-Barneo J, Pardal R, Ortega-Saenz P 2001 Cellular mechanism of oxygen sensing. Annu Rev Physiol 63:259–287

Maines MD 1997 The heme oxygenase system: a regulator of second messenger gases. Annu Rev Pharmacol Toxicol 37:517–554

Patel AJ, Honore E 2001 Molecular physiology of oxygen-sensitive channels. Eur Respir J 18:221–227

Peers C 1990 Hypoxic suppression of K^+ currents in type-I carotid-body cells-selective effect on the Ca^{2+}-activated K^+ current. Neurosci Lett 119:253–256
Peers C, Kemp PJ 2001 Acute oxygen sensing: diverse but convergent mechanisms in airway and arterial chemoreceptors. Respir Res 2:145–149
Peng W, Hoidal JR, Farrukh IS 1999 Role of a novel KCa opener in regulating K^+ channels of hypoxic human pulmonary vascular cells. Am J Respir Cell Mol Biol 20:737–745
Prabhakar NR 1999 NO and CO as second messengers in oxygen sensing in the carotid body. Respir Physiol 115:161–168
Prabhakar NR, Dinerman JL, Agani FH, Snyder SH 1995 Carbon monoxide: a role in carotid body chemoreception. Proc Natl Acad Sci USA 92:1994–1997
Riesco-Fagundo AM, Perez-Garcia MT, Gonzalez C, Lopez-Lopez JR 2001 O_2 modulates large-conductance Ca^{2+}-dependent K^+ channels of rat chemoreceptor cells by a membrane-restricted and CO-sensitive mechanism. Circ Res 89:430–436
Tang XD, Xu R, Reynolds MF, Garcia ML, Heinemann SH, Hoshi T 2003 Haem can bind to and inhibit mammalian calcium-dependent Slo1 BK channels. Nature 425:531–535
Thompson RJ, Nurse CA 1998 Anoxia differentially modulates multiple K^+ currents and depolarizes neonatal rat adrenal chromaffin cells. J Physiol 512:421–434
Tseng-Crank J, Foster CD, Krause JD et al 1994 Cloning, expression, and distribution of functionally distinct Ca^{2+}-activated K^+ channel isoforms from human brain. Neuron 13:1315–1330
Verma A, Hirsch DJ, Glatt CE, Ronnett GV, Snyder SH 1993 Carbon monoxide: a putative neural messenger. Science 259:381–384
Wang R, Wu L 1997 The chemical modification of KCa channels by carbon monoxide in vascular smooth muscle cells. J Biol Chem 272:8222–8226
Williams SE, Wootton P, Mason HS et al 2004a Hemoxygenase-2 is an oxygen sensor for a calcium-sensitive potassium channel. Science 306:2093–2097
Williams SE, Wootton P, Mason HS, Iles DE, Peers C, Kemp PJ 2004b siRNA knock-down of gamma-glutamyl transpeptidase does not affect hypoxic K^+ channel inhibition. Biochem Biophys Res Commun 314:63–68
Wood LS, Vogeli G 1997 Mutations and deletions within S8-S9 interdomain region abolish complementation of N- and C-terminal domains of Ca^{2+}-activated K^+ (BK) channels. Biochem Biophys Res Comm 240:623–628
Wyatt CN, Peers C 1995 Ca^{2+}-activated K^+ channels in isolated type-I cells of the neonatal rat carotid-body. J Physiol 483:559–565
Wyatt CN, Wright C, Bee D, Peers C 1995 O_2-sensitive K^+ currents in carotid-body chemoreceptor cells from normoxic and chronically hypoxic rats and their roles in hypoxic chemotransduction. Proc Natl Acad Sci USA 92:295–299

DISCUSSION

Chandel: The important consideration with any oxygen sensor is to think about the K_m with respect to oxygen, and whether this fits the physiological system under study. What is the K_m of these haem oxygenases.

Kemp: I am fairly convinced that the K_m for oxygen for HO-2 has not been measured. The K_m for HO-1 is between 10 and 30 Torr. These two enzymes are quite different, so I wouldn't like to extrapolate those data.

Chandel: Are the HO-2 knockout mice viable?

Kemp: Yes.

Chandel: So what happens in these mice?

Kemp: Nanduri Prabhakar has some data on this.

Prabhakar: In 1995, we reported HO-2 expression in glomus cells of the carotid bodies. Zinc protoporphyrin-9, an inhibitor of HO-2 stimulated the carotid body sensory activity in a dose-dependent manner resembling the stimulatory effects of hypoxia, and the effects of HO-2 inhibitor could be reversed with exogenous administration of CO (Prabhakar et al 1995). With regard to the affinity of O_2 for HO-2, I recall a publication from Dr Ikeda-Saito's group in Cleveland reporting a rather high affinity of O_2 to HO-2, perhaps much greater than myoglobin (Migita et al 1998).

Kemp: So you saw hyperexcitability of the sensory discharge?

Prabhakar: Yes. We saw increased sensory activity following HO-2 inhibitor. As regards to HO-2 knockout animals, Dr Tonegawa's group at MIT developed these mice, and they are functionally viable. Recently, a Japanese group reported blunted hypoxic sensitivity in these mice. We found absence of ventilatory acclimatization to chronic hypoxia in HO-2 mice. I suspect that there might be complex interaction between CO from HO-2 and nitric oxide (NO) from NOS in intact carotid bodies.

Duchen: These cells are producing bilirubin and CO under normoxic conditions, and this changes when they become hypoxic. All of these things have absorption spectra. Would they turn up on Helmut Acker's absorption spectra?

Acker: If we had an isolated spectrum we could fit it into our spectral deconvolution and see whether it contributes or not.

Kemp: Presumably we could test the hypothesis that your shifted haem A3 is due to CO binding. We could do this in cells with or without the siRNA knockout of HO-2, so there is no CO production.

Duchen: Isn't that the wrong way round? You would have a decrease in CO during hypoxia, whereas you are looking at an increase.

Acker: It might be the question of a ligand. The unbinding of a ligand might also change.

Kemp: I think it would change the other way.

Archer: I have a question about NO from another angle. NO does exactly that to BK Ca^{2+} channels. The more open a channel is the larger its closing effect when an inhibitor is given. How did you show that it doesn't block the CO effect, and do you know from your 2D gels that NOS does not co-associate with the BK Ca^{2+} channels?

Kemp: HEK 293 cells express eNOS, and eNOS is not co-localized. This shows that what we are looking at is not a non-specific protein–protein interaction.

Archer: Have you looked at protein kinase G (PKG) sensitivity?

Kemp: No. I am expecting that PKG will not be co-localized with the channel in an excised inside–out patch. We never saw it from the proteomics. The mass spectrometry data from the trypsin mass mapping never produced with any probability at all masses of fragments consistent with PKG. I am assuming that the potential convergence of NO and CO pathways would not be membrane delimited.

Kummer: The open probability of the channel under hypoxia was always the same, regardless of the condition: siRNA treated or not. It always came down to the same level under hypoxia. What changed was the baseline under normoxia. In these siRNA-treated cells you showed us data only after stimulation with NADPH and haem presumably under unphysiologically high concentrations.

Kemp: Actually, they are not unphysiologically high. Haem is at 1 nM.

Kummer: What happens in the siRNA-treated cells when there is no NADPH added?

Kemp: When we knocked out the HO-2 protein, NADPH plus haem at any concentration did not activate the channel.

Kummer: But what happens when you just compare normoxia and hypoxia, if you take siRNA-treated cells, expose them to normoxia and hypoxia without adding anything?

Kemp: There is no hypoxic inhibition. The normoxic level and hypoxic level are the same when HO-2 is knocked out.

Buckler: So what would account for the oxygen sensing you see in excised patch when there is no substrate for haem oxygenase?

Kemp: I have been thinking hard about this. What is the nature of the component that isn't haem oxygenase sensitive? And why does it go away when you remove haem oxygenase. The answer to the second could be that haem oxygenase binding to the channel confers a baseline activity to the BK channel. In other words, it is required to be colocalized for the usual background activity to be seen. When HO-2 is removed completely, this basal, relatively high activity drops practically to nothing. HO-2, which is ubiquitously expressed, is required for normal BK function. In answer to the first part, the nature of the haem oxygenase insensitive component, I have no idea. One thing I am certain of though is that conformational coupling of some sort is required for anything other than a *P* open of 0.001.

Buckler: You didn't show summary data so we could see what the relative effects are. In relation to the type 1 cell, what proportion of sensitivity to hypoxia can be ascribed to the intrinsic oxygen sensitivity, as opposed to the oxygen sensitivity conferred by CO production?

Kemp: Inhibition in the absence of HO-2 substrates is around 20–40%. In the presence, it is 60–80%.

Buckler: Your published data on type 1 cells suggest that there isn't much activation of the channel by haem oxygenase substrates at all (Williams et al 2004).

Kemp: There was evident stimulation. What you are quoting are mean data from carotid body cells, which are inherently very noisy. If you go from normoxia with substrates to hypoxia with substrates, what you see in every patch is a 50–60% inhibition.

Duchen: Does NADPH itself have any effect on the channel?

Kemp: We did the experiments in both directions. We added NADPH on its own, followed by haem, and then haem followed by NADPH. We needed coapplication of both to see an effect. There are interesting effects at higher concentrations of haem, which may or may not be consistent with Tang's paper showing haem inhibition (Tang et al 2003).

López-Barneo: What kind of β subunit are you using in your heterologous maxi K expression experiments?

Kemp: The $\beta 1$ subunit.

López-Barneo: This is not expressed in glomus cells.

Kemp: No, but the haem oxygenase activation is.

López-Barneo: Do you think that $\beta 1$ is needed?

Kemp: No. We have done similar studies showing that we get NADPH plus haem activation in the α-only expressing cell. The β subunit is neither required for oxygen sensitivity nor is it apparently involved in bridging the protein–protein interaction between the α subunit and HO-2. I still have no good evidence that the HO-2 and the α subunit of BK are directly coupled. There could be any number of peptide and protein intermediates in that interaction. However, I am fairly sure that the β subunit is not required.

López-Barneo: There are some old data by Gonzalez's group showing that the O_2-sensitive K^+ channel that we initially described in the rabbit carotid body is also modulated by CO. This would imply that this mechanism is actively involved in regulating the K^+ channel. If this is a mechanism for O_2 sensing, don't you think it is a bit complicated? Oxygen is converted to CO, which leads to the question of how CO is sensed

Kemp: I might design it differently!

Murphy: My question is related to this: is there a model for what the CO is doing?

Kemp: Yes, the big question is how CO modulates the channel activity. I'd like some advice on how we could look at that. The obvious route is to discover whether it needs another haem-containing intermediate. This would really begin to get complicated. Is there a CO-specific binding motif? I doubt it. Is it binding via a second haem-containing protein partner? We can do the light activation spectrum of this. This is something we need to do.

Sylvester: The haem oxygenase reaction also produces molecular iron. Is that playing some role?

Kemp: This is something else I have been thinking about. Is the iron that is being produced interacting with any of the other potential sensors? It's a good question. We know that bilirubin (BR) is also a channel activator. Not only this, BR plus CO have a more than additive effect. My guess is that although CO is a short-lived gas, BR/biliverdin reactions are just as fast.

Sylvester: I have a rather philosophical question. In your proteomics approach, you started with a 2D gel, determined a region of interest and apparently immediately threw a strike by selecting haem oxygenase. Can you give us some guidance as to how you achieved such perspicacity?

Kemp: What we do is run a 2D gel, get a pipette tip, and cut every single band out. When you realise that the awful MALDI-TOFF machine you have can't detect many of the small spots, you then run 1D gels and cut bands out. When you have some information from the 1D band, you go back. We had no robotics, the proteomics facility was slow, so we took the areas of interest which represented some of our first hits in terms of pI and relative molecular weight, and took them back to the MALDI to go through the mass spec system again and again. In a perfect world you would run 12 gels with all the different controls, put it on a robotic spot picker and go away for three days. You come back, it having been through a tandem MS machine, and are supplied with plenty of structural data. This gives a better chance. What happened is that I had a list of about 3000 potential proteins from that single 2D gel. We sat down for about eight afternoons and looked at which ones might be oxygen sensitive. The first one we picked was γ-glutamyl transpeptidase, because there was some precedence in the literature that this might be involved in oxygen sensing, and while this proved to be involved in BK channel regulation it was not involved in oxygen sensitivity. The next one down the list was HO-2, and this was our 'bingo' moment.

Schumacker: Is HO-2 a mixed function oxygenase? Does it put two oxygen atoms into the substrate or just one?

Kemp: One.

Schumacker: Then it generates superoxide. Is it possible that superoxide generated by flux through the enzyme system is actually the signalling molecule that is acting? You could test that in your system if you added a superoxide dismutase (SOD) or an oxygen radical scavenger.

Kemp: That would be very interesting and easy enough to do in our cell free system.

Chandel: The superoxide would go down, because the haem oxygenase is not as functional under hypoxia.

Schumacker: So there would be a decrease in reactive oxygen species (ROS) there.

Gurney: The BKα subunit has many splice variants, but only one or two are inhibited by hypoxia. Does this help you in understanding where the binding sites might be?

Kemp: We are trying to figure out the upstream event of the protein–protein interaction. The natural splice variants give us a good start. Sandile Williams has C-terminal constructs, each with a different splice variant. We are now trying to solubilize these and then co-crystallize with HO-2.

Gurney: Is the splice variant that you are using in your cells the same as the carotid body one?

Kemp: I don't know. It is not clear to me which splice variant is there.

Duchen: One thing that has worried me a bit is that this is a Ca^{2+}-dependent K^+ channel, and it is also voltage dependent. Do you know how the Ca^{2+} and voltage dependence are modulated by these processes?

Kemp: We don't know anything about how Ca^{2+} and voltage impact on haem oxygenase activation. If you do a current–voltage relationship, CO activates at every activating voltage. We know a little bit about the haem oxygenase-insensitive component.

Duchen: You seem to have a situation where you have a channel which is going to reduce its open probability in response to hypoxia, and as a consequence intracellular Ca^{2+} goes up and the cells depolarise. These are both going to increase the open probability of the channel again. Is your haem oxygenase altering the Ca^{2+} or voltage dependence of the channel?

Kemp: It's a key question which we've not addressed.

Peers: We think that hypoxia is changing the Ca^{2+} sensitivity.

Kemp: Only in the HO-2-independent component.

Peers: We haven't looked in the presence of substrates, but early studies show that if the channel is driven by voltage alone, it is not oxygen sensitive.

Kemp: It's the Ca^{2+} titration curve that shifts in hypoxia.

References

Migita CT, Matera KM, Ikeda-Saito M et al 1998 The oxygen and carbon monoxide reactions of heme oxygenase. J Biol Chem 273:945–949

Prabhakar NR, Dinerman JL, Agani FH, Snyder SH 1995 Carbon monoxide: a role in carotid body chemoreception. Proc Natl Acad Sci USA 92:1994–1997

Tang XD, Xu R, Reynolds MF, Garcia ML, Heinemann SH, Hoshi T 2003 Haem can bind to and inhibit mammalian calcium-dependent Slo1 BK channels. Nature 425:531–535

Williams SE, Wootton P, Mason HS et al 2004 Hemoxygenase-2 is an oxygen sensor for a calcium-sensitive potassium channel. Science 306:2093–2097

A central role for oxygen-sensitive K^+ channels and mitochondria in the specialized oxygen-sensing system

Stephen L. Archer, Evangelos D. Michelakis*, Bernard Thébaud†, Sebastien Bonnet*, Rohit Moudgil*, Xi-Chen Wu* and E. Kenneth Weir‡

*Department of Medicine (Cardiology), Department of Physiology and the Vascular Biology Group, University of Alberta, Edmonton, Canada, *Department of Medicine (Cardiology) and the Vascular Biology Group, University of Alberta, †Department of Pediatrics and the Vascular Biology Group, University of Alberta, Canada and ‡Department of Medicine (Cardiology), Minneapolis Veterans Affairs Medical Center, USA*

Abstract. Mammals possess a specialized O_2-sensing system (SOS), which compensates for encounters with hypoxia that occur during development, disease, and at altitude. Consisting of the resistance pulmonary arteries (PA), ductus arteriosus, carotid body, neuroepithelial body, systemic arteries, fetal adrenomedullary cell and fetoplacental arteries, the SOS optimizes O_2-uptake and delivery. Hypoxic pulmonary vasoconstriction (HPV), a vasomotor response of resistance PAs to alveolar hypoxia, optimizes ventilation/perfusion matching and systemic pO2. Though modulated by the endothelium, HPV's core mechanism resides in the smooth muscle cell (SMC). The Redox Theory proposes that HPV results from the coordinated action of a redox *sensor* (proximal mitochondrial electron transport chain) which generates a diffusible *mediator* (a reactive O_2 species, ROS) that regulates *effector* proteins (voltage-gated K_v channels). Hypoxic withdrawal of ROS inhibits $K_v1.5$ and $K_v2.1$, depolarizes PASMCs, activates voltage-gated Ca^{2+} channels, increasing Ca^{2+} influx and causing vasoconstriction. Hypoxia's effect on ROS (decrease vs. increase) and the molecular origins of ROS (mitochondria vs. NADPH oxidase) remains controversial. Distal to this pathway, Rho kinase regulates the contractile apparatus' sensitivity to Ca^{2+}. Also, a role for cADP ribose as a redox-regulated mediator of intracellular Ca^{2+} release has been proposed. Despite tissue heterogeneity in the SOS's output (vasomotion versus neurosecretion), O_2-sensitive K^+ channels constitute a conserved effector mechanism. Disorders of the O_2-sensing may contribute to diseases, such as pulmonary hypertension.

2005 Signalling pathways in acute oxygen sensing. Wiley, Chichester (Novartis Foundation Symposium 272) p 157–175

Hypoxic pulmonary vasoconstriction (HPV) is a homeostatic vasomotor response of small, muscular 'resistance' pulmonary arteries (PA) to alveolar hypoxia. HPV actively diverts perfusion to optimally ventilated lung segments. The central mechanism of HPV involves a redox-based O_2 sensor (e.g. mitochondria) which, during normoxia, tonically generates a diffusible redox mediator (e.g. H_2O_2). During hypoxia the reactive O_2 species (ROS) are withdrawn. The resulting reduced redox environment and/or loss of the ROS inhibits certain voltage-gated K^+ channels (K_v) in pulmonary artery smooth muscle cells (PASMCs). Upstream of this pathway, the magnitude of HPV is impacted by the endothelium (e.g. endothelin, nitric oxide); downstream it is modulated by Rho kinase through sensitization of the contractile apparatus to Ca^{2+}. This overview, extracted from a recent review (Moudgil et al 2005), presents a critical appraisal of the Redox Theory for the mechanism of HPV (Figs. 1–2).

Historical perspective

The first observation of HPV was made by Bradford and Dean in 1894 (see (Moudgil et al 2005). Fifty years later von Euler and Liljestrand recognized the potential homeostatic role of HPV in optimizing systemic pO2, noting that it 'increases the blood flow to better aerated lung areas, which leads to improved conditions for the utilization of alveolar air'.

Properties of HPV in humans

Research defining the properties of HPV in humans offers practical definitions of physiological hypoxia and delineates the fundamental characteristics of HPV. Results from reductionist models should be judged against these studies. At the summit of Mount Everest the inspired and arterial pO2 are both ~43 mmHg with a $PaCO_2$ of ~11 mmHg, and an arterial pH of ~7.53 (Malconian et al 1993). This serves as a practical definition of 'physiologically tolerable' hypoxia. To go higher (in altitude) or lower (in pO2) is to court death; yet much of the molecular science of HPV is conducted at pO2s below 30 mmHg with no measure of pH or pCO_2. Refocusing on 'physiologically tolerable' hypoxia would reduce confusion in the HPV literature.

HPV results in a maximum increase in pulmonary vascular resistance (PVR) within 15 min (Bindslev et al 1985). Moderate hypoxia (12.5% O_2, pO2 ~50 mmHg) more than doubles PVR and HPV is not potentiated by repeated hypoxic challenges; nor does it decay with hours of sustained hypoxia (Carlsson et al 1985). In healthy volunteers, 8 hours of hypoxia increases PVR from 1 to 3 Wood Units within 2 hours; thereafter PVR plateaus. Systemic vascular resistance (SVR) decreases in parallel, falling from 19 to 14 Wood Units. PVR reverts to normoxic levels upon return to breathing 20% O_2 (Dorrington et al 1997). The intrinsic nature of these opposing responses to hypoxia is recapitulated in the isolated, serially perfused, lung-kidney model (Michelakis et al 2002a).

In humans, HPV is significantly impaired by hyperventilation and the resulting respiratory alkalosis; however, HPV is not reduced by inhibition of eicosanoids synthesis or endothelin receptor antagonism (reviewed in Moudgil et al 2005), consistent with endothelium-independence of the response. A role for ion channels in HPV is implicated by *in vivo* studies of patients with chronic obstructive pulmonary disease (COPD). HPV is attenuated 53% in COPD patients by nifedipine, an L-type, voltage-gated Ca^{2+} channel blocker, at doses that don't reduce SVR (Burghuber 1987).

HPV in animal models

HPV is primarily elicited by alveolar, not mixed venous, pO2 (see Moudgil et al 2005). HPV is intrinsic to the resistance segment of the PA, specifically to the PASMCs of intrapulmonary resistance pulmonary arteries (PAs) (<200 μm) that directly experience alveolar pO2 (Kato & Staub 1966). HPV can be demonstrated in isolated lungs, PA rings denuded of endothelium (Archer et al 2004) and even isolated PASMCs from resistance PAs (Madden et al 1985). Furthermore, the magnitude of HPV increases in inverse proportion to the PA diameter (Shirai et al 1986). The opposing effects of hypoxia on pulmonary and systemic circulations indicate that HPV cannot be explained by endocrine or paracrine vasoconstrictors that have concordant constrictor effects on the pulmonary and systemic circulation (e.g. endothelin, leukotrienes, etc). We believe that the restricted occurrence of hypoxic constriction to intrapulmonary PAs (and to a lesser extent veins) reflects the localized abundance of HPV's molecular apparatus in these segments (Archer et al 2004).

In contrast to the clarity of the *in vivo* and isolated lung research, there is disagreement about the characteristics of HPV in PA rings. While we find that HPV causes a rapid constriction in resistance PA rings that gradually plateaus and is sustained (Archer et al 2001), much as occurs *in vivo*, several groups find that hypoxia elicits a biphasic response consisting of an immediate, endothelium-independent constriction which peaks in ~10 min (phase I) and a subsequent slowly developing endothelium-dependent contraction peaking at ~30–40 min (phase II) (Ward & Robertson 1995). The basis for the discrepancy is uncertain but likely relates to tissue selection and handling.

Role of HPV in health and disease

This is reviewed in Moudgil et al (2005). In acute atelectasis, PVR increases within 15 min to a maximal value by 1 hour and optimizes systemic pO2. HPV is also important in the transitional circulation. HPV actively reduces blood flow in the fetal lung *in utero* and conversely, the reversal of HPV contributes to the fall in PVR that occurs in the transitional circulation at birth. Thus HPV has a pivotal role to play in developmental and adult human physiology. HPV has been implicated in the pathophysiology of high altitude pulmonary oedema (HAPE), the pathogenesis of which entails a high pulmonary artery pressure (PAP) and inflammation, leading to increased capillary pressure and pulmonary oedema. Drugs that reduce HPV such as nifedipine and inhaled nitric oxide lower PAP and improve gas exchange in HAPE. However, under other circumstances, suppression of HPV can be detrimental. HPV is often suppressed in bacterial pneumonia, contributing to ventilation/perfusion (V/Q) mismatch and systemic hypoxemia. Clinically, HPV is exploited in single-lung anaesthesia to minimize blood flow to a lung that is intentionally made hypoxic to create a dry operative field.

Mechanisms of O_2 sensing in HPV

The redox mechanism for HPV, proposed initially in 1986 (Archer et al 1986) (Fig. 1), and refined subsequently (Fig. 2) (reviewed in Moudgil et al 2005) indicates that a drop in alveolar O_2 decreases production of a redox mediator virtually instantaneously (<30 s). The mediator, in turn, alters the function of one or more effector proteins, which ultimately increases cytosolic Ca^{2+}, thereby activating the PASMCs' contractile machinery. There are two potential cytochrome-based O_2-sensitive sources of ROS which are candidate O_2 sensors, the mitochondrial electron transport chain (ETC) and a family of classical and novel nicotinamide adenine dinucleotide phosphate (NADPH) oxidases (Phox and Nox). The final common pathway of all forms of pulmonary vasoconstriction involves activation of the contractile apparatus. Thus actin and myosin, and their regulatory systems, such as Rho kinase, are very important determinants of the magnitude of HPV. However, since they appear not to be unique to the pulmonary circulation and act downstream from the 'O_2-sensing unit', they are not discussed in this mini-review (see Moudgil et al 2005).

Mitochondria as O_2 sensors

The mitochondria's role as a primary site for O_2 consumption during aerobic metabolism makes it an ideal candidate for early detection of changes in pO2. The evidence suggesting that the mitochondrial ETC could be important in O_2 sensing is based in part on the concordant effects of certain ETC inhibitors and hypoxia. Inhibitors of the mitochondrial ETC and oxidative phosphorylation cause pulmonary vasoconstriction in isolated blood-perfused lungs (Rounds & McMurtry 1981). Moreover, inhibitors of complex I (rotenone) and complex III (antimycin A), but not inhibitors of complex IV (cyanide), mimic hypoxia's haemodynamic effects, causing pulmonary vasoconstriction, and dilatation of the human ductus

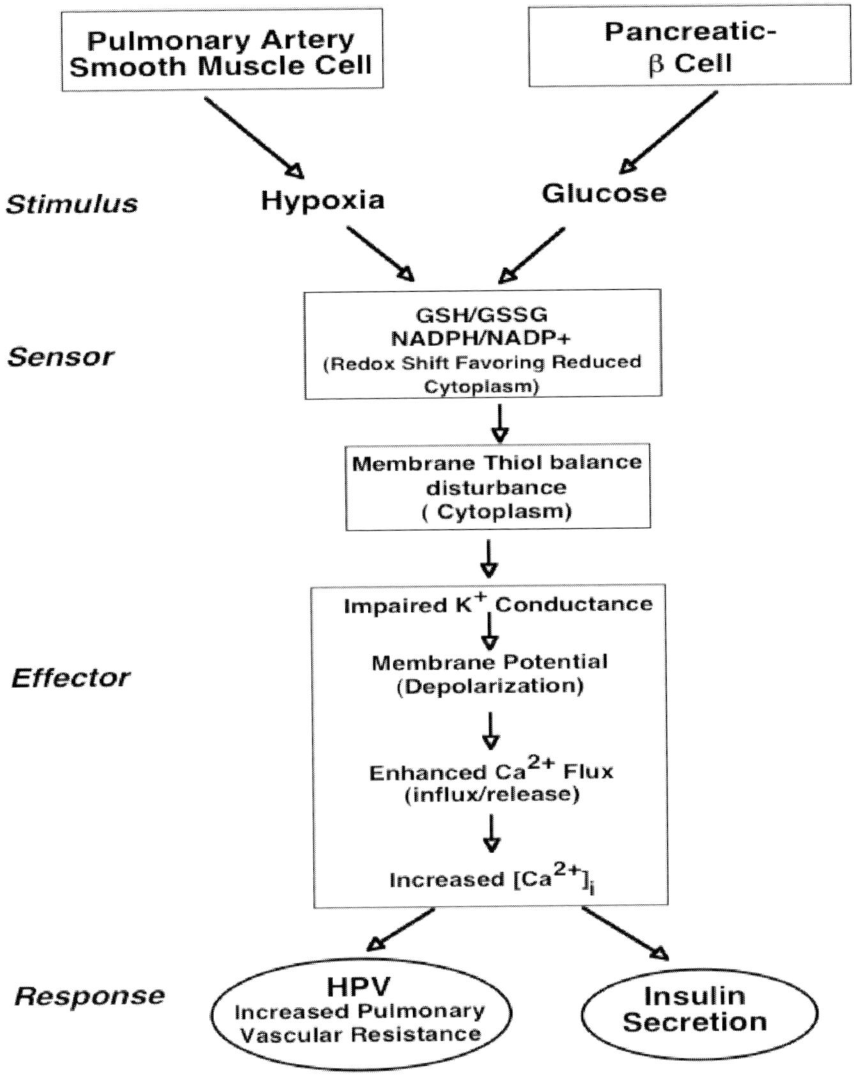

FIG. 1. 1986 Initial redox model for HPV. The concept is that redox state, determined by metabolism (in the mitochondria and elsewhere) regulates sulfhydryl balance in the membranes. This controls PASMC K^+ flux and membrane potential and thereby regulates the L-type Ca^{2+} channel and thus cytosolic Ca^{2+}. This pathway was also proposed as a means of regulating insulin secretion in pancreatic β cells (Modified from Archer et al 1986).

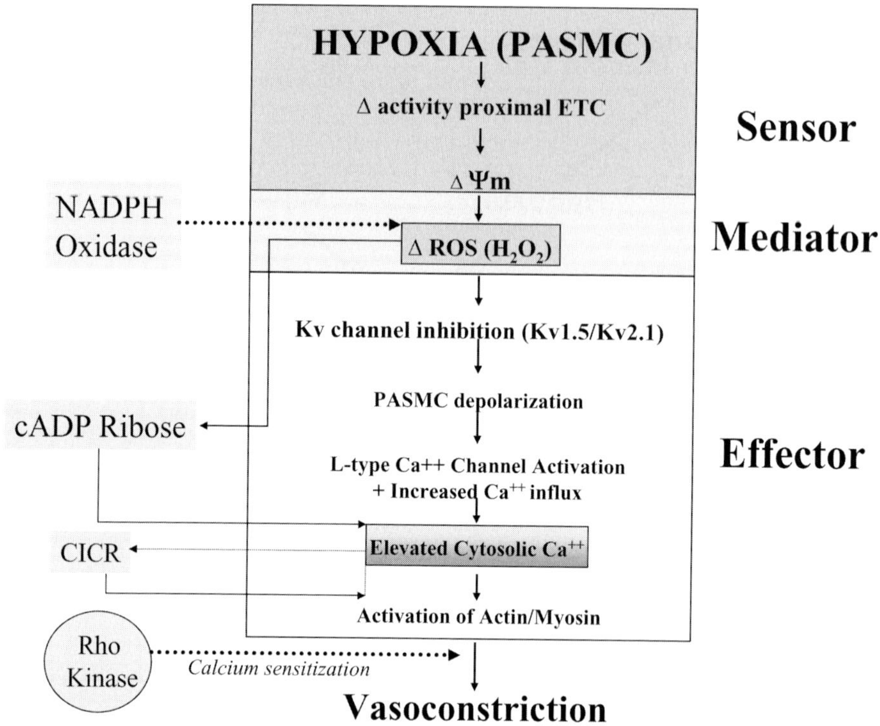

FIG. 2. Updated redox theory of HPV. In this model the mitochondrial ETC senses hypoxia and decreases ROS production. A decrease in ROS (superoxide anion and/or H_2O_2) inhibits the effector, $K_v1.5$ and $K_v2.1$ channels. K_v channels depolarize the membrane and activate Ca^{2+} entry via L-type Ca^{2+} channels. Other theories, which are not mutually exclusive, include NADPH oxidase as a sensor and source of ROS, cyclic adenosine diphosphate ribose (cADPR) as a redox-activated mediator that causes intracellular Ca^{2+} release by activating ryanodine receptors in the SR. In addition, activation of Rho kinase, may enhance HPV by sensitizing the contractile apparatus to Ca^{2+}.

arteriosus (DA) (Michelakis et al 2002b) and systemic circulation (Michelakis et al 2002a). Rotenone and antimycin not only inhibit subsequent acute HPV but their constrictor potency, like that of hypoxia, is selectively suppressed in chronic hypoxic pulmonary hypertension (CH-PHT), a condition in which constriction to most stimuli is enhanced (Reeve et al 2001). The parallels between hypoxia and ETC inhibitors are also evident in nonvascular tissue. ETC inhibitors mimic hypoxia's effects on the carotid body (Duchen & Biscoe 1992). The initial phase of the pulmonary vasoconstriction induced by both metabolic inhibitors and by hypoxia appears to occur without ATP depletion. ATP is preserved in HPV induced by moderate hypoxia; conversely, anoxia, which does deplete ATP, results in only transient pulmonary vasoconstriction followed by pulmonary vasodilatation (Buescher et al 1991). Our group finds that rotenone and antimycin reduce lung and PA ROS production in a manner analogous to that of physiological hypoxia (Archer et al 1993, 1999, Michelakis et al 2002a). However, the effects of hypoxia and mitochondrial inhibitors in the pulmonary circulation are controversial, as is discussed subsequently. Other groups have similarly concluded that mitochondria are important in HPV, albeit disagreeing on the precise signalling mediator (Waypa et al 2001, 2002).

On a cellular level, hypoxia inhibits K^+ current and depolarizes PASMCs, but does not have this effect in systemic arterial SMCs (Post et al 1992, Yuan et al 1993). Like hypoxia, rotenone and antimycin A (and some other ETC inhibitors) inhibit whole-cell K^+ current in PASMCs (Archer et al 1993). As would be predicted from the redox theory, the ETC inhibitors, which dilate the DA, increase I_K in human DA SMCs (Michelakis et al 2002b).

The PASMC's O_2-sensing mechanism appears to relate to the small proportion of total electron flux involving unpaired electrons and the associated production of intramitochondrial ROS. Teleologically, this uncoupled electron flow provides an early warning of potential risk to the upstream supply of electrons and their originating reducing equivalents which could limit ATP production. The decreased ROS signals initiation of HPV to restore homeostasis.

The ETC is comprised of four mega-complexes, each with many subunits, that mediate transfer of electrons down a redox potential gradient through series of carriers resulting in final acceptance of the electron by O_2 producing ATP, with water as an end product. In the course of electron transport, H^+ ions are translocated creating a proton gradient that generates the mitochondria's extremely negative $\Delta\Psi m$. This potential energy is used to produce ATP. Complexes I and III produce most of the mitochondrial ROS (Duchen 1999), although there is tissue heterogeneity in which complex predominates. Complex I, which is inhibited by the isoflavinoid rotenone, is an NADH ubiquinone oxidoreductase where NADH is oxidized to NAD^+ as it transfers two electrons to ubiquinone. Complex III is a ubiquinol-cytochrome c oxidoreductase that transfers electrons from $FADH_2$ to cytochrome b, and subsequently to the final electron acceptor, cytochrome c. Cyanide-sensitive complex IV, the cytochrome oxidase, is composed of cytochromes a and a3. It accepts electrons from reduced cytochrome c and passes them to O_2. Inhibition of a complex proximal to the site of ROS production should decrease ROS production (as we observe with rotenone and antimycin); conversely, distal inhibition would disrupt electron flow, diverting them to react with O_2 and generate ROS, a phenomenon we have observed with cyanide (Archer et al 1993). This suggests the site of critical ROS generation is near complex III.

To detoxify the mitochondrial ROS, the mitochondria uniquely express an inducible superoxide dismutase isoform (manganese superoxide dismutase, SOD2 or MnSOD). MnSOD transforms toxic superoxide radicals to H_2O_2. We hypothesize that H_2O_2 is a diffusible redox mediator connecting the sensor (mitochondria) to the effector (K^+ channel), linking redox state and vascular tone (Michelakis et al 2002a). In this redox theory the low normoxic pulmonary vascular tone is due to tonic generation of ROS which maintains the O_2-sensitive K_v channels in PASMCs oxidized and open (thereby relaxing the PASMC) (Fig. 2). Conversely, HPV results from the withdrawal of these ROS and K_v channel closure. Differences in mitochondrial ETC function and ROS generation appear to account for the observed heterogeneity in MnSOD expression in PA versus renal artery mitochondria, levels of ROS and MnSOD being far higher in PASMCs (Michelakis et al 2002a). In general, MnSOD levels are inversely proportional to mitochondrial ROS production.

NADPH oxidase or Nox as O_2 sensors

Another potential O_2 sensor is NAD(P)H oxidase or one of the Nox isoforms. NAD(P)H oxidase is a flavocytochrome that contains two membrane-bound subunits, (p22phox and gp91phox) and two cytosolic proteins (p47phox and p67phox). NADPH oxidase (or a variant that preferentially uses NADH as a substrate) produces ROS in proportion to the pO2 and has been suggested as an O_2 sensor (Mohazzab-H & Wolin 1994). Production of ROS by Nox isoforms has been demonstrated in cells derived from several O_2-sensitive tissues including the neuroepithelial body (NEB), PASMCs and carotid body. Much of the evidence favouring NADPH oxidase as an O_2 sensor derives from experiments using diphenylene iodonium (DPI). DPI does inhibit NADPH oxidase and, in several respects, mimics hypoxia. After causing slight vasoconstriction, DPI decreases subsequent pressor responses to hypoxia (reviewed in Moudgil et al 2005). Moreover, it reduces normoxic ROS production in the PA and inhibits PASMC I_K. Unfortunately, DPI nonspecifically inhibits flavoprotein-containing enzymes, including nitric oxide synthase and complex I of the mitochondrial ETC (reviewed in Moudgil et al 2005).

Mice deficient in the 91-phox-containing NADPH oxidase (Nox-2) manifest dramatically lower normoxic lung ROS production than back-crossed control mice (Archer et al 1999). However, HPV and the O_2-sensitive portion of PASMC I_K are preserved (Archer et al 1999). Moreover, rotenone constriction is preserved or enhanced, consistent with the mitochondrial O_2-sensor hypothesis (Archer et al 1999). Preserved O_2 sensing has also been reported in the type 1 cell of the carotid body from these mice (Roy et al 2000). Overall, these findings argue against the classical NADPH oxidase system as an O_2 sensor in HPV, although they do not exclude a role for other novel oxidases. Moreover, since the Nox is simply another cytochrome-based electron shuttle, it may combine with a mitochondria signal to regulate K^+ channels and vascular tone.

Effector mechanisms of HPV

K^+ *channels*. Inhibitors of K_v channels, such as 4-aminopyridine (4-AP) (Hasunuma et al 1991, Post et al 1992, Yuan 1995) or correolide, cause pulmonary vasoconstriction (Hasunuma et al 1991, Archer et al 2004). Hypoxia inhibits I_K and depolarizes E_M in canine PASMCs, but not renal arterial SMCs (Post et al 1992). This was independently confirmed (Yuan et al 1993) and is now widely accepted. In the last decade significant advances have been made in the molecular identification of the relevant O_2-sensitive K^+ channels.

K^+ channels are proteins consisting of four transmembrane-bound α subunits and four regulatory β subunits. The ionic pore, which determines the channel's intrinsic conductance and ionic specificity, is created by the formation of tetramers of α subunits. The K_v channels also have, in their S4 region, a voltage sensor. β subunits associate with many K^+ channels and alter their expression and kinetics. There are several types of K^+ channel α subunits, including K_v, inward rectifier (Kir) and two-pore, acid sensitive K^+ channel (TASK).

How do K^+ channels control pulmonary vascular tone? In PASMCs, as in most vascular SMCs, closure of K_v channels decreases the tonic efflux of K^+ that occurs down the intra/extracellular concentration gradient (145/5 mM). Channel closure renders the cell interior relatively more positive (depolarized). At EMs positive to ~ -30 mV, the open probability of L-type voltage-gated Ca^{2+} channels increases and intracellular Ca^{2+} concentrations rise. K^+ channel inhibition increases both cytosolic Ca^{2+} and K^+, which not only promotes contraction, but also induces a phenotype that favours proliferation and suppression of apoptosis. Thus, regulation of K^+ channel activity, and the subsequent regulation of Ca^{2+}, may be important to maintaining both low PVR and vascular morphology.

Eleven K_v channel families have been identified, each with multiple isoforms. A variety of putative O_2-sensitive channels exist in the PASMC (K_v1.2, K_v1.5, K_v2.1, K_v3.1b and K_v9.3) (see review Moudgil et al 2005). Certain K^+ channels are specially suited to O_2 sensing, by virtue of possessing key cysteine and methionine groups. Reduction or oxidation of sulfhydryl residues in these channels by a redox mediator, such as ROS, can cause conformational changes in the channel, thereby altering pore function and channel gating.

Amongst the many types of K^+ channels, only certain K_v subtypes (e.g. K_v1.5 and K_v2.1) appear to be effectors of HPV (see review (Moudgil et al 2005). In PASMCs, oxidants increase I_K (e.g. H_2O_2, diamide and oxidized glutathione); in contrast, reducing agents (e.g. reduced glutathione) and agents that facilitate electron shuttling (e.g. duroquinone) inhibit I_K (reviewed in Moudgil et al 2005). In accordance with their electrophysiological effects, oxidants (e.g. diamide) dilate the pulmonary circulation (mimicking O_2) while reducing agents mimic hypoxia; an analogous situation occurs in the DA. Hypoxia and redox agents may alter the function of K^+ channels directly, through modulation of levels of electron donors (NADH) or by modulating ROS production.

The role of $K_v1.5$ and $K_v2.1$ in HPV can best be seen in two models in which HPV is selectively suppressed, the chronic hypoxic pulmonary hypertension (CH-PHT) model and the $K_v1.5$ knockout mouse. Suppression of acute HPV in CH-PHT results, in part, from loss of $K_v1.5$ expression with concordant suppression of O_2-sensitive I_K (Smirnov et al 1994, Reeve et al 2001). Human PA-derived $K_v1.5$ adenoviral gene transfer restores K_v expression, O_2-sensitive I_K and HPV (Pozeg et al 2003). Mice with targeted $K_v1.5$ deletions have impaired HPV and reduced PASMC O_2-sensitive I_K (Archer et al 2001). Additionally, preconstriction of normal resistance PAs with the K_v blocker 4-AP eliminates subsequent HPV (Archer et al 2004). Furthermore, the strong parallels between constriction to correolide, a selective $K_v1.x$ channel inhibitor (Archer et al 2004), and hypoxia suggests that a $K_v1.x$ channel contributes to the mechanism of HPV (Archer et al 2004). In patch clamp studies, intracellular dialysis of PASMCs with anti-$K_v1.5$ antibodies (immunoelectropharmacology) depolarizes resistance, but not conduit, PASMCs. Indeed, the combination of anti-$K_v1.5$ and anti-$K_v2.1$, causes a large depolarization and virtually eliminates hypoxic depolarization (Archer et al 1998, 2004). Of the many K_v channels expressed in PAs, only $K_v1.5$ protein is expressed more abundantly in the resistance vs. conduit PAs, the vascular segment which most strongly manifests HPV (Archer et al 2004). Finally, heterologously expressed human $K_v1.5$, cloned from normal PA, generates an outward K_v current that is active near the resting E_M of −65 mV in Chinese Hamster Ovary (CHO) cells and which is inhibited by hypoxia (Archer et al 2004). Thus, the consequence of enriched $K_v1.5$ expression in resistance PAs is relative hyperpolarization of resistance vs. conduit PASMCs (~−60 mV vs. ~−35 mV) and a unique O_2 sensitivity that underlies the localized enhancement of HPV in resistance PAs.

Controversies regarding the role of ROS in HPV

Even among groups that agree that there is a redox basis for HPV and that the mitochondria are the source of signalling ROS there remains controversy as to the nature of the effects of hypoxia on ROS generation (decreased versus increased).

Hypoxia decreases ROS

During alveolar hypoxia, we and others find that ROS production falls in proportion to the inspired O_2 (Paky et al 1987, Archer et al 1989a, Archer et al 1993). Consequently, K^+ channels are reduced and inhibited, depolarizing E_M (Reeve et al 1995). Higher normoxic levels of ROS cause physiological oxidation of K^+ channel in PASMCs. In support of this view we measured PA ROS production using three different detection methods (lucigenin-enhanced chemiluminescence, AmplexRed H_2O_2 assay and 2′, 7′-dichlorodihydrofluorescein diacetate [DCF]). In endothelium-denuded, resistance PA rings, a concordant decrease in ROS levels during hypoxia or exposure to ETC inhibitors was detected by all three methods (Michelakis et al 2002a). Likewise, Paky et al (1993) found that hypoxia reduces lung ROS production, as measured using lucigenin. They noted that the SOD-mimetic, tiron or antimycin A decreased lung ROS. Likewise, Freeman and Crapo noted that ROS production in rat lungs or isolated lung mitochondria increases in direct proportion to pO2. Wolin's group also suggested that it is the withdrawal of normoxic ROS (H_2O_2) that elicits HPV, although they attribute the source of the ROS to NADH oxidase (Mohazzab et al 1995). In summary, we find that regardless of the technique of ROS measurement, both in isolated lungs and isolated PA rings, hypoxia consistently and rapidly decreases ROS production prior to the onset of HPV (reviewed in Moudgil et al 2005). This fall in ROS is not an artifact of vasoconstriction as the vasoconstrictors, angiotensin II and KCl have no effect (Archer et al 1993, 1999). An indirect validation of the chemiluminescence technique for measuring ROS is that it shows the expected results, consistent with those found with electron spin resonance, when used in a myocardial ischaemia-reperfusion model. The use of lucigenin in the Langendorff heart model of ischaemia reperfusion reveals ROS suppression during ischaemia and overshoot ROS production, beyond normoxic levels, during reperfusion (Henry et al 1990). Thus in both the heart and the lung, low ambient pO2 translates into reduced levels of ROS. If one excludes anoxia, reperfusion or prolonged ischaemia this is a robust finding.

Hypoxia increases ROS

In contrast, Marshall et al (1996) found an increase in ROS during hypoxia in isolated PA myocytes, although they attributed this to NADPH oxidase. In experiments using both isolated perfused lungs and cultured PASMCs, the Schumacker group showed that hypoxia increased intracellular DCF fluorescence in cultured PASMC and this was blocked by the complex III inhibitor myxothiazol (Waypa et al 2001). Many of their measurements of HPV are conducted in isolated PASMCs (passage number unspecified but >6 days in culture) using a surrogate for PVR (SMC shortening) (Waypa et al 2001). Since cultured cells rapidly lose ion channels and also have diminished O_2 sensitivity, at least in the DA (Michelakis et al 2002b), we believe that conclusions about O_2 sensing should be derived from experiments in freshly isolated tissue (when possible).

While agreeing that PASMC mitochondria function as O_2 sensors, they concluded that increased ROS production triggers HPV. Supporting their conclusions, they observed that a variety of antioxidants (pyrrolidinedithiocarbamate, ebselen and diethyldithiocarbamate) abolished HPV (Waypa et al 2001). They measured

ROS in cultured PASMCs using DCF. DCF is a suboptimal ROS probe, as discussed in a recent review on controversies in HPV (Moudgil et al 2005). However, it is unlikely that the use of cultured cells and DCF are the only explanations for the divergent findings amongst groups because Liu et al recently reported that hypoxia increases ROS (measured by DCF fluorescence), and 'tended' to increase ROS (using lucigenin-enhanced chemiluminescence), and variably increased ROS (measured by electron paramagnetic resonance spectra) (Liu et al 2003). Like the Schumacker group, they found that SOD and catalase reduced HPV in porcine PAs. The latter finding contradicts our previous report that liposomal delivery of SOD plus catalase enhances HPV (Archer et al 1989b).

Points of agreement and likely causes for disagreements

Both camps conclude that ROS produced by the proximal ETC signals the haemodynamic response to hypoxia and ETC inhibitors. Moreover, the isolated lung data from the Schumacker group agree with our observations (i.e. rotenone increases normoxic pressure and inhibits subsequent HPV whereas cyanide does not) (Waypa et al 2001, 2002). The Schumacker group finds that proximal ETC inhibitors (DPI 10 μM), rotenone (50 ng/ml), and myxothiazole (50 ng/ml) inhibit HPV; whereas antimycin A (1 ng/ml) does not. However, they acknowledge that antimycin A at 10 ng/ml does cause pulmonary vasoconstriction and inhibits HPV (Waypa et al 2001), consistent with our findings (Archer et al 1993). They discount the antimycin A effects because it also inhibits constriction to U46619 (Waypa et al 2001). Interestingly myxothiazole does not cause pulmonary vasoconstriction, which they interpret as indicating that it targets the relevant ROS generator mediating HPV. Thus their conclusions about the site of ROS generation are based on the assumption that the relevant 'sensor' complex(es) when blocked will selectively inhibit HPV but will not mimic hypoxia by increasing PA pressure. In contrast, we assume that blocking the sensor should both elicit vasoconstriction and inhibition of subsequent HPV. Since these metabolic inhibitors eventually do lower ATP levels, perhaps over the 30 min that the Schumacker group observed contraction in their cultured cell model, the confounding effects of ATP depletion may have occurred. The resulting KATP channel activation could promote relaxation and might explain the observed reduction in U46619 contraction. In our studies, where ROS and PVR were measured simultaneously the changes in PVR and ROS occurred within seconds/minutes of administration of hypoxia or ETC inhibitors and ROS levels fell before constriction commenced (as expected if the ROS are indeed a signal).

It is likely that the discord in the literature relates to divergence in the techniques employed to measure ROS, differences in the timing of the measurement, variations in incubation conditions of the ROS assay (pH, pO2 and pCO_2) and differences in the precise tissue preparations studied. We advocate using multiple independent measurement techniques (luminol, lucigenin, L-012, Amplex red) to measure ROS in freshly isolated resistance PA rings (Michelakis et al 2002a) and isolated perfused lungs (see Moudgil et al 2005).

In conclusion, the opinion of these two groups that HPV is sensed by mitochondria, that the sensor is in the proximal ETC, that the mediator is a ROS, and that the mechanism resides in the PASMC are more concordant than discordant. The REDOX model (Fig. 2) offers a comprehensive explanation for HPV and is relevant to studies of a specialized O_2-sensing system (SOS) in other organs. Although there are undoubtedly deficiencies in the detail of the model and areas of controversy remain, this Redox Hypothesis is testable and we encourage its widespread evaluation.

Acknowledgements

Drs Thébaud, Michelakis and Archer are supported by the Canada Foundation for Innovation, the Alberta Heart and Stroke Foundation (H&SF), the Alberta Heritage Foundation for Medical Research (AHFMR) and the Canadian Institutes for Health Research (CIHR). Dr. Archer is a Canada Research Chair in Oxygen Sensing and Translational Cardiovascular Research and is supported by NIH-RO1-HL071115. Dr. Michelakis is a Canada Research Chair in Pulmonary Hypertension. Dr Thébaud is supported by the Stollery Children's Hospital Foundation. E. K. Weir is supported by VA Merit Review Funding and NHLBI Grant RO1 HL-65322. The authors apologize to the many researchers whose work was used to write this paper but could not be cited due to page constraints.

References

Archer SL, Will JA, Weir EK 1986 Redox status in the control of pulmonary vascular tone. Herz 11:127–141

Archer SL, Nelson DP, Weir EK 1989a Simultaneous measurement of O_2 radicals and pulmonary vascular reactivity in rat lung. J Appl Physiol 67:1903–1911

Archer SL, Peterson D, Nelson DP et al 1989b Oxygen radicals and antioxidant enzymes alter pulmonary vascular reactivity in the rat lung. J Appl Physiol 66:102–111

Archer SL, Huang J, Henry T, Peterson D, Weir EK 1993 A redox-based O_2 sensor in rat pulmonary vasculature. Circ Res 73:1100–1112

Archer SL, Reeve HL, Michelakis E et al 1999 O_2 sensing is preserved in mice lacking the gp91 phox subunit of NADPH oxidase. Proc Natl Acad Sci USA 96:7944–7949

Archer SL, Souil E, Dinh-Xuan AT et al 1998 Molecular identification of the role of voltage-gated K^+ channels, $K_v1.5$ and $K_v2.1$, in hypoxic pulmonary vasoconstriction and control of resting membrane potential in rat pulmonary artery myocytes. J Clin Invest 101:2319–2330

Archer SL, London B, Hampl V et al 2001 Impairment of hypoxic pulmonary vasoconstriction in mice lacking the voltage-gated potassium channel $K_v1.5$. FASEB J 15:1801–1803

Archer SL, Wu XC, Thebaud B et al 2004 Preferential expression and function of voltage-gated, O_2-sensitive K^+ channels in resistance pulmonary arteries explains regional heterogeneity in hypoxic pulmonary vasoconstriction. Ionic diversity in smooth muscle cells. Circ Res 95:308–318

Bindslev L, Jolin A, Hedenstierna G, Baehrendtz S, Santesson J 1985 Hypoxic pulmonary vasoconstriction in the human lung: effect of repeated hypoxic challenges during anesthesia. Anesthesiology 62:621–625
Buescher P, Perse D, Pillai R et al 1991 Energy state and vasomotor tone in hypoxic pig lungs. J Appl Physiol 70:1874–1881
Burghuber OC 1987 Nifedipine attenuates acute hypoxic pulmonary vasoconstriction in patients with chronic obstructive pulmonary disease. Respiration 52:86–93
Carlsson AJ, Bindslev L, Santesson J, Gottlieb I, Hedenstierna G 1985 Hypoxic pulmonary vasoconstriction in the human lung: the effect of prolonged unilateral hypoxic challenge during anaesthesia. Acta Anaesthesiol Scand 29:346–351
Dorrington KL, Clar C, Young JD et al 1997 Time course of the human pulmonary vascular response to 8 hours of isocapnic hypoxia. Am J Physiol 273:H1126–1134
Duchen MR 1999 Contributions of mitochondria to animal physiology: from homeostatic sensor to calcium signalling and cell death. J Physiol (Lond) 516:1–17
Duchen MR, Biscoe TJ 1992 Relative mitochondrial membrane potential and [Ca2+]i in type I cells isolated from the rabbit carotid body. J Physiol 450:33–61
Hasunuma K, Rodman D, McMurtry I 1991 Effects of K^+ channel blockers on vascular tone in the perfused rat lung. Am Rev Respir Dis 144:884–887
Henry T, Archer SL, Nelson D, Weir EK, From AHL 1990 Enhanced chemiluminescence as a measure of oxygen-derived free radical generation during ischemia and reperfusion. Circ Res 167:1453–1461
Kato M, Staub N 1966 Response of small pulmonary arteries to unilobar alveolar hypoxia and hypercapnia. Circ Res 19:426–440
Liu JQ, Sham JS, Shimoda LA, Kuppusamy P, Sylvester JT 2003 Hypoxic constriction and reactive oxygen species in porcine distal pulmonary arteries. Am J Physiol Lung Cell Mol Physiol 285:L322–333
Madden JA, Dawson CA, Harder DR 1985 Hypoxia-induced activation in small isolated pulmonary arteries from the cat. J Appl Physiol 59:113–118
Malconian MK, Rock PB, Reeves JT, Cymerman A, Houston CS 1993 Operation Everest II: gas tensions in expired air and arterial blood at extreme altitude. Aviat Space Environ Med 64: 37–42
Marshall C, Mamary AJ, Verhoeven AJ, Marshall BE 1996 Pulmonary artery NADPH-oxidase is activated in hypoxic pulmonary vasoconstriction. Am J Respir Cell Mol Biol 15:633–644
Michelakis E, Hampl V, Nsair A et al 2002a Diversity in mitochondrial function explains differences in vascular oxygen sensing. Circ Res 90:1307–1315
Michelakis ED, Rebeyka I, Wu X et al 2002b O_2 sensing in the human ductus arteriosus: regulation of voltage-gated K^+ channels in smooth muscle cells by a mitochondrial redox sensor. Circ Res 91:478–486
Mohazzab KM, Fayngersh RP, Kaminski PM, Wolin MS 1995 Potential role of NADH oxidoreductase-derived reactive O_2 species in calf pulmonary arterial PO_2-elicited responses. Am J Physiol 269:L637–644
Mohazzab-H K, Wolin M 1994 Properties of a superoxide anion-generating microsomal NADH oxidoreductase, a potential pulmonary artery PO_2 sensor. Am J Physiol 267:L823–L831
Moudgil R, Michelakis ED, Archer SL 2005 Hypoxic pulmonary vasoconstriction. J Appl Physiol 98:390–403
Paky A, Farrukh I, Michael J, Gurtner G 1987 Basal superoxide production in the isolated perfused rabbit lung. Fed Proc 46:995
Paky A, Michael JR, Burke-Wolin T, Wolin MS, Gurtner GH 1993 Endogenous prodution of superoxide by rabbit lungs: effects of hypoxia or metabolic inhibitors. J Appl Physiol 74:2868–2874
Post JM, Hume JR, Archer SL, Weir EK 1992 Direct role for potassium channel inhibition in hypoxic pulmonary vasoconstriction. Am J Physiol 262:C882–890

Pozeg ZI, Michelakis ED, McMurtry MS et al 2003 In vivo gene transfer of the O_2-sensitive potassium channel $K_v1.5$ reduces pulmonary hypertension and restores hypoxic pulmonary vasoconstriction in chronically hypoxic rats. Circulation 107:2037–2044
Reeve HL, Michelakis E, Nelson DP, Weir EK, Archer SL 2001 Alterations in a redox oxygen sensing mechanism in chronic hypoxia. J Appl Physiol 90:2249–2256
Reeve HL, Weir EK, Nelson DP, Peterson DA, Archer SL 1995 Opposing effects of oxidants and antioxidants on K^+ channel activity and tone in rat vascular tissue. Exp Physiol 80:825–834
Rounds S, McMurtry I 1981 Inhibitors of oxidative ATP production cause transient vasoconstriction and block subsequent pressor responses in rat lungs. Circ Res 48:393–400
Roy A, Rozanov C, Mokashi A et al 2000 Mice lacking in gp91 phox subunit of NAD(P)H oxidase showed glomus cell [Ca(2 +)](i) and respiratory responses to hypoxia. Brain Res 872:188–193
Shirai M, Sada K, Ninomiya I 1986 Effects of regional alveolar hypoxia and hypercapnia on small pulmonary vessels in cats. J Appl Physiol 61:440–448
Smirnov SV, Robertson TP, Ward JP, Aaronson PI 1994 Chronic hypoxia is associated with reduced delayed rectifier K^+ current in rat pulmonary artery muscle cells. Am J Physiol 266:H365–370
Ward JP, Robertson TP 1995 The role of the endothelium in hypoxic pulmonary vasoconstriction. Exp Physiol 80:793–801
Waypa GB, Chandel NS, Schumacker PT 2001 Model for hypoxic pulmonary vasoconstriction involving mitochondrial oxygen sensing. Circ Res 88:1259–1266
Waypa GB, Marks JD, Mack MM et al 2002 Mitochondrial reactive oxygen species trigger calcium increases during hypoxia in pulmonary arterial myocytes. Circ Res 91:719–726
Yuan X-J 1995 Voltage gated K^+ currents regulate resting membrane potential and $[Ca^{2+}]i$ in pulmonary artery myocytes. Circ Research 77:370–378
Yuan X-J, Goldman W, Tod M, Rubin L, Blaustein M 1993 Hypoxia reduces potassium currents in cultured rat pulmonary but not mesenteric arterial myocytes. Am J Physiol 264:L116–L123

DISCUSSION

Gurney: I have a question relating to the 4-aminopyridine (4-AP) effect on hypoxic pulmonary vasoconstriction (HPV). You showed that 4-AP inhibits it, but there are three other papers showing that 4-AP enhances HPV in lungs and isolated vessels (Hasunuma et al 1991, Barman 1998, Archer et al 1996). One of these papers is yours back in 1996. Does it decrease or increase?

Archer: When you are challenged to do an experiment to prove a point you lay it out so that the conditions are right! In this experiment we made sure we had a maximal constriction from 4-AP, which was not the case in previous work. Previously we showed that it enhanced HPV, but this is dependent on the dose of 4-AP given. This was also in isolated lung where the amount of 4-AP we could give was limited by the tendency for oedema with too much 4-AP. We have consistently shown that if we give 4-AP we get almost identical constriction to oxygen. It is a little bit larger, suggesting that oxygen is a subset of K_v constriction in the human and rabbit ductus. I admit that in any given experiment there is variability, but most of the time they are about the same size and correolide is a subset.

Gurney: Have you tested the effect of correolide on HPV?

Archer: No. We haven't given it to an isolated lung. In the pulmonary artery (PA) rings it causes constriction itself.

Gurney: The selectivity of correolide isn't really known yet.

Archer: There are about three papers that suggest that red coral-derived toxin, which is available from Merck if you beg, blocks K_v1x families. It blocks this new ion channel class which is not related to any known channel. There are correolide binding sites: about seven amino acids confer coreolide binding sensitivity, and this mystery channel we have discovered has three of those.

Gurney: What does this mystery channel look like? Is it like a $K_v1.5$?

Archer: It is a four-transmembrane domain channel that we got out of collagen 1a. It seems to be a different open reading frame from within that gene that we picked up because it has some similarity to $K_v1.5$. We were looking for splice variants and found something very different from what we were expecting. One of the confounding factors is that CHO cells, which everyone says have no ion channels, have mRNA for almost every K_v channel and they also have some of the protein. This is a big confounding factor when we assess these channels.

Murphy: Has anyone carried out site-directed mutagenesis on the cysteine residues on the Ca^{2+} channels to see if they are essential?

Archer: Work by Ruppersberg et al (1991) shows that the cysteines were important in the redox gating of channels. There is pretty general agreement that these channels open and close with oxidation and reduction. Our own data and Ken Weir's data show that the ductus constricts with oxygen and PAs dilate. In both cases H_2O_2 goes up as pO_2 goes up, but the responses of the K^+ channels are opposing (Michelakis et al 2002).

Murphy: If, for example, you replace the cysteine with alanine, do you lose this effect?

Archer: We haven't done this.

López-Barneo: In the $K_v1.5$ knockout, the ductus works fine. Hypoxic vasoconstriction is abolished. Is hypoxic relaxation of the ductus maintained?

Archer: We haven't looked at this. It should be decreased, although in the course of pulmonary artery dissections we have never noticed a patent ductus arteriosus (DA). However, we have shown that $K_v1.5$ or $K_v2.1$ gene therapy *ex vivo* will help make preterm ductuses constrict to oxygen more vigorously (Thebaud et al 2004).

López-Barneo: I have a question concerning your mitochondrial hypothesis. How can you explain the lack of effect of cyanide in your preparation? It should be doing something very similar to antimycin A in terms of mitochondrial function. You have been reporting similar effects between rotenone and hypoxia in different preparations. Does antimycin A also follow this?

Archer: By and large, yes. Cyanide in our hands does cause pulmonary vasoconstriction, but unlike the constriction caused by rotenone, antimycin or hypoxia, this can be removed by superoxide dismutase (SOD) and was associated with an increase in reactive oxygen species (ROS) production (Archer et al 1993). Other groups have reported this. We think in theory this has something to do with cyanide causing a downstream block; if you block distally the electrons are already committed to the chain and there is probably autooxidation and ROS generation. The reviewer of the last paper we wrote on this subject insisted that it was not possible that cyanide would not act like the other inhibitors. So we took H1 cells which are beating cardiac cells and compared the mitochondrial function and sensitivity of cyanide in these with our ductus cells. We found that they were very different: for a start, the heart has far more mitochondria/volume and their mitochondrial membrane potential is more sensitive to cyanide. If we gave millimolar doses of cyanide, perhaps it might mimic hypoxia. It is not effective at the micromolar dose at which rotenone and antimycin A are active.

Ward: We only get complete block of electron transport with millimolar cyanide.

Archer: This is a weakness of our data. We assumed, perhaps incorrectly, years ago that if we gave millimolar cyanide we got a dead preparation and so didn't pursue this dose range. Ten times the dose we are using does affect mitochondrial membrane potential.

Aaronson: You showed in your chronic hypoxic model that if you put $K_v1.5$ back, the HPV came back. What about the oxidant signalling? This won't come back.

Archer: It doesn't come back completely.

Aaronson: Does it come back at all?

Archer: There is a paper we did with Helen Reeve as first author in 2001 in which we looked at this (Reeve et al 2001). We found that in chronic hypoxia both the signalling and the channel response are dysfunctional. I don't know whether $K_v1.5$ replacement therapy restored HPV (despite abnormal ROS production) because of the degree of overexpression of the channel (Pozeg et al 2003). I agree, though, that this is an apparent inconsistency. Since we show that in chronic hypoxia rotenone doesn't cause much pulmonary vasoconstriction or K^+ channel inhibition, this implies that the mitochondria are dysfunctional as O_2 sensors in chronic hypoxia. But then we give $K_v1.5$ and the problem goes away, this is somewhat consistent. In the Reeve et al (2001) paper we certainly show that there is a redox change. If you are a true believer you could argue that there still is some signal.

Schumacker: As I understand your model the oxidant signal that is produced as a function of pO2 is due to a non-specific interaction between molecular oxygen and the sources of electrons from the electron transport chain (ETC). It varies with pO2 from ambient conditions: above ambient conditions, to hyperoxia and down to anoxia in a kind of linear way. Looking at the tracings of your chemoluminescent probe, though, it looked as if the change in oxidant stress or oxidant signal from 21% down to 1% was less than the change from 1% to 0%. Is it a non-linear measurement?

Archer: ROS production isn't linear with pO2; however ROS production is quantitatively greater as pO2 increases from 20–100 mmHg. It certainly is not linear as you go up to hyperoxia. If we give 95% oxygen we get a 10% increase in chemiluminescence measured from the whole lung over what we see with 20% oxygen. The relationship between pO2 and ROS production is tighter as FiO_2 decreases from 20% to 0%. Parenthetically, the chemiluminescence technique does detect xanthine/xanthine oxidase ROS production and the other standard ROS-inducing stimuli (such as phorbol ester induced activation of NADPH oxidase). We have used it in the ischaemia reperfusion model and ROS production does what you'd expect it to. It goes to zero when there is ischaemia and overshoots with reperfusion. Moreover, in the NADPH oxidase (phos 91) knockout mice, ROS production as measured by enhanced or unenhanced chemiluminescence is significantly suppressed. We have also measured PA ROS production without any luminescence enhancer. There is a technique called unenhanced chemiluminescence in which the signal is supposed to be singlet oxygen. In extreme darkness this can be detected. The NOS knockouts are different from the wild-type even without any lucigenin on board.

Schumacker: Is it the lucigenin technique that is non-linear, or is it the process of ROS production?

Archer: I don't know.

Murphy: It could be that the mitochondrial membrane potential is lower with very low oxygen tension, thereby decreasing ROS production.

Chandel: From 1–20% oxygen you predict that there might be a pO2 effect but no effect on the redox state of the electron transport carriers. Below 1% cytochrome oxidase starts to back up, so there is a decline in pO2 and also a robust redox effect that is not present at 1% or above.

Murphy: Above 20% it could be that you are just using up any radical-producing intermediate.

Archer: The pulmonary artery mitochondria have more MnSOD than other blood vessels, whereas their ETC complex expression is not as dramatically different (Michelakis et al 2002).

Duchen: The pulmonary arteries themselves are producing more radicals at rest than the renal arteries. According to the JC-1 data the mitochondria are a bit depolarized in the pulmonary as compared to the renal artery cells. You would expect that free radical production should be greater with greater mitochondrial membrane potential.

Archer: On the contrary, in the PASMCs we find that hyperpolarized mitochondria (measured using JC-1 or TMRM) make fewer ROS, and dichloroacetate (which turns on pyruvate dehydrogenase by blocking PDH kinase) depolarizes mitochondria and enhances ROS production (McMurtry et al 2004).

References

Archer SL, Huang J, Henry T et al 1993 A redox based oxygen sensor in rat pulmonary vasculature. Circ Res 73:1100–1112

Archer SL, Huang JMC, Reeve HL et al 1996 Differential distribution of electrophysiologically distinct myocytes in conduit and resistance arteries determines their response to nitric oxide and hypoxia. Circ Res 78:431–442

Barman SA 1998 Potassium channels modulate hypoxic pulmonary vasoconstriction. Am J Physiol 275:L64–L70

Hasunuma K, Rodman DM, McMurtry IF 1991 Effects of K+ channel blockers on vascular tone in the perfused rat lung. Am Rev Respir Dis 144:884–887

McMurtry MS, Bonnet S, Wu X et al 2004 Dichloroacetate prevents and reverses pulmonary hypertension by inducing pulmonary artery smooth muscle cell apoptosis. Circ Res 95: 830–840

Michelakis ED, Rebeyka I, Wu X et al 2002 O_2 sensing in the human ductus arteriosus: regulation of voltage-gated K^+ channels in smooth muscle cells by a mitochondrial redox sensor. Circ Res 91:478–486

Pozeg ZI, Michelakis ED, McMurtry MS et al 2003 In vivo gene transfer of the O_2-sensitive potassium channel $K_v1.5$ reduces pulmonary hypertension and restores hypoxic pulmonary vasoconstriction in chronically hypoxic rats. Circulation 107:2037–2044

Reeve HL, Michelakis E, Nelson DP et al 2001 Alterations in a redox oxygen sensing mechanism in chronic hypoxia. J Appl Physiol 90:2249–2256

Ruppersberg J, Stocker M, Pongs O et al 1991 Regulation of fast inactivation of cloned mammalian 1K(A) channels by cysteine oxidation. Nature 352:711–714

Thebaud B, Michelakis ED, Wu XC et al 2004 Oxygen-sensitive K_v channel gene transfer confers oxygen responsiveness to preterm rabbit and remodeled human ductus arteriosus: implications for infants with patent ductus arteriosus. Circulation 110:1372–1379

Role for mitochondrial reactive oxygen species in hypoxic pulmonary vasoconstriction

Gregory B. Waypa and Paul T. Schumacker[1]

Department of Pediatrics, Northwestern University Feinberg School of Medicine, Chicago, Illinois 60611, USA

Abstract. Recently, the mitochondria have become the focus of attention as the site of O_2 sensing underlying hypoxic pulmonary vasoconstriction (HPV). From a teleological standpoint, it is reasonable that the organelle where most of the cellular O_2 is consumed would also be the site of O_2 sensing. Originally, it was proposed that a drop in pO2 decreases the rate of mitochondrial reactive oxygen species (ROS) generation resulting in a decrease in oxidant stress and an accumulation of reducing equivalents, thus causing the inhibition of voltage-dependent K^+ channels, membrane depolarization, and the influx of calcium through voltage-gated (L-type) Ca^{2+} channels. Recently, a new model has emerged that suggests hypoxia triggers a paradoxical increase in a mitochondrial-induced ROS signal. The resulting shift of the cytosol to an oxidized state triggers the release of intracellular Ca^{2+} stores, recruitment of Ca^{2+} channels in the plasma membrane, and activation of contraction. Below we will discuss the aspects of this novel model of O_2 sensing and its applicability to the HPV response.

2005 Signalling pathways in acute oxygen sensing. Wiley, Chichester (Novartis Foundation Symposium 272) p 176–195

When challenged with alveolar hypoxia, the pulmonary arteries (PA) constrict, a response aptly termed hypoxic pulmonary vasoconstriction (HPV). This physiological response was first described in 1946 by von Euler and Liljestrand, who found that PA pressure increased when feline lungs were ventilated with low O_2 mixtures, and decreased when ventilated with pure O_2 (von Euler & Liljestrand 1946). They speculated that this hypoxia-induced change in pulmonary vascular resistance could help to improve gas exchange efficiency by diverting blood flow away from poorly ventilated lung regions and towards areas with better oxygenation. Hypoxia-induced arterial constriction is unique to the pulmonary circulation; in the systemic circula-

[1]This paper was presented at the symposium by Paul T. Schumacker, to whom correspondence should be addressed.

tion, tissue hypoxia cause arterioles to dilate, resulting in a restoration of blood flow and oxygen supply to the affected region (Michelakis et al 2002, Weir et al 1997).

Since the discovery of HPV, many investigators have characterized the pulmonary vascular response to hypoxia, and the importance of this response in health and disease has become apparent. However, an important and still unresolved question relates to the underlying mechanism by which vascular cells detect a decrease in O_2 tension and convert this into a biological signal. That said, it has been well established that the HPV response is retained in excised lungs and in rings of pulmonary artery (Rounds & McMurtry 1981, McMurtry 1984, Archer et al 1993, Grimminger et al 1995, Weissmann et al 1998, 1995, Marshall et al 1996, Zhao et al 1996), as well as isolated PA smooth muscle cells (Zhang et al 1997, Waypa et al 2001). Furthermore, removing the endothelium from PA vessels partially diminishes, but does not fully attenuate the HPV response (Archer et al 1998). This would suggest that PA myocytes possess an O_2 sensor and the signalling mechanisms necessary to activate a contractile response, while the endothelium plays an amplification role in HPV.

The HPV response is biphasic (Dipp & Evans 2001, Dipp et al 2001, Leach et al 2001, Robertson et al 1995, 2000b, 2001), and consists of an initial transient constriction response (phase 1) followed by a slow sustained constriction (phase 2). Phase 1 is referred to as the acute phase because it begins within seconds after hypoxic challenge and appears to involve the release of Ca^{2+} from the intracellular stores followed by the opening of voltage-gated L-type Ca^{2+} channels and/or store operated Ca^{2+} channels (Archer et al 1998, Dipp et al 2001, Morio & McMurtry 2002, Dipp & Evans 2001, Snetkov et al 2003, Wang et al 2004, Sweeney & Yuan 2000, Robertson et al 2000b). Phase 1 also occurs independent of the endothelium (Zhang et al 1997). By contrast, phase 2 does not appear until minutes after hypoxic challenge, but it can last for hours to days during chronic hypoxia. And unlike phase 1, phase 2 requires an intact endothelium, which suggests that endothelium-derived promoting factors may be involved (Liu et al 2001, Aaronson et al 2002, Ward & Robertson 1995, Dipp et al 2001, Leach et al 1994, 2001, Lazor et al 1996).

Searching for the O_2 sensor

Because of the nature of hypoxia, it makes sense for one to look at HPV as a physiological response to the hypoxia-induced changes in the redox environment of the cell. In this regard, early studies suggested that hypoxia decreases extracellular oxidant levels, based on measurements using lucigenin or luminol chemiluminescence in perfused lungs (Archer et al 1989, 1993, Mohazzab & Wolin 1994a). More recent studies in endothelium-denuded rings of distal PA also show a decrease in reactive oxygen species (ROS) release using chemiluminescence, Amplex Red, and DCF (2′,7′-dichlorodihydrofluorescein diacetate) fluorescence during hypoxia

(Michelakis et al 2002). These findings suggested that during normoxia a tonic level of ROS generation occurs possibly through a constitutively active NADPH oxidase system (Archer et al 1986, Archer & Michelakis 2002). It is presumed that this system would respond to a fall in its substrate, O_2, by decreasing its production of superoxide. However, when mice with targeted deletion of the gp^{91} phox subunit of NADPH oxidase were shown to retain their response to hypoxia (Archer et al 1999), that system was ruled out. It is conceivable that alternative isoforms (Nox-1 or Nox-4) to the gp^{91} (Nox-2) subunit are involved (Lassegue et al 2001), but definitive studies testing this hypothesis have not yet appeared.

As an alternative, Archer and colleagues proposed a model in which hypoxia decreases mitochondrial electron transport, thus decreasing ROS production which is a by-product of cellular respiration (Archer et al 1993, Michelakis et al 2002) (Fig. 1A). This idea arose from their finding that the mitochondrial electron transport inhibitors antimycin A and rotenone mimicked the hypoxic response, presumably by decreasing tonic mitochondrial ROS generation (Archer et al 1993, Archer & Michelakis 2002). From a teleological standpoint, it is reasonable that the organelle where most of the cellular O_2 is consumed would also be the site of O_2 sensing.

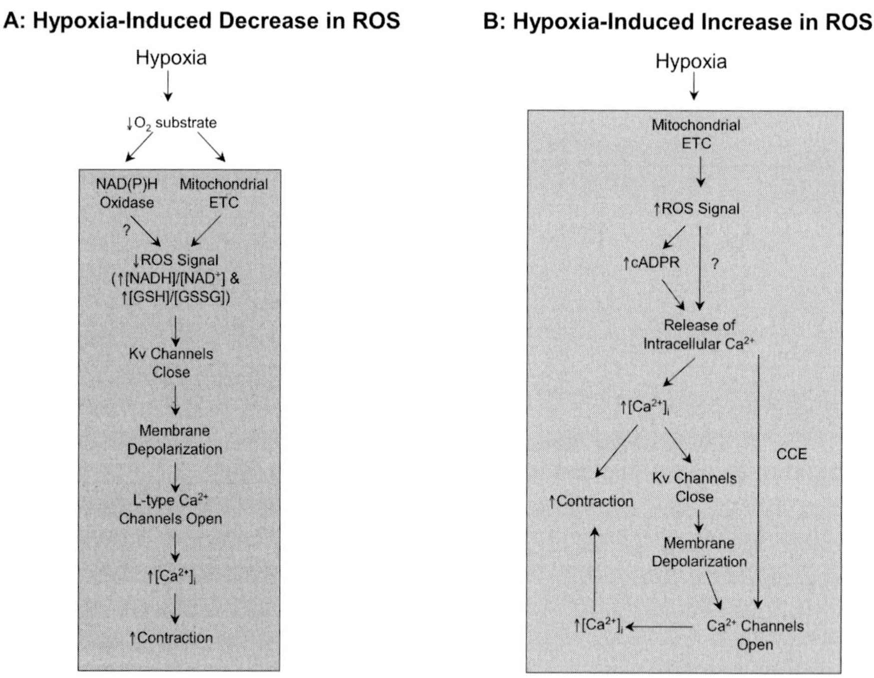

FIG. 1. Two contradictory models of O_2 sensing underlying the HPV response. (A) HPV is a result of hypoxia-induced decrease in ROS signalling. (B): HPV is a result of a paradoxical increase in ROS signalling. Reproduced from Waypa & Schumacker (2005), with permission from the American Physiological Society.

Archer and colleagues suggested that the decrease in NADH utilization by mitochondria would shift the cytosolic redox balance to a more reduced state (Weir & Archer 1995). Such a redox shift was postulated to cause an inhibition of redox-sensitive K_v channel subunits (Rettig et al 1994), resulting in membrane depolarization, leading to the opening of L-type, voltage-gated Ca^{2+} channels and subsequent myocyte contraction. In support of this hypothesis, hypoxia was reported to cause inhibition of steady-state K^+ currents and membrane depolarization in rat PA smooth muscle cells (Olschewski et al 2004, 2002b). In addition, reducing agents caused decreases in K^+ currents and constriction in pulmonary artery smooth muscle cells (Olschewski et al 2002a, Reeve et al 1995, Weir et al 2002, Olschewski et al 2004) while diamide, a chemical that oxidizes sulfhydryls in the cell, caused vasodilation in the isolated perfused lung (Weir et al 1985). However, this sole necessity for K_v channel inhibition during HPV has been met with criticism. Sham et al (2000) and Hasunuma et al (1991) demonstrated that pharmacological pre-inhibition of K_v channels had no effect on HPV and in some cases it augmented the response. While these studies don't rule out the involvement of K_v channels in the HPV response, they suggest that K_v channel inhibition alone cannot account for the response. Thus, K_v channels may represent a downstream amplification step that contributes to the HPV response.

The model of hypoxia-induced decreases in ROS was further questioned when it was recognized that autoxidation and other methodological limitations may have interfered with the ability of chemiluminescence measurements to accurately detect intracellular oxidant stress (Spasojevic et al 2000). Furthermore, other investigators report that the mitochondrial inhibitor rotenone decreases ROS production while antimycin A augments oxidant generation (Turrens et al 1985, Chen et al 2003, Chandel et al 2000b, Cowan et al 2003, Garcia-Ruiz et al 1997, Mohazzab & Wolin 1994b). Therefore, if the Archer model is to be believed, (Archer et al 1993, Michelakis et al 2002), rotenone should have mimicked HPV while antimycin A should have caused vasodilation; this was not the case. Other investigators have demonstrated that antioxidants attenuate the HPV response in intact lungs, isolated PA vessels and isolated PA smooth muscle cells and do not cause a sustained increase in pulmonary vascular tone as suggested by the Archer model (Waypa et al 2002, 2001, Liu et al 2003). The basis for these contradictory findings may involve the methodology and site of action for the various agents. Still, these contradictory observations bring into question the idea that hypoxia-induced decreases in ROS trigger HPV.

Evidence implicating an increase in ROS signalling during HPV

An alternative to the hypoxia-induced decrease in ROS signalling was initially suggested by Marshall et al (1996) who measured by chemiluminescence a paradoxical increase in ROS production in isolated PA smooth muscle cells (Fig. 1B). Soon after, other studies suggested that hypoxia triggers an increase in ROS signalling in pulmonary artery smooth muscle cells (PASMCs), as evidenced by an increase in oxidation of the intracellular probe 2′,7′-dichlorofluorescein-diacetate (DCFH) (Waypa et al 2001, Killilea et al 2000, Liu et al 2003). However, those results have been challenged because DCFH oxidation is non-specific (Sham 2002, Sylvester 2001), and the sites of its oxidation in the cell are not known. However, lucigenin-derived chemiluminescence and electron paramagnetic resonance (EPR) spectrometry have also detected increases in radical production during hypoxia (Liu et al 2003). Furthermore, exogenous H_2O_2 mimics HPV in PASMCs and in isolated lungs (Waypa et al 2002, 2001), although it should be noted that H_2O_2 can also release Ca^{2+} from mitochondria (Richter et al 1995) and also potentially trigger non-specific effects (Jin et al 1991, Rhoades et al 1990). Nevertheless, cell permeable antioxidants have been shown to block the HPV response without affecting alternative methods of inducing contraction (Waypa et al 2002, 2001, Liu et al 2003). In addition, over-expression of catalase in the cytosol was found to attenuate both hypoxia induced and H_2O_2-induced increases in cytosolic Ca^{2+} (Waypa et al 2001). Recently, we have taken steps to critically evaluate the role of ROS in the HPV response through the use of a redox-sensitive cytosolic FRET (fluorescence resonance energy transfer) sensor expressed in PA myocytes (Waypa & Schumacker 2004). Superfusion of the cells with hypoxic media results in an increase in the FRET ratio indicating an increase in ROS signalling. Preincubation of the cells with either the antioxidants pyrrolidinedithiocarbamate or N-acetyl-L-cysteine or over-expression of glutathione peroxide in the cytosol attenuates the hypoxia-induced increase in the FRET ratio. This along with the previously mentioned data supports the idea that an increase rather than a decrease in a ROS signalling occurs during HPV.

To determine the source of the hypoxia-induced ROS signal, we again look to the mitochondria where it has been shown that the complex III inhibitor myxothiazol attenuates the increase in ROS signalling during hypoxia (Waypa et al 2001) (Fig. 2). Further evidence of mitochondrial participation comes from studies using other site-specific mitochondrial inhibitors. These studies suggest that HPV requires electron transport in the proximal (but not the distal) region of the electron transport chain (ETC) (Waypa et al 2002, 2001, Leach et al 2001, Weissmann et al 2003). When the ETC was inhibited at complex I by rotenone or diphenylene iodonium (DPI), HPV was abrogated (Weissmann et al 2003, Waypa et al 2002, 2001) (Fig. 2). By contrast, inhibitors of the distal region of the ETC (antimycin A or cyanide) do not abolish HPV, which suggests that a fully functional ETC is not required for HPV (Waypa et al 2002, 2001, Leach et al 2001). Rounds and McMurtry were the first to show that antimycin A elicited vasoconstriction in normoxic lungs (Rounds & McMurtry 1981). Although that study preceded the ROS hypothesis, it now seems clear that antimycin A prolongs the lifetime of ubisemiquinone at the Qo site in complex III, enhancing the opportunity for superoxide generation and

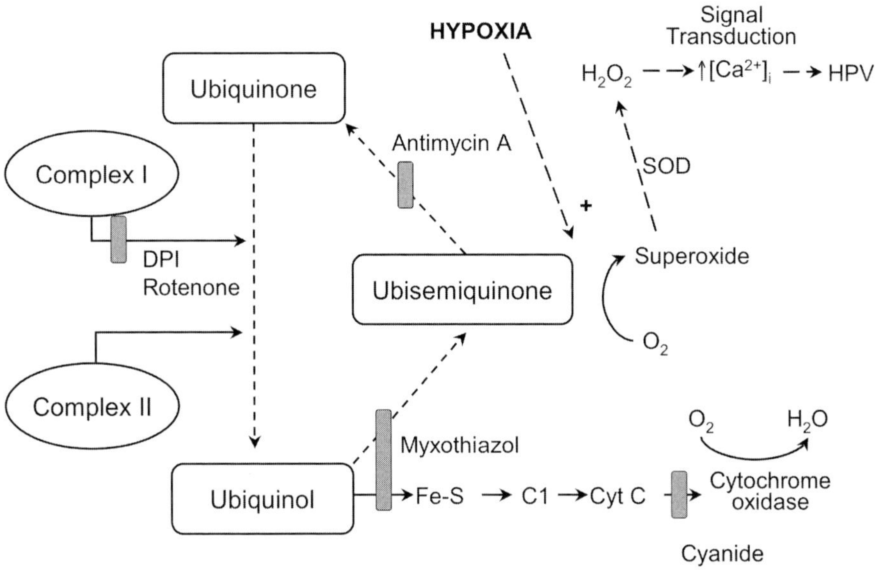

FIG. 2. Mitochondria function as the O_2 sensor during hypoxia by increasing ROS generation at complex III. This increase in mitochondrial ROS results in an increase in $[Ca^{2+}]_i$ and finally HPV. Mitochondrial inhibitors that act upstream of ubisemiquinone, DPI, rotenone and myxothiazol attenuate HPV, whereas mitochondrial inhibitors that act downstream of ubisemiquinone, antimycin A and cyanide have either no effect or augment HPV. Reproduced from Waypa & Schumacker (2002), with permission from the American Physiological Society.

contraction during normoxia. Cyanide, which inhibits the terminal cytochrome oxidase in the ETC, increases ROS production at complex III, induces constriction during normoxia (Waypa et al 2002, 2001) and triggers $[Ca^{2+}]_i$ increases in PASMC (Waypa et al 2002, 2003a). This response was blocked by myxothiazol, which prevents electron flow into complex III, and by over-expression of catalase (Waypa et al 2002), which augments the scavenging of H_2O_2. Interestingly, Archer et al (1993) also found that cyanide induced vasoconstriction during normoxia and augmented HPV. In an interesting set of experiments, Leach et al (2001) extended the model of hypoxia-induced mitochondrial ROS by demonstrating that rotenone blocked hypoxic contraction of PA vessels and then restoring the HPV response by using succinate to shuttle electrons into complex III via complex II, bypassing the rotenone inhibition at complex I. Unfortunately a weakness of many of these studies is their reliance on pharmacological inhibitors, which may have unexpected effects. However according to this model, ρ^0-PASMCs which lack a functional ETC (King & Attardi 1989) should be incapable of generating ROS during hypoxia (Chandel & Schumacker 1999). Interestingly, these cells lost their hypoxic response

(Waypa et al 2001). Collectively, these findings are consistent with a role for increased mitochondrial ROS production in HPV.

Which mitochondrial complexes contribute to ROS generation during hypoxia?

During respiration, reducing equivalents generated in glycolysis or the Krebs cycle are passed along the ETC (Fig. 3). At complexes I, III and IV, protons are extruded from the mitochondrial matrix thus generating an electrochemical gradient ($\Delta\Psi m$) across the inner mitochondrial membrane, which is used by F_0F_1 ATP synthase to generate ATP. Most of the O_2 consumed is reduced to H_2O at complex IV (cytochrome oxidase). However single electrons can escape from the various transport proteins upstream of complex IV to be captured by O_2 resulting in the generation of superoxide (Chance & Williams 1955). Normally thought of as toxic by-products of the ETC, superoxide and its degradation product H_2O_2 have gained interest as participants in signal transduction (Asencio et al 2003, Boveris et al 1972). Though the precise mechanism of ROS generation is not fully understood (Kushnareva et al 2002), electrons have the potential to escape from the complex I flavoprotein (FMN) allowing for their capture by O_2 (Turrens & Boveris 1980). On the other hand, Paddenberg et al (2003) have suggested that complex II may function in reverse as a fumarate reductase resulting in an increase in ROS production during hypoxia. However, additional studies are required to more fully understand this potential mechanism.

Complex III has received the greatest attention in terms of its potential for generating ROS (Boveris et al 1976, Chen et al 2003, Rana et al 2000, Turrens et al 1985). During respiration, ubiquinol carries two electrons obtained from either complex I or II and binds to complex III at the Qo site. After transferring one of its electrons to the Rieske iron-sulfur protein, the free radical ubisemiquinone is formed (Chen et al 2003, Iwata et al 1998, Moghaddas et al 2003, Sun & Trumpower 2003). Normally, ubisemiquinone transfers its remaining electron to the b cytochromes in complex III. However, this remaining electron can be captured by O_2 to yield superoxide. This would explain how myxothiazol, which inhibits the binding of ubiquinol to the Qo site, acts to prevent the generation of ubisemiquinone, which is the source of superoxide. Furthermore, other 'upstream' inhibitors such as rotenone and DPI limit HPV by also preventing the formation of ubisemiquinone. Finally, complex IV (cytochrome oxidase) does not appear to be the site of ROS generation (Fabian & Palmer 1998), due to the high-affinity trapping of O_2 at the binuclear centre (Verkhovsky et al 1996). This would explain why the complex IV inhibitor cyanide does not inhibit HPV. However, cyanide does induce the HPV response during normoxia (Waypa et al 2002, 2001, Archer et al 1993) most likely by causing electrons to backup in complexes I, II and III, causing

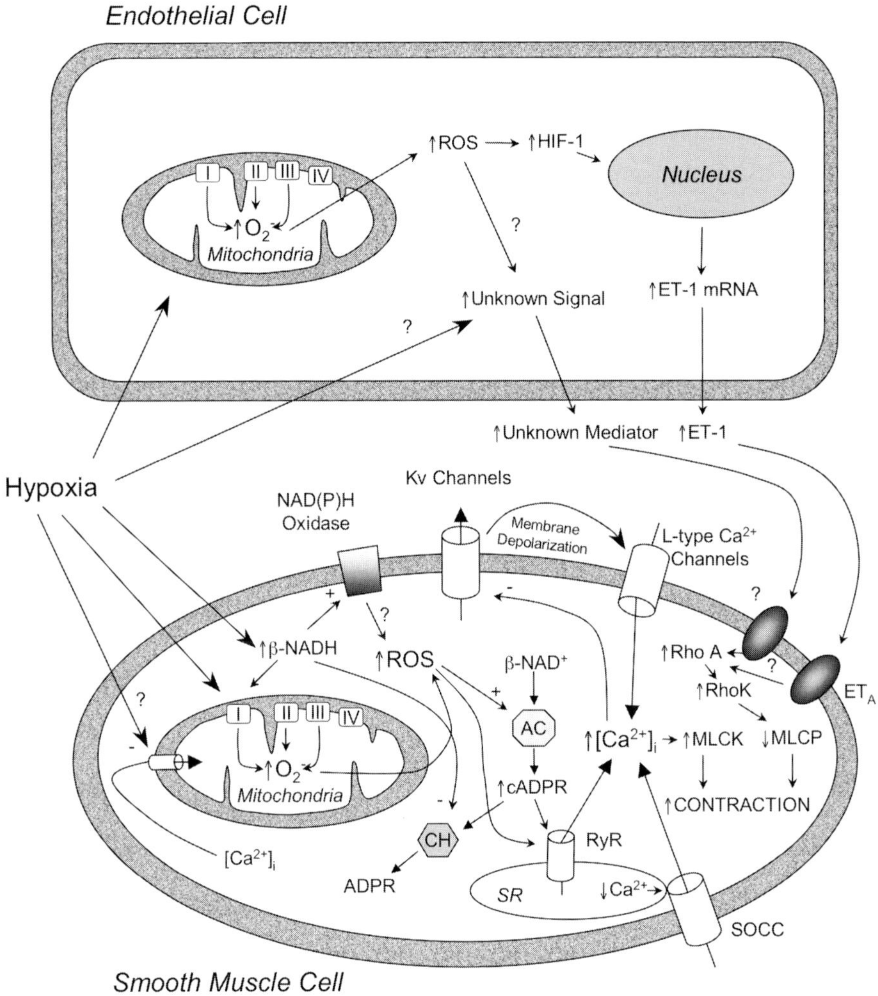

FIG. 3. Possible mechanisms underlying the biphasic constriction (phase 1 and 2) of the pulmonary arteries in response to hypoxia. ROS, reactive oxygen species; HIF-1, hypoxia-inducible factor 1; ET-1, endothelin 1; ETA, endothelin receptor subtype A; RhoK, Rho kinase; MLCK, myosin light chain kinase; MLCP, myosin light chain phosphatase; AC, ADP-ribosyl cyclase; CH, cADPR hydrolase; SR, sarcoplasmic reticulum; SOCC, store operated Ca^{2+} channel; RyR, ryanodine receptor. Reproduced from Waypa & Schumacker (2005), with permission from the American Physiological Society.

those sites to become fully reduced and increasing the generation of superoxide from reduced flavin groups.

Role of ROS signalling in phase 1 of the HPV response

Phase 1 contraction appears to result from the release of Ca^{2+} from intracellular sources (Morio & McMurtry 2002, Dipp & Evans 2001, Dipp et al 2001), followed by the entry of extracellular Ca^{2+} through voltage-dependent, store-operated, and/or receptor-regulated Ca^{2+} channels (Archer et al 2000, Kang et al 2002, 2003, Wang et al 2004, Robertson et al 2000b). Evans and Dipp, have proposed that phase 1 may reflect sarcoplasmic reticulum (SR) Ca^{2+} release through stimulation of ryanodine receptors via generation of the Ca^{2+} mobilizing β-NAD^+ metabolite, cyclic ADP ribose (cADPR) (Dipp & Evans 2001, Evans & Dipp 2002). cADPR accumulation during hypoxia appears to be dependent on hypoxia-induced increases in β-NADH levels (Wilson et al 2001). Presently, it is not known how β-NADH may modulate cADPR production, but it is possible that cADPR accumulation may be the result of an increase in β-NADH-dependent ROS signalling (Kumasaka et al 1999). β-NADH feeds into complex I of the ETC, resulting in the mitochondrial complexes to become reduced and allowing for the generation of superoxide during hypoxia. By this mechanism, cADPR-dependent Ca^{2+} release from the SR could occur as a downstream consequence of aforementioned hypoxia-induced increases in ROS.

The hypoxia-induced Ca^{2+} release from the SR is then thought to trigger capacitative Ca^{2+} entry (CCE) and/or voltage-dependent Ca^{2+} entry, thus increasing $[Ca^{2+}]_i$ and therefore amplifying phase 1 tension generation (Snetkov et al 2003, Wang et al 2004, Sweeney & Yuan 2000, Robertson et al 2000b) (Fig. 1B). Of the two regulators of Ca^{2+} entry, CCE via store operated Ca^{2+} channels recently emerged as a potentially important means to increase $[Ca^{2+}]_i$ during HPV when inhibition of extracellular Ca^{2+} entry with La^{3+} was shown to attenuate the rise in $[Ca^{2+}]_i$ and phase 1 constriction (Robertson et al 2000b). Alternatively, the release of Ca^{2+} from internal stores could potentially inhibit K_v channels, resulting in membrane depolarization and the opening of voltage-gated L-type Ca^{2+} channels leading to increased $[Ca^{2+}]_i$. When voltage-dependent Ca^{2+} entry was blocked by the L-type Ca^{2+} channel inhibitor nifedipine, phase 1 of the HPV response was partially attenuated (Robertson et al 2000b). These results suggest that voltage-dependent Ca^{2+} entry and CCE may have a synergistic relationship during phase 1 of the HPV response.

Role of ROS signalling in phase 2 of the HPV response

Current research suggests that the pulmonary endothelium is critical for phase 2 of HPV. Removal of the endothelium significantly attenuates phase 2 constriction in isolated pulmonary vessels though it plateaus slightly above baseline and remains at this level throughout the period of hypoxia (Archer et al 1998, Dipp & Evans 2001, Wilson et al 2001, Robertson et al 1995, Ward & Robertson 1995, Leach et al 2001, Dipp et al 2001, Liu et al 2001). Interestingly, Robertson et al (2003) demonstrated that the removal of the endothelium had no effect on $[Ca^{2+}]_i$ in pulmonary smooth muscle cells at any point during HPV. Furthermore, PA constriction during phase 2 is dependent on a small rise in $[Ca^{2+}]_i$ which is independent of extracellular Ca^{2+} entry (Dipp et al 2001, Robertson et al 2000b, Jin et al 1992). This suggests that phase 2 constriction results from endothelial-derived factors that augment the sensitivity of the smooth muscle cell to the hypoxia-induced rise in $[Ca^{2+}]_i$ (Aaronson et al 2002, Shimoda et al 2002). This led Dipp and colleagues to propose that continued hypoxia-induced cADPR synthesis during phase 2 could trigger the tonic release of low levels of Ca^{2+} from the SR through constant stimulation of the ryanodine receptor (Dipp & Evans 2001). Depletion of Ca^{2+} from intracellular stores as well as inhibition of ryanodine-sensitive Ca^{2+} release by the cADPR antagonist, 8-bromo-cADPR, abolishes tension generation during phase 2 of the HPV response, in support of this idea (Robertson et al 2000b, Dipp & Evans 2001).

Increased sensitivity to $[Ca^{2+}]_i$ in smooth muscle cells is regulated by transduction pathways such as RhoA kinase. With this in mind, the RhoA kinase antagonist Y-27632 was shown to significantly attenuate the HPV response during phases 1 and 2 (Robertson et al 2000a, Wang et al 2001, 2003b, Fagan et al 2004). Smooth muscle contraction is a balance between the actions of Ca^{2+} /calmodulin-activated-myosin light chain (MLC) kinase (contraction) and MLC phosphatase (relaxation) on MLC phosphorylation. Hypoxia has been shown to activate RhoA in PA smooth muscle and endothelial cells, leading to the activation of RhoA kinase (Robertson et al 2000a, Wang et al 2001, 2003b) which in turn inactivates MLC phosphatase through phosphorylation of its myosin binding unit (Somlyo & Somlyo 2003). This shifts the balance to an increase in MLC phosphorylation, thus augmenting tension generation at a given level of $[Ca^{2+}]_i$ and MLC kinase activity.

Endothelin (ET-1) is a 21 amino acid peptide secreted by endothelial cells, and is considered a contributing mediator of phase 2 (Shimoda et al 2000). Hypoxia has been shown to increase ET-1 expression and secretion from cultured endothelium (Hu et al 1998, Yamashita et al 2001). Moreover, BQ-123, an ET-1 receptor A (ETA) antagonist, attenuates phase 2 of HPV in isolated vessels and whole lungs (Liu et al 2003, Sato et al 2000). This would suggest that the endothelium possesses an O_2 sensor, which detects the fall in O_2 and activates transcription of ET-1. Hu et al (1988) found that the proximal promoter of the ET-1 gene contains a hypoxia-inducible factor 1 (HIF-1) binding site (HRE) on the antisense DNA strand, positioned at –118 to –125 bp upstream from the transcription start site. Hypoxia-inducible factor (HIF)-1 is the principal O_2-responsive factor underlying increased expression of glycolytic enzymes, glucose transporters, and other genes during hypoxia (Semenza 2000b, a, 2001). Interestingly, mitochondrial ROS have been proposed to trigger the stabilization of HIF-1α during hypoxia in hepatocytes (Chandel et al 1998, 2000a) and umbilical vein endothelial cells have been shown to increase mitochondrial ROS signalling during hypoxia (Pearlstein et al 2002). This suggests the possibility that the same mechanism of O_2 sensing may function in PA myocytes and endothelium, with the sensor activating cell-specific responses that, in combination, result in the HPV response. ET-1 can increase Ca^{2+} -sensitivity via a RhoA/RhoA kinase pathway in the basilar artery of rabbits (Miao et al 2002), yet whether a similar mechanism is present in the pulmonary arteries is not known.

The possibility that endothelium-derived factors in addition to ET-1 could exist was suggested by studies involving a candidate that was partially purified from hypoxic, isolated perfused rat lungs using stepwise acetonitrile elution from SEP-Pak C_{18} cartridges (Robertson et al 2001). This factor caused a slow and sustained constriction of pulmonary arteries without a corresponding increase in $[Ca^{2+}]_i$, and was unaffected by combined treatment with BQ-123 and BQ-788 (a ET_B receptor antagonist) (Robertson et al 2001). This led Robertson et al (2001, 2003) to suggest that this endothelium-derived factor demonstrated Ca^{2+} sensitizing properties yet it was not ET-1. At this time, further research is needed to determine the identity of this putative mediator, how hypoxia triggers its release, and whether a link exists between this hypoxia-induced, endothelium-derived factor and the hypoxia-induced RhoA-dependent increase in Ca^{2+} -sensitivity.

Concluding remarks

Based on current research, there is increasing evidence that a paradoxical increase in mitochondria-dependent ROS signalling is involved, which initiates a chain of events resulting in both an increase in $[Ca^{2+}]_i$ in PA smooth muscle cells and an endothelium-derived increase in Ca^{2+} sensitivity (Fig. 3). This model is attractive given that the mitochondria are the primary consumers of oxygen in the cell. However, definitive experiments are still needed to verify hypoxia-induced ROS signalling. In that regard, preliminary reports show that redox-sensitive cytosolic FRET (fluorescence resonance energy transfer) sensors detect an increase in oxidant stress in PA myocytes during hypoxia (Waypa & Schumacker 2004). Additional studies are needed to assess the subcellular sources of these ROS, and the various compartments where oxidants or reductant signals regulate Ca^{2+} and Ca^{2+} sensitivity. It is possible that hypoxia could cause a decrease in ROS production from NADH oxidase, while it causes a simultaneous increase in ROS release from the mitochondrial ETC thus explaining the early results of Archer and colleagues (Archer et al 1989, 1993, Mohazzab & Wolin 1994a, Michelakis et al 2002). Finally, many of the present data are dependent on problematic pharmacological agents. Therefore, given the availability of genetic tools to modulate ROS production and

ROS scavenging, we remain optimistic that more definitive information regarding this important mechanism will soon be forthcoming.

Acknowledgements

Supported by NHLBI Grants HL66315 and HL35440 and AHA Grant 0235457Z.

References

Aaronson PI, Robertson TP, Ward JP 2002 Endothelium-derived mediators and hypoxic pulmonary vasoconstriction. Respir Physiolo Neurobiol 132:107–120

Archer SL, Will JA, Weir EK 1986 Redox status in the control of pulmonary vascular tone. Herz 11:127–141

Archer SL, Nelson DP, Weir EK 1989 Simultaneous measurement of O_2 radicals and pulmonary vascular reactivity in rat lung. J Appl Physiol 67:1903–1911

Archer SL, Huang J, Henry T, Peterson D, Weir EK 1993 A redox-based O_2 sensor in rat pulmonary vasculature. Circ Res 73:1100–1112

Archer SL, Souil E, Dinh-Xuan AT et al 1998 Molecular identification of the role of voltage-gated K^+ channels, $K_v1.5$ and $K_v2.1$, in hypoxic pulmonary vasoconstriction and control of resting membrane potential in rat pulmonary artery myocytes. J Clin Invest 101:2319–2330

Archer SL, Reeve HL, Michelakis E et al 1999 O_2 sensing is preserved in mice lacking the gp91 phox subunit of NADPH oxidase. Proc Natl Acad Sci U S A 96:7944–7949

Archer SL, Weir EK, Reeve HL, Michelakis E 2000 Molecular identification of O2 sensors and O2-sensitive potassium channels in the pulmonary circulation. Adv Exp Med Biol 475:219–240

Archer S, Michelakis E 2002 The mechanism(s) of hypoxic pulmonary vasoconstriction: potassium channels, redox O(2) sensors, and controversies. News Physiol Sci 17:131–137

Asencio C, Rodriguez-Aguilera JC, Ruiz-Ferrer M, Vela J, Navas P 2003 Silencing of ubiquinone biosynthesis genes extends life span in Caenorhabditis elegans. Faseb J 17:1135–1137

Boveris A, Oshino N, Chance B 1972 The cellular production of hydrogen peroxide. Biochem J 128:617–630

Boveris A, Cadenas E, Stoppani AO 1976 Role of ubiquinone in the mitochondrial generation of hydrogen peroxide. Biochem J 156:435–444

Chance B, Williams GR 1955 Respiratory enzymes in oxidative phosphorylation. I. Kinetics of oxygen utilization. J Biol Chem 217:383–393

Chandel NS, Maltepe E, Goldwasser E et al 1998 Mitochondrial reactive oxygen species trigger hypoxia-induced transcription. Proc Natl Acad Sci U S A 95:11715–11720

Chandel NS, Schumacker PT 1999 Cells depleted of mitochondrial DNA (rho0) yield insight into physiological mechanisms. FEBS Lett 454:173–176

Chandel NS, McClintock DS, Feliciano CE et al 2000a Reactive oxygen species generated at mitochondrial complex III stabilize hypoxia-inducible factor-1alpha during hypoxia: a mechanism of O_2 sensing. J Biol Chem 275:25130–25138

Chandel NS, Trzyna WC, McClintock DS, Schumacker PT 2000b Role of oxidants in NF-kappa B activation and TNF-alpha gene transcription induced by hypoxia and endotoxin. J Immunol 165:1013–1021

Chen Q, Vazquez EJ, Moghaddas S, Hoppel CL, Lesnefsky EJ 2003 Production of reactive oxygen species by mitochondria: central role of complex III. J Biol Chem 278:36027–36031

Cowan DB, Jones M, Garcia LM et al 2003 Hypoxia and stretch regulate intercellular communication in vascular smooth muscle cells through reactive oxygen species formation. Arterioscler Thromb Vasc Biol 23:1754–1760

Dipp M, Evans AM 2001 Cyclic ADP-ribose is the primary trigger for hypoxic pulmonary vasoconstriction in the rat lung in situ. Circ Res 89:77–83

Dipp M, Nye PC, Evans AM 2001 Hypoxic release of calcium from the sarcoplasmic reticulum of pulmonary artery smooth muscle. Am J Physiol Lung Cell Mol Physiol 281:L318–325

Evans AM, Dipp M 2002 Hypoxic pulmonary vasoconstriction: cyclic adenosine diphosphate-ribose, smooth muscle Ca(2+) stores and the endothelium. Respir Physiolo Neurobiol 132:3–15

Fabian M, Palmer G 1998 Hydrogen peroxide is not released following reaction of cyanide with several catalytically important derivatives of cytochrome c oxidase. FEBS Lett 422:1–4

Fagan KA, Oka M, Bauer NR et al 2004 Attenuation of acute hypoxic pulmonary vasoconstriction and hypoxic pulmonary hypertension in mice by inhibition of Rho-kinase. Am J Physiol Lung Cell Mol Physiol 287:656–664

Garcia-Ruiz C, Colell A, Mari M, Morales A, Fernandez-Checa JC 1997 Direct effect of ceramide on the mitochondrial electron transport chain leads to generation of reactive oxygen species. Role of mitochondrial glutathione. J Biol Chem 272:11369–11377

Grimminger F, Weissmann N, Spriestersbach R et al 1995 Effects of NADPH oxidase inhibitors on hypoxic vasoconstriction in buffer-perfused rabbit lungs. Am J Physiol 268:L747–752

Hasunuma K, Rodman DM, McMurtry IF 1991 Effects of K+ channel blockers on vascular tone in the perfused rat lung. Am Rev Respir Dis 144:884–887

Hu J, Discher DJ, Bishopric NH, Webster KA 1998 Hypoxia regulates expression of the endothelin-1 gene through a proximal hypoxia-inducible factor-1 binding site on the antisense strand. Biochem Biophys Res Commun 245:894–899

Iwata S, Lee JW, Okada K et al 1998 Complete structure of the 11-subunit bovine mitochondrial cytochrome bc1 complex. Science 281:64–71

Jin N, Packer CS, Rhoades RA 1991 Reactive oxygen-mediated contraction in pulmonary arterial smooth muscle: cellular mechanisms. Can J Physiol Pharmacol 69:383–288

Jin N, Packer CS, Rhoades RA 1992 Pulmonary arterial hypoxic contraction: signal transduction. Am J Physiol 263:L73–78

Kang TM, Park MK, Uhm DY 2002 Characterization of hypoxia-induced $[Ca2+]_i$ rise in rabbit pulmonary arterial smooth muscle cells. Life Sci 70:2321–2333

Kang TM, Park MK, Uhm DY 2003 Effects of hypoxia and mitochondrial inhibition on the capacitative calcium entry in rabbit pulmonary arterial smooth muscle cells. Life Sci 72:1467–1479

Killilea DW, Hester R, Balczon R, Babal P, Gillespie MN 2000 Free radical production in hypoxic pulmonary artery smooth muscle cells. Am J Physiol Lung Cell Mol Physiol 279:L408–412

King MP, Attardi G 1989 Human cells lacking mtDNA: repopulation with exogenous mitochondria by complementation. Science 246:500–503

Kumasaka S, Shoji H, Okabe E 1999 Novel mechanisms involved in superoxide anion radical-triggered Ca2+ release from cardiac sarcoplasmic reticulum linked to cyclic ADP-ribose stimulation. Antioxid Redox Signal 1:55–69

Kushnareva Y, Murphy AN, Andreyev A 2002 Complex I-mediated reactive oxygen species generation: modulation by cytochrome c and NAD(P)+ oxidation-reduction state. Biochem J 368:545–553

Lassegue B, Sorescu D, Szocs K et al 2001 Novel gp^{91}(phox) homologues in vascular smooth muscle cells : nox1 mediates angiotensin II-induced superoxide formation and redox-sensitive signaling pathways. Circ Res 88:888–894

Lazor R, Feihl F, Waeber B, Kucera P, Perret C 1996 Endothelin-1 does not mediate the endothelium-dependent hypoxic contractions of small pulmonary arteries in rats. Chest 110:189–197

Leach RM, Robertson TP, Twort CH, Ward JP 1994 Hypoxic vasoconstriction in rat pulmonary and mesenteric arteries. Am J Physiol 266:L223–231

Leach RM, Hill HM, Snetkov VA, Robertson TP, Ward JP 2001 Divergent roles of glycolysis and the mitochondrial electron transport chain in hypoxic pulmonary vasoconstriction of the rat: identity of the hypoxic sensor. J Physiol London 536:211–224

Liu Q, Sham JS, Shimoda LA, Sylvester JT 2001 Hypoxic constriction of porcine distal pulmonary arteries: endothelium and endothelin dependence. Am J Physiol Lung Cell Mol Physiol 280:L856–865

Liu JQ, Sham JS, Shimoda LA, Kuppusamy P, Sylvester JT 2003 Hypoxic constriction and reactive oxygen species in porcine distal pulmonary arteries. Am J Physiol Lung Cell Mol Physiol 285:L322–333

Marshall C, Mamary AJ, Verhoeven AJ, Marshall BE 1996 Pulmonary artery NADPH-oxidase is activated in hypoxic pulmonary vasoconstriction. Am J Respir Cell Mol Biol 15:633–644

McMurtry IF 1984 Angiotensin is not required for hypoxic constriction in salt solution-perfused rat lungs. J Appl Physiol 56:375–380

Miao L, Dai Y, Zhang J 2002 Mechanism of RhoA/Rho kinase activation in endothelin-1- induced contraction in rabbit basilar artery. Am J Physiol Heart Circ Physiol 283:H983–989

Michelakis ED, Hampl V, Nsair A et al 2002 Diversity in mitochondrial function explains differences in vascular oxygen sensing. Circ Res 90:1307–1315

Moghaddas S, Hoppel CL, Lesnefsky EJ 2003 Aging defect at the QO site of complex III augments oxyradical production in rat heart interfibrillar mitochondria. Arch Biochem Biophys 414:59–66

Mohazzab KM, Wolin MS 1994a Properties of a superoxide anion-generating microsomal NADH oxidoreductase, a potential pulmonary artery PO_2 sensor. Am J Physiol 267:L823–831

Mohazzab KM, Wolin MS 1994b Sites of superoxide anion production detected by lucigenin in calf pulmonary artery smooth muscle. Am J Physiol 267:L815–822

Morio Y, McMurtry IF 2002 Ca(2+) release from ryanodine-sensitive store contributes to mechanism of hypoxic vasoconstriction in rat lungs. J Appl Physiol 92:527–534

Olschewski A, Hong Z, Linden BC et al 2002a Contribution of the K(Ca) channel to membrane potential and O_2 sensitivity is decreased in an ovine PPHN model. Am J Physiol Lung Cell Mol Physiol 283:L1103–1109

Olschewski A, Hong Z, Nelson DP, Weir EK 2002b Graded response of K(+) current, membrane potential, and [Ca(2+)](i) to hypoxia in pulmonary arterial smooth muscle. Am J Physiol Lung Cell Mol Physiol 283:L1143–1150

Olschewski A, Hong Z, Peterson DA et al 2004 Opposite effects of redox status on membrane potential, cytosolic calcium, and tone in pulmonary arteries and ductus arteriosus. Am J Physiol Lung Cell Mol Physiol 286:L15–22

Paddenberg R, Ishaq B, Goldenberg A et al 2003 Essential role of complex II of the respiratory chain in hypoxia-induced ROS generation in the pulmonary vasculature. Am J Physiol Lung Cell Mol Physiol 284:L710–719

Pearlstein DP, Ali MH, Mungai PT et al 2002 Role of mitochondrial oxidant generation in endothelial cell responses to hypoxia. Arterioscler Thromb Vasc Biol 22:566–73

Rana M, de Coo I, Diaz F, Smeets H, Moraes CT 2000 An out-of-frame cytochrome b gene deletion from a patient with parkinsonism is associated with impaired complex III assembly and an increase in free radical production. Ann Neurol 48:774–781

Reeve HL, Weir EK, Nelson DP, Peterson DA, Archer SL 1995 Opposing effects of oxidants and antioxidants on K+ channel activity and tone in rat vascular tissue. Exp Physiol 80:825–834

Rettig J, Heinemann SH, Wunder F et al 1994 Inactivation properties of voltage-gated K+ channels altered by presence of beta-subunit. Nature 369:289–294

Rhoades RA, Packer CS, Roepke DA, Jin N, Meiss RA 1990 Reactive oxygen species alter contractile properties of pulmonary arterial smooth muscle. Can J Physiol Pharmacol 68:1581–1589

Richter C, Gogvadze V, Laffranchi R et al 1995 Oxidants in mitochondria: from physiology to diseases. Biochim Biophys Acta 1271:67–74

Robertson TP, Aaronson PI, Ward JP 1995 Hypoxic vasoconstriction and intracellular Ca^{2+} in pulmonary arteries: evidence for PKC-independent Ca^{2+} sensitization. Am J Physiol 268: H301–307

Robertson TP, Dipp M, Ward JP, Aaronson PI, Evans AM 2000a Inhibition of sustained hypoxic vasoconstriction by Y-27632 in isolated intrapulmonary arteries and perfused lung of the rat. Br J Pharmacol 131:5–9

Robertson TP, Hague D, Aaronson PI, Ward JP 2000b Voltage-independent calcium entry in hypoxic pulmonary vasoconstriction of intrapulmonary arteries of the rat. J Physiol London 525 Pt 3:669–680

Robertson TP, Aaronson PI, Ward JP 2003 Ca^{2+}-sensitization during sustained hypoxic pulmonary vasoconstriction is endothelium-dependent. Am J Physiol Lung Cell Mol Physiol 284:L1121–1126

Robertson TP, Ward JP, Aaronson PI 2001 Hypoxia induces the release of a pulmonary-selective, Ca(2+)-sensitising, vasoconstrictor from the perfused rat lung. Cardiovasc Res 50:145–150

Rounds S, McMurtry IF 1981 Inhibitors of oxidative ATP production cause transient vasoconstriction and block subsequent pressor responses in rat lungs. Circ Res 48:393–400

Sato K, Morio Y, Morris KG, Rodman DM, McMurtry IF 2000 Mechanism of hypoxic pulmonary vasoconstriction involves ET(A) receptor-mediated inhibition of K(ATP) channel. Am J Physiol Lung Cell Mol Physiol 278:L434–442

Semenza GL 2000a HIF-1: mediator of physiological and pathophysiological responses to hypoxia. J Appl Physiol 88:1474–1480

Semenza GL 2000b Surviving ischemia: adaptive responses mediated by hypoxia-inducible factor 1. J Clin Invest 106:809–812

Semenza GL 2001 HIF-1, O(2), and the 3 PHDs: how animal cells signal hypoxia to the nucleus. Cell 107:1–3

Sham JS, Crenshaw BR, Jr., Deng LH, Shimoda LA, Sylvester JT 2000 Effects of hypoxia in porcine pulmonary arterial myocytes: roles of K(V) channel and endothelin-1. Am J Physiol Lung Cell Mol Physiol 279:L262–272

Sham JS 2002 Hypoxic pulmonary vasoconstriction: ups and downs of reactive oxygen species. Circ Res 91:649–651

Shimoda LA, Sylvester JT, Sham JS 2000 Mobilization of intracellular Ca(2+) by endothelin-1 in rat intrapulmonary arterial smooth muscle cells. Am J Physiol Lung Cell Mol Physiol 278:L157–164

Shimoda LA, Sham JS, Liu Q, Sylvester JT 2002 Acute and chronic hypoxic pulmonary vasoconstriction: a central role for endothelin-1? Respir Physiolo Neurobiol 132:93–106

Snetkov VA, Aaronson PI, Ward JP, Knock GA, Robertson TP 2003 Capacitative calcium entry as a pulmonary specific vasoconstrictor mechanism in small muscular arteries of the rat. Br J Pharmacol 140:97–106

Somlyo AP, Somlyo AV 2003 Ca^{2+} sensitivity of smooth muscle and nonmuscle myosin II: modulated by G proteins, kinases, and myosin phosphatase. Physiol Rev 83:1325–1358

Spasojevic I, Liochev SI, Fridovich I 2000 Lucigenin: redox potential in aqueous media and redox cycling with O-(2) production. Arch Biochem Biophys 373:447–450

Sun J, Trumpower BL 2003 Superoxide anion generation by the cytochrome bc1 complex. Arch Biochem Biophys 419:198–206

Sweeney M, Yuan JX 2000 Hypoxic pulmonary vasoconstriction: role of voltage-gated potassium channels. Respir Res 1:40–48

Sylvester JT 2001 Hypoxic pulmonary vasoconstriction: a radical view. Circ Res 88:1228–30

Turrens JF, Boveris A 1980 Generation of superoxide anion by the NADH dehydrogenase of bovine heart mitochondria. Biochem J 191:421–427

Turrens JF, Alexandre A, Lehninger AL 1985 Ubisemiquinone is the electron donor for superoxide formation by complex III of heart mitochondria. Arch Biochem Biophys 237: 408–414

Verkhovsky MI, Morgan JE, Puustein A, Wikstrom M 1996 Kinetic trapping of oxygen in cell respiration. Nature 380:268–270

von Euler U, Liljestrand G 1946 Observations on the pulmonary arterial blood pressure of the cat. Acta Physiol Scand 12:301–320

Wang Z, Jin N, Ganguli S et al 2001 Rho-kinase activation is involved in hypoxia-induced pulmonary vasoconstriction. Am J Respir Cell Mol Biol 25:628–635

Wang YX, Zheng YM, Abdullaev I, Kotlikoff MI 2003a Metabolic inhibition with cyanide induces calcium release in pulmonary artery myocytes and Xenopus oocytes. Am J Physiol Cell Physiol 284:C378–388

Wang Z, Lanner MC, Jin N et al 2003b Hypoxia inhibits myosin phosphatase in pulmonary arterial smooth muscle cells: Role of Rho-kinase. Am J Respir Cell Mol Biol 29:465–471

Wang J, Shimoda LA, Sylvester JT 2004 Capacitative calcium entry and TRPC channel proteins are expressed in rat distal pulmonary arterial smooth muscle. Am J Physiol Lung Cell Mol Physiol 286:L848–858

Ward JP, Robertson TP 1995 The role of the endothelium in hypoxic pulmonary vasoconstriction. Exp Physiol 80:793–801

Waypa GB, Chandel NS, Schumacker PT 2001 Model for hypoxic pulmonary vasoconstriction involving mitochondrial oxygen sensing. Circ Res 88:1259–1266

Waypa GB, Marks JD, Mack MM et al 2002 Mitochondrial reactive oxygen species trigger calcium increases during hypoxia in pulmonary arterial myocytes. Circ Res 91:719–726

Waypa GB, Schumacker PT 2002 O(2) sensing in hypoxic pulmonary vasoconstriction: the mitochondrial door re-opens. Respir Physiolo Neurobiol 132:81–91

Waypa GB, Schumacker PT 2004 Hypoxia results in increased oxidative stress in pulmonary arterial smooth muscle cells. American Journal of Respiratory and Critical Care Medicine 169:A167

Waypa GB, Schumacker PT 2005 Hypoxic pulmonary vasoconstriction: redox events in oxygen sensing. J Appl Physiol 98:404–414

Weir EK, Eaton JW, Chesler E 1985 Redox status and pulmonary vascular reactivity. Chest 88:249S–252S

Weir EK, Archer SL 1995 The mechanism of acute hypoxic pulmonary vasoconstriction: the tale of two channels. Faseb J 9:183–189

Weir EK, Reeve HL, Cornfield DN et al 1997 Diversity of response in vascular smooth muscle cells to changes in oxygen tension. Kidney Int 51:462–166

Weir EK, Hong Z, Porter VA, Reeve HL 2002 Redox signaling in oxygen sensing by vessels. Respir Physiol Neurobiol 132:121–130

Weissmann N, Grimminger F, Walmrath D, Seeger W 1995 Hypoxic vasoconstriction in buffer-perfused rabbit lungs. Respir Physiol 100:159–169

Weissmann N, Grimminger F, Voswinckel R, Conzen J, Seeger W 1998 Nitro blue tetrazolium inhibits but does not mimic hypoxic vasoconstriction in isolated rabbit lungs. Am J Physiol 274:L721–727

Weissmann N, Ebert N, Ahrens M et al 2003 Effects of mitochondrial inhibitors and uncouplers on hypoxic vasoconstriction in rabbit lungs. Am J Respir Cell Mol Biol 29:721–32

Wilson HL, Dipp M, Thomas JM et al 2001 Adp-ribosyl cyclase and cyclic ADP-ribose hydrolase act as a redox sensor. a primary role for cyclic ADP-ribose in hypoxic pulmonary vasoconstriction. J Biol Chem 276:11180–11188

Yamashita K, Discher DJ, Hu J, Bishopric NH, Webster KA 2001 Molecular regulation of the endothelin-1 gene by hypoxia. Contributions of hypoxia-inducible factor-1, activator protein-1, GATA-2, AND p300/CBP. J Biol Chem 276:12645–12653

Zhang F, Carson RC, Zhang H, Gibson G, Thomas HM, 3rd 1997 Pulmonary artery smooth muscle cell $[Ca2+]_i$ and contraction: responses to diphenyleneiodonium and hypoxia. Am J Physiol 273:L603–611

Zhao Y, Rhoades RA, Packer CS 1996 Hypoxia-induced pulmonary arterial contraction appears to be dependent on myosin light chain phosphorylation. Am J Physiol 271:L768–774

DISCUSSION

Murphy: I was curious about the rationale that complex II is the site of the ROS production. It is quite possible that it is, but it could also be that by altering complex III you are decreasing the mitochondrial membrane potential, and that this decrease is responsible for blocking the ROS production.

Schumacker: This wouldn't explain why we are able to measure an increase in the radical adducts in a submitochondrial particle where we are supplying electrons into complex III via succinate in the presence of rotenone and antimycin A. This doesn't rule out the possibility that there could also be ROS coming from complex I.

Murphy: I would be wary of taking that as absolute proof.

Schumacker: I agree. In fact, the HSP FRET data show that there is some increase with hypoxia, even when we have done what we think is a substantial knockdown of complex III or knockout of cytochrome c. This indicates that perhaps there are other sites within the mitochondria that could generate ROS.

Murphy: With the FRET probe, are you assuming that H_2O_2 comes out and then directly reacts with the thiol, or do you think it goes via the glutathione pool or via oxidation of thirodoxin, or a combination of all of these?

Schumacker: I don't know. I see the thiol pool as an environment where they are at equilibrium, at least after some time. Something that comes along and pulls this environment in an oxidizing direction is going to be reflected in a change in the overall thiol redox status. The thiol pool is being oxidized.

Murphy: To play devil's advocate, I could say that what you are looking at is not necessarily through H_2O_2, but it is something that comes from the mitochondria which changes the cytosolic glutathione/glutathione disulfide ratio.

Kummer: You have beautiful data describing what is going on. Can you also explain why? What makes the difference if you have a little bit more or less oxygen in the system?

Schumacker: It is counterintuitive. Why should less oxygen (a substrate for superoxide production) produce more ROS? There are two possible explanations. First, if oxygen were somehow interacting with this subunit in the mitochondrial membrane such that at lower oxygen concentrations there is a change: even though there is less oxygen there, it has better access to the semiquinone, which is the source of the electrons that produce superoxide. If this were the case, maybe less oxygen could paradoxically increase the production of superoxide. The alternative hypothesis is that there is some oxygen-induced change in the structure of the protein such that the lifetime of the semiquinone is increased by some amount. The longer the lifetime of the semiquinone radical, the greater the opportunity for oxygen to steal an electron from it, thus resulting in superoxide.

Kummer: Oxygen must then bind somewhere to this complex at a site where no electrons flow through.

Schumacker: I would be hesitant to use the word 'binding' because this would imply a specific receptor for oxygen. Maybe it is possible that oxygen, which is fairly lipophilic, could change the fluidity in the membrane by interacting with membrane phospholipids. It appears that oxygen is somehow interacting with this complex in the membrane, such that the tendency to make superoxide is increased. We don't understand what this mechanism is, but we really want to find out.

Archer: I noticed in those three cells where you used the FRET reporter, the signal went up but never came back down with reoxygenation, whereas in the myocardial cells it went up and down. What property of the reporter wasn't reversible? Aren't you surprised that a heart cell, a pulmonary artery cell, and an 'anonymous manipulated cell' would all have the same response to hypoxia? Should O_2-sensitive ROS production not be different in specialized O_2-sensitive tissues.

Schumacker: Those are both good questions. There are two factors in the reversibility. One has to do with how quickly the thiol pool redox status returns to the pre-hypoxia state. If this wasn't instantaneous, if it requires some glutathione reductase activity, then perhaps it takes a little while to return. The second aspect of restoration of the FRET to control conditions has to do with refolding of the protein, which might involve a chaperone. In the cardiac myocytes where we made the cells ischaemic and then reperfused, upon reperfusion there is a big reoxygenation ROS burst. I am suspicious of the rapid drop in the FRET ratio that we see upon reoxygenation, because this is when a lot of cells start dying. I don't know whether this represents oxidation of the probe or a recovery from oxidative stress. To your second question, whether I am surprised that cells with different oxygen-sensing functions behave similarly, I don't find this surprising. If this is a sensing mechanism that works to control a variety of transcriptional and post-translational responses to hypoxia, it would make great sense from a parsimonious viewpoint to have an O_2 sensor that is easily reproduced from cell to cell, and for different cells to use the same oxidant signal to mediate different responses by expressing different signalling molecules, causing increases in Ca^{2+} in some cells and stabilization of transcription factors in other cells. It is a little too much to imply that this is the only O_2 sensor: it is quite clear that it is not. But if it is playing a role as an O_2 sensor, mitochondria tend to be more similar among tissues than they are different. It would make great sense if this system, which is preserved from yeasts to mammals, could be a route of signalling.

Harris: I'd like to explore the superoxide dismutase (SOD) story in more detail. It is a bit surprising that expressing mitochondrial SOD or cytosolic SOD made no difference. What happens if you inhibit these?

Schumacker: If most of the superoxide is dismuted by the SOD that is constitutively expressed there, then adding more superoxide will shorten the lifetime of superoxide that is floating around because it will be more quickly dismuted. It won't result in a net increase in the production of hydrogen peroxide, because this is limited ultimately by the rate that superoxide is generated. If there is lots of constitutive SOD around and you are generating superoxide at some rate, and then you double the amount of SOD present, you will decrease the lifetime of superoxide but you won't increase the concentration of hydrogen peroxide.

Harris: We have inhibited SOD by copper chelation and this has no effect at all on HIF generation in normoxia or mild hypoxia.

Schumacker: Using dihydroethidium as a sensor for superoxide, we have made cells ischaemic and then blocked copper/zinc SOD pharmacologically with a chelator. We see an increase in the superoxide signal from this dihydroethidium in response to SOD inhibition. This observation is compatible with the idea that there is an increase in oxidant production during the hypoxia/ischaemia, and there is less H_2O_2 and more superoxide if you block the dismutation of superoxide.

Harris: We have done exactly the same experiment and found that when the free radicals go up by a dye method in normoxia, SOD inhibition could help this go up with no effect on HIF expression. I would have expected HIF to go up, unless it is localization: perhaps the localization of the generation of free radicals is important, or perhaps the HIF has to be near the mitochondria. What is the diffusion distance in the cell of H_2O_2? How far can it travel before it can hit something and have an effect?

Schumacker: I don't know.

Harris: I was wondering whether the localization was important.

Schumacker: I have a couple of thoughts. If you are blocking SOD1 and there is still SOD2 activity, perhaps this is enough.

Harris: Then the diffusion distance would have to be quite short for this to work.

Schumacker: The diffusion distances are submicron.

Sylvester: Where is prolyl hydroxylase located?

Harris: People have thought that there is a mitochondrial and cytoplasmic localization. Other studies suggest that all three are in the cytosol.

Weissmann: In terms of hypoxic vasoconstriction, can you speculate what is downstream of H_2O_2? How does the increase in H_2O_2 induce the vasoconstrictor response?

Schumacker: That is a good question: I don't really have an answer to that. We saw some data earlier today which suggest that oxidants might be playing a role in triggering release of Ca^{2+} from intracellular stores. We have heard other models, also.

It is possible that H_2O_2 acts to trigger release from intracellular stores and perhaps there is capacitative Ca^{2+} entry (CCE) after that.

Ward: There is also a fair amount of evidence that various ROS can directly or indirectly activate a variety of members of the TRP superfamily of non-selective cation channels, so potentially causing Ca^{2+} influx. Direct evidence for this is however a bit loose.

Schumacker: The issue here is that although everyone uses exogenous oxidants, what we are learning is that much of the oxidant signalling may be compartmentalized, or at least localized. So it may be difficult to mimic with exogenous oxidants the physiological effects seen when oxidants are produced at some complex and then diffused to nearby targets. It is hard to extrapolate the effects of localized oxidant signalling to the level of oxidant production in the intact tissue, and it is difficult to apply oxidants to the intact tissue and expect the responses to mimic those produced by endogenous oxidant signalling systems. Moreover, it is hard to know what the concentration of H_2O_2 is in the cell where it is acting on its target.

Aaronson: Is your construct going to be sensitive enough to do imaging of ROS distribution within the cell, in a dynamic way?

Schumacker: I don't think so, partly because what it is responding to is a change in the thiol redox status. I think it is a change of the thiol redox status in the pool, in its environment. This might not follow in a dynamic way the wave of oxidant stress, in the same way you could follow a wave of Ca^{2+} as it propagated through an excited cell.

Duchen: None of the fluorescent free radical indicators provide decent spatial signals. They all have reasons why they partition into mitochondria or bind to DNA, for example, giving a misleading result.

Archer: Equally importantly, many of the fluorescent ROS measuring assays (such as dihydroethidium) have a signal that will dynamically decrease. The signal tends to rise with time (and oxidation) but when there is less oxidant stress this integrative measure of oxidation does not go down much and may only increase less rapidly (making it poorly suited for measuring rapid changes in ROS).

Schumacker: If you have a normoxic cell in a tissue culture incubator under ambient 5% CO_2 conditions, if you add a thiol reductant to this and the sensor had previously existed at a state of partial oxidation, you should be able to further reduce the thiols and cause a decrease in the FRET ratio. We have tried these experiments but we can't ever get a decrease in it. We think this is because the GSH to GSSG ratio in a typical cell is about 300:1, and under these circumstances virtually every thiol in the cytosol is completely reduced.

Peet: Do you have any thoughts on whether the reactive oxygen species might also regulate HIF transcriptional activity?

Schumacker: That's a good question, but no.

Hypoxic pulmonary vasoconstriction—triggered by an increase in reactive oxygen species?

Norbert Weissmann, Ralph T. Schermuly, Hossein A. Ghofrani, Jörg Hänze, Parag Goyal, Friedrich Grimminger and Werner Seeger

Department of Internal Medicine, University of Giessen Lung Center (UGLC), Justus-Liebig-University Giessen, Klinikstrasse 36, 35392 Giessen, Germany

Abstract. Hypoxic pulmonary vasoconstriction (HPV) is an essential mechanism of the lung that matches perfusion to ventilation in order to optimize pulmonary gas exchange. Despite intensive research, the underlying mechanism has not yet been fully elucidated. Reactive oxygen species (ROS) have been proposed as key mediators of HPV. However, there is ongoing discussion as to whether ROS really contribute to HPV regulation and if so, whether an increase or a decrease in ROS occurs during alveolar hypoxia. In this overview, we summarize our data that have led us to conclude that alveolar hypoxia induces an increase in superoxide and subsequently H_2O_2, and thus elicits HPV. This conclusion is drawn from investigations employing various inhibitors that interfere with ROS in isolated buffer-perfused rabbit lungs challenged with 10-minute periods of alveolar hypoxia. Targeting possible sources of a hypoxia-induced increase in ROS, our data are only partially in accordance with the hypothesis that mitochondria are the hypoxia-dependent ROS generators, and suggest NADPH oxidases as an alternative source. From measurements of intracellular and exhaled H_2O_2, we hypothesize that total lung ROS release is reduced in alveolar hypoxia, but that in specialized cells or sub-cellular structures an increased ROS release may occur, triggering HPV.

2005 Signalling pathways in acute oxygen sensing. Wiley, Chichester (Novartis Foundation Symposium 272) p 196–213

Hypoxic pulmonary vasoconstriction (HPV) matches lung perfusion to ventilation in order to optimize pulmonary gas exchange (Fishman 1976, Staub 1985). The modern era of HPV research started with the description of this mechanism by von Euler and Liljestrand in 1946 (von Euler 1946) although even earlier reports observed that alveolar hypoxia induces an increase in pulmonary artery pressure. Since then, considerable effort has been spent to identify the cell(s) responsible for O_2 sensing, the sensor mechanism(s), and the pathway(s) of signal transduction leading to contraction of the vascular smooth muscle cells in the precapillary resistance vessels, which are proposed to be the predominant site of HPV (Hillier et al 1997, Staub 1985). The precise mechanism of HPV, however, has not been elucidated yet.

For more than a decade, a variety of investigations have focused on the role of reactive oxygen species (ROS) in the regulation of HPV (Archer et al 1993, 1995, 1999, Marshall et al 1996, Mohazzab et al 1995, Paddenberg et al 2003a,b, Thomas et al 1991, Waypa et al 2001, 2002, Weissmann et al 2000, 2001a, 2003, Wilson et al 2001). These investigations proposed a role for ROS in the regulation of HPV. However, among these studies there is dissent on the question of whether ROS are up- or down-regulated during alveolar hypoxia and if such an up- or down-regulation contributes to the regulation of HPV. ROS are good candidates for the regulation of HPV because they can be generated rapidly and are short-lived. Indeed, HPV was shown to be a very fast reaction, occurring within seconds after onset of alveolar hypoxia (Jensen et al 1992, Weissmann et al 1995). Furthermore, HPV is fully reversible after reversal of alveolar hypoxia, also within seconds. This rapid response demands a rapidly controlled pathway for its regulation. The first evidence for a role for ROS was provided in elegant experiments by Archer, Weir and colleagues based on luminol-enhanced chemiluminescence measurements in isolated rat lungs (Archer et al 1993). This study proposed that a decrease in mitochondrial ROS release with subsequent shift in redox couples (e.g. GSH : GSSG, NADH : NAD), leading to inhibition of voltage-gated K^+ channels, triggers HPV. Investigations from the same laboratory later provided further evidence for a decrease in ROS as a trigger for HPV (Archer et al 1995). This hypothesis was recently supported by Archer and Michelakis comparing ROS formation in pulmonary and renal arteries with a variety of different methods (Michelakis et al 2002). In line with a decrease of lung ROS, Mohazzabh and colleagues provided evidence for an NADH-oxidoreductase as the pulmonary oxygen sensor, which, via superoxide, generates H_2O_2. This H_2O_2 then stimulates a guanylate cyclase to release vasodilatating cGMP. The NO pathway synergistically contributes to cGMP synthesis via guanylate cyclase. According to this model there is tonic stimulation of cGMP formation during normoxia. Under hypoxic conditions, the release of H_2O_2 by the oxidoreductase is decreased, and the subsequent loss of vasodilatation contributes to HPV (Mohazzab et al 1995, Monaco & Burke-Wolin 1995).

In contrast to the idea of decreased ROS release as a trigger for HPV, other investigations suggested that an increase in ROS is the underlying mechanism of HPV. The first evidence in support of this hypothesis was provided by the investigation of Marshall et al (1996) that identified an activation of a NADPH oxidase in hypoxia, which then increased the release of ROS. Subsequently, the investigations of Waypa et al (2001, 2002) suggested that hypoxia-induced increases in ROS release from mitochondrial complex III was the underlying mechanism of HPV. In their investigations, evidence was based not only on the use of inhibitors of the

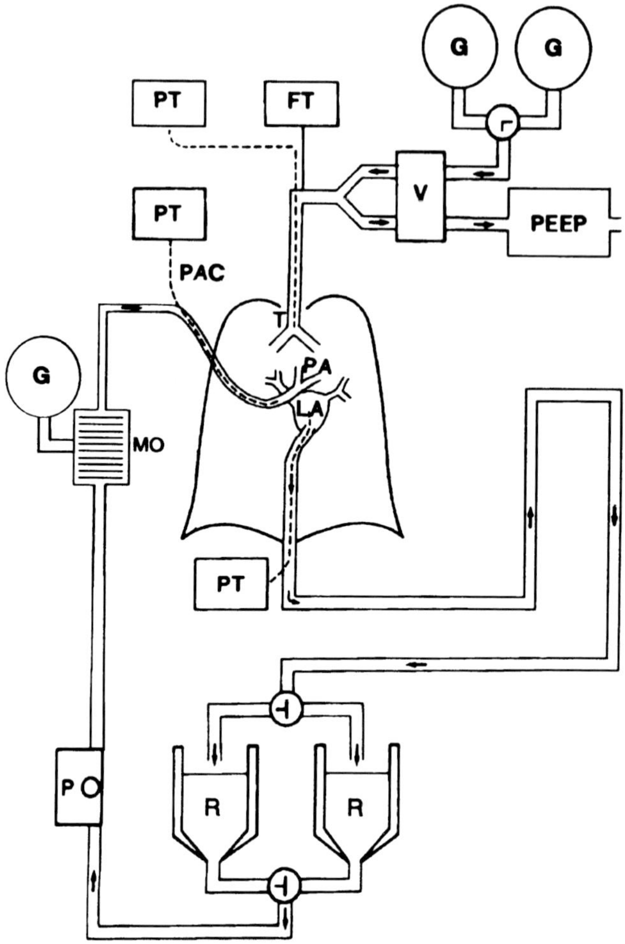

FIG. 1. Schematic description of the isolated rabbit lung model. (Modified with permission from Weissmann et al 1995.) Lungs are perfused with Krebs-Henseleit buffer at constant flow and are ventilated at constant volume with positive end-expiratory pressure (PEEP). For measurement of pulmonary artery pressure a small diameter catheter (PAC) is placed in the pulmonary artery (PA). A membrane oxygenator (MO) may be integrated to control mixed venous gas partial pressures. Left ventricular pressure is set at 2 mmHg. The direction of gas and buffer flow is indicated by arrows. Inhibitors or activators of different biochemical pathway can be added to the buffer reservoir or into the pulmonary arterial line. FT, force transducer (for measurement of organ weight); G, gas supply for the membrane oxygenator and for normoxic as well as hypoxic ventilation; LA, left atrium; P, perfusion pump; PT pressure transducer; R, buffer reservoir; V, ventilator.

mitochondrial electron transport chain, superoxide release, and measurement of ROS with 2′,7′-dichlorofluorescein diacetate, but also on the fact that pulmonary artery cells lacking mitochondria displayed an attenuated response to hypoxia, but not to stimulation with the thromboxane mimetic U46619 (Waypa et al 2001). This concept that complex III of the mitochondrial electron transport chain acts as a pulmonary oxygen sensor was also supported by a study of Leach et al (2001). Recently, another concept suggesting mitochondrial complex II as a potential oxygen sensor was proposed, which also provided evidence for an increase in ROS in hypoxia in pulmonary artery smooth muscle cells (Paddenberg et al 2003a,b). Further evidence for increased ROS generation in hypoxia was provide by Liu et al (2003).

Against this background, here we summarize our evidence for the concept of increased ROS as a trigger for HPV. These investigations are related primarily to the use of inhibitors that interfere with ROS generation and metabolism.

Methods

For quantification of the strength of HPV, we use an isolated buffer-perfused and ventilated rabbit lung model (Weissmann et al 1995). This set-up allows for the investigation of the basic mechanism of HPV by maintaining natural cell–cell contacts, without humoral or neural interference by the body. Explanted lungs are perfused with Krebs-Henseleit buffer at constant flow and are ventilated with either 21% or 3% of oxygen under normobaric conditions (Weissmann et al 1995). The ventilator gas contains 5.3% CO_2, balanced with N_2. By addition of $NaHCO_3$ to the buffer fluid, the pH is maintained at 7.35–7.40. As the perfusion rate (150 ml/min) and venous pressure are constant, pulmonary artery pressure directly reflects the strength of HPV. In this set-up hypoxic vasoconstrictor responses are very reproducible, if 10 min hypoxic ventilation manoeuvres are alternated with 15 min periods of normoxic ventilation (Weissmann et al 1995, 1998a,b, 1999, 2000, 2003). Addition of inhibitors or activators of distinct signalling pathways to the perfusate fluid 5 min before each hypoxic ventilation manoeuvre allows the investigation of the effects of these agents on the strength of HPV. To elucidate pathways specific for the regulation of HPV, the effect of these agents on non-hypoxic induced vasoconstrictions are quantified. In these experiments, repetitive bolus applications (every 25 min) of the thromboxane mimetic U46619 or angiotensin II into the pulmonary artery are performed in a time schedule corresponding to that of the hypoxic experiments (Grimminger et al 1995a, Weissmann et al 1998a 1999, 2000, 2003). In these studies with U46619- or angiotensin II-induced vasoconstrictions, the lungs are continuously ventilated with normoxic gas. It is important to note that all of our investigations focused on the maximal strength of HPV during 10 min phases of hypoxic ventilation, because mechanisms of hypoxic vasoconstriction may differ between short-term (min) and long-term (hours) hypoxic ventilation (Weissmann et al 2001a,b).

Results and discussion

The role of the superoxide-hydrogen peroxide axis for the regulation of HPV

There is ongoing discussion concerning whether ROS are involved in the regulation of HPV, and if so, whether an up- or a down-regulation of ROS triggers HPV. In our isolated rabbit lung model, the contribution of ROS to HPV regulation was explored by the addition of various inhibitors that interfere with ROS metabolism, to the perfusate fluid. For interpretation of the data it is worth mentioning what effects might be expected when interfering with ROS:

- If a down-regulation of ROS is the underlying mechanism of HPV, then inhibitors of ROS release/generation should inhibit HPV and should mimic the hypoxic response under normoxic conditions.
- If an up-regulation of ROS is the underlying mechanism of HPV, then inhibitors of ROS release/generation should inhibit HPV without mimicking the hypoxic response under normoxic conditions.

We investigated the lung response to the following agents interfering with ROS:

(1) Nitroblue tetrazolium (NBT), an agent that scavenges superoxide and prevents its dismutation to H_2O_2. This agent thus should decrease superoxide, and subsequently the H_2O_2 concentration in the lung (Weissmann et al 1998a) (Fig. 2).
(2) Tiron, an agent that scavenges superoxide but allows the formation of H_2O_2. This agent should not change lung superoxide and H_2O_2 concentrations, assuming that the endogenous superoxide dismutase is already maximally effective for dismutation of superoxide (Weissmann et al 1998a) (Fig. 2).
(3) Triethylenetetramine (TETA), an agent inhibiting superoxide dismutases. This agent should increase the superoxide and decrease the H_2O_2 concentration (Weissmann et al 2000) (Fig. 2).
(4) Deferoxamine (DFO), an iron chelator that should decrease hydroxy-radical formation via the Fenton reaction (Fig. 2).

Establishing dose–response curves for these agents in the isolated, perfused and ventilated rabbit lung, we demonstrated that all of these agents dose-dependently inhibited HPV without mimicking hypoxia (i.e. that they decreased or blocked HPV without increasing normoxic vascular tone) (example see Fig. 3). Moreover, all of these agents specifically inhibited HPV as they exhibited negligible effects on the strength of the U46619-induced vasoconstrictions. To avoid a possible masking of HPV mimicry under normoxic conditions by endogenously produced vasodilators,

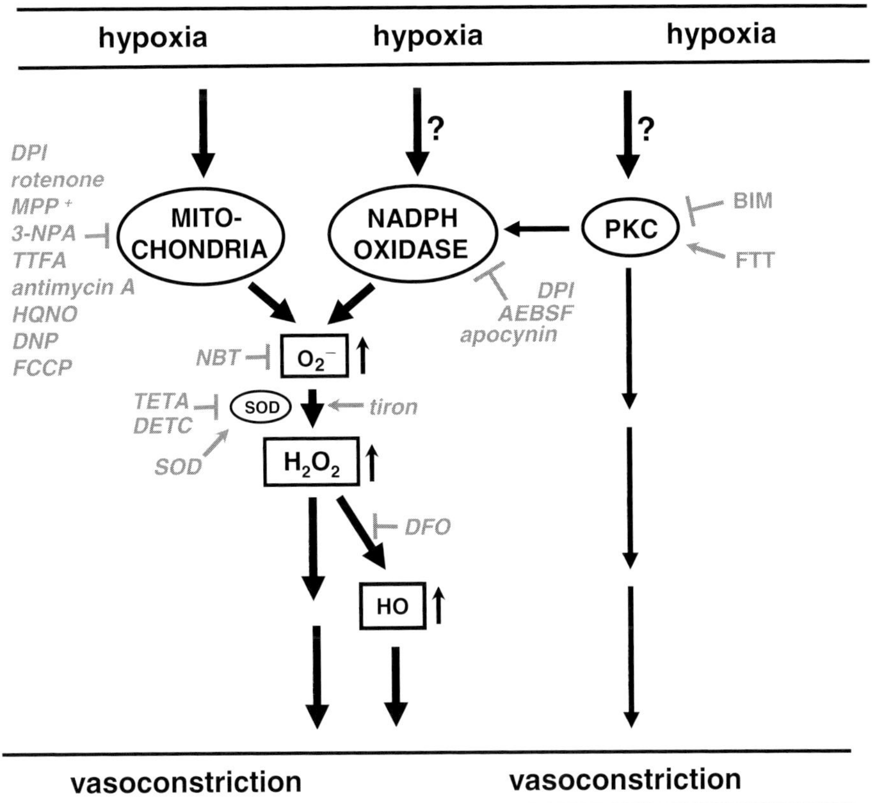

FIG. 2. Suggested pathways for the regulation of hypoxic pulmonary vasoconstriction (HPV) by reactive oxygen species. The pathways shown represent our conclusion from investigations in isolated buffer-perfused and ventilated lungs by use of various inhibitors (⊥) or activators (→) of the indicated pathways. Mitochondria or NADPH oxidases are suggested as oxygen sensors. Protein kinase C (PKC) may also play a role in this process. The model proposes a subcellular increase in lung superoxide generation during alveolar hypoxia with a subsequent increase in H_2O_2 concentration which (maybe via OH radicals) serves as a trigger for HPV. Inhibitors and activators are shown in grey. AEBSF, 4-(2-aminoethyl)-benzenesulfonyl fluoride; BIM, bisindolylmaleimide I; DETC, diethyldithiocarbamic acid; DFO, deferoxamine; DPI, diphenyleneiodonium; FCCP, carbonyl cyanide 4-(trifluoromethoxy) phenylhydrazone; FTT, farnesylthiotriazole; HQNO, 2-heptyl-4-hydroxyquinoline-N-oxide; DNP, 2,4-dinitrophenol NBT; MPP^+, 1-methyl-4-phenylpyridinium iodide; NBT, nitroblue tetrazolium; 3-NPA, 3-nitropropionic acid; SOD, superoxide dismutase; TTFA, 2-thenoyltrifluoroacetone.

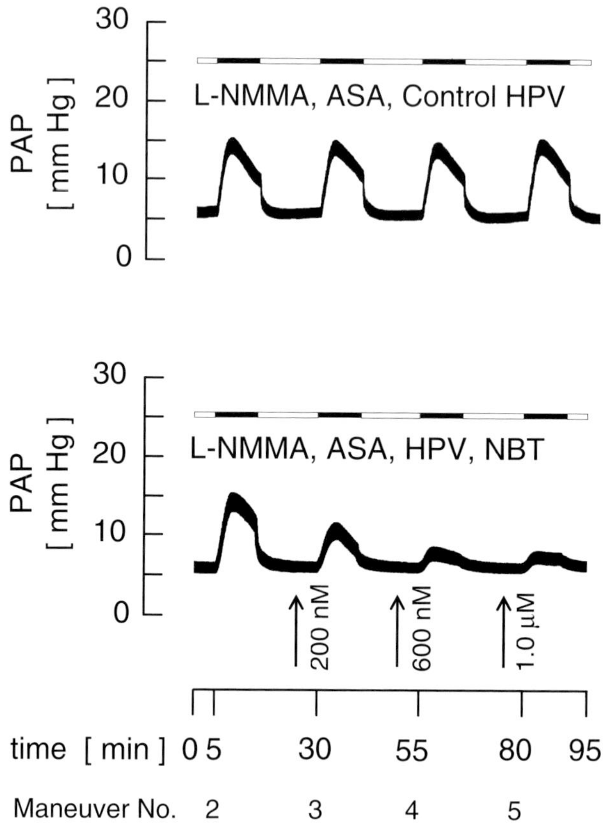

FIG. 3. Original registrations of repetitive hypoxic ventilation manoeuvres (21% O_2, normobar), alternating with normoxic ventilation (3% O_2, normobar). (Reprinted from Weissmann et al 1998a, with permission from the American Physiological Society.) Experiments were performed in the absence (control) and presence of NBT. Arrows indicate the addition of NBT to the perfusate, cumulative concentrations are related to the recirculating perfusion fluid. Both experiments were performed in the presence of 400 μM N^G-monomethyl-L-arginine (L-NMMA) and 1 mM acetylsalicylic acid (ASA), admixed to the buffer at the onset of the recirculating perfusion. (PAP, pulmonary artery pressure). Black bars indicate hypoxic ventilation manoeuvres (3% O_2), white bars normoxic ventilation manoeuvres (21% O_2).

parallel experiments were carried out with prior pharmacological blockage of the lung nitric oxide (NO) and/or prostacyclin systems by application of 400 μM N^G-monomethyl-L-arginine (L-NMMA) and 1 mM acetylsalicylic acid to the perfusate fluid. Even under these conditions, neither NBT, nor TETA, tiron, or DFO mimicked HPV (NO increases reactivity to HPV, but it is not involved in the basic mechanism of HPV). The effects of acetylsalicylic acid and metabolites of arachidonic acid with respect to lipoxy- and cyclooxygenase pathways have previously been demonstrated not to be involved in the regulation of HPV (Grimminger et al 1995b, Weissmann et al 1998b).

Thus we conclude that an increase in superoxide, and subsequently H_2O_2 (followed by generation of hydroxyl radicals), can trigger HPV. If not, then these agents should have mimicked HPV under normoxic conditions. The results that we obtained with TETA are particularly relevant to this conclusion, since TETA should increase superoxide, but decrease H_2O_2. Our data indicate that an increase in superoxide, and the subsequent increase in H_2O_2, can elicit HPV. This concept is supported by the fact that a second SOD-inhibitor, diethyldithiocarbamic acid (DETC) decreased HPV without mimicking hypoxia, and that application of exogenous SOD did not affect HPV or normoxic vascular tone (Weissmann et al 2000).

Possible sources of the hypoxia-induced ROS generation

NADPH oxidases. To investigate potential sources of lung ROS generation, we initially focused on NADPH oxidases. Applying the flavoprotein and thereby NADPH oxidase inhibitor diphenyleneiodonium (DPI, Fig. 2) we demonstrated that DPI inhibited HPV without inhibition of U46619- or angiotensin II-induced vasoconstriction (Grimminger et al 1995c). These experiments were also performed in the presence of the NO inhibitor L-NMMA, as we were able to prove that DPI also inhibits lung NO production. Under these experimental conditions, interference with the lung NO pathway was avoided. The DPI turned out to be a specific inhibitor that did not mimic HPV under normoxic conditions. As DPI is a non-specific inhibitor of NADPH oxidases, we then applied two structurally unrelated NADPH oxidase inhibitors, apocynin and 4-(2-aminoethyl)-benzenesulfonyl fluoride (AEBSF) (Grimminger et al 1995c, Weissmann et al 2000) (Fig. 2). Apocynin turned out to be a non-specific inhibitor of HPV, also blocking the strength of U46619-induced vasoconstriction, with a corresponding dose response profile. Apocynin did not alter normoxic vascular tone. However, because of its non-specific vasodilatory effects, no conclusions can be drawn concerning a prospective role for NADPH oxidases in the regulation of HPV from the apocynin experiments. In contrast, AEBSF was a specific inhibitor of HPV compared to its effect on non-hypoxia induced vasoconstrictions. Furthermore, AEBSF did not mimic the hypoxic response under normoxic conditions. Apart from the inhibitory effect of AEBSF on NADPH oxidases (which we demonstrated in parallel experiments with isolated alveolar macrophages) (Weissmann et al 2000), this compound is known as a protease inhibitor. We were able to exclude a role for proteases in the regulation of HPV in parallel experiments that demonstrated that protease inhibitors like aprotinin and phenylmethylsulfonyl fluoride (PMSF) had no effect on HPV, even when applied at very high concentrations (Weissmann et al 2000).

In summary, our experiments suggest that NADPH oxidases may function as pulmonary oxygen sensors. The fact that the inhibition of NADPH oxidases inhibited HPV but did not mimic the hypoxic response under normoxic conditions further supports the hypothesis that an increase in superoxide, the primary metabolite generated by NADPH oxidases from molecular oxygen, is a trigger for HPV.

NADPH oxidases are multiprotein complexes. The classical phagocytic type of NADPH oxidase consists of two membrane bound subunits, $gp91^{phox}$ and $p22^{phox}$, as well as the cytosolic subunits $p40^{phox}$, $p47^{phox}$ and $p67^{phox}$. Superoxide production by this complex is induced by an assembly of the cytosolic and the membrane-bound subunits. Such an assembly can be induced by phosphorylation of at least one of these subunits, $p47^{phox}$ (Bokoch & Knaus 2003). Moreover, Rac GTPase is also involved in the activation process. Although it has been shown recently that $gp91^{phox}$-deficient mice have a normal response to acute hypoxia (Archer et al 1999), this investigation could not exclude that the 'low output' isoforms of the leukocytic NADPH oxidase are involved in the oxygen sensing process.

NADPH oxidases can be activated by phosphorylation of the p47 subunit, e.g. via a protein kinase C (PKC)-mediated pathway. In line with the concept of increased NADPH oxidase-derived superoxide production as an initial step in the oxygen-sensing process underlying HPV, we were able to demonstrate (1) that the PKC inhibitor bisindolylmaleimide I (BIM I) selectively inhibited HPV compared to non-hypoxia-induced vasoconstrictions and (2) that this agent did not alter normoxic vascular tone, even when NO and PGI_2 synthesis were blocked (Weissmann et al 1999). These data are supported by the fact that the PKC activator farnesylthiotriazole (FTT) induced a vasoconstriction under normoxic conditions, concomitant with an inhibition of HPV (Weissmann et al 1999). However, when considering the role of PKC, one should bear in mind that this enzyme family has multiple, NADPH oxidase-independent cellular effects.

Mitochondria. Although the application of three structurally unrelated NADPH oxidase inhibitors in our study support a role for these enzyme systems in hypoxia-regulated ROS release, their lack of specificity must be considered. For example, DPI can also inhibit mitochondrial complex I, although such effects have not been described for AEBSF, with the exception of its protease-inhibitory capacity. This effect of DPI is consistent with a possible role for mitochondria as the oxygen sensors underlying HPV. Mitochondria have long been thought to be involved in redox changes underlying the regulation of HPV (Archer et al 1993, Rounds & McMurtry 1981). However, in recent years a new concept emerged strongly suggesting that a mitochondria-derived increase, rather than a decrease, in superoxide release is the initial trigger for HPV. In several elegant studies evidence was provided that an increased electron leak at complex III of the mitochondrial electron transport chain, upstream of the binding site for antimycin A, occurs under hypoxic conditions (Waypa et al 2001, 2002). The concept that increased ROS may be a trigger for HPV is well in line with our findings. However, our investigations, employing different agents interfering with mitochondrial electron transport, do not fully support the hypothesis that an increased leak of electrons at complex III of the mitochondrial electron transport chain is the source of this increase in lung ROS formation. We employed four inhibitors that blocked mitochondrial electron transport upstream of complex III [rotenone, 1-methyl-4-phenylpyridinium iodide (MPP^+), myxothiazole, 2-thenoyltrifluoroacetone (TTFA), and 3-nitropropionic acid (3-NPA)]. These agents are expected to decrease the electron leak at complex III. We additionally used two complex III inhibitors [antimycin A and 2-heptyl-4-hydroxyquinoline-N-oxide (HQNO)], which are expected to increase the electron leak at complex III. Moreover, we applied two uncouplers [(2,4-dinitrophenol (DNP) and carbonyl cyanide 4-(trifluoromethoxy) phenylhydrazone (FCCP)] which are thought to increase electron flow through the mitochondrial respiratory chain and thus decrease electron leakage at complex III (Weissmann et al 2003). If an increased electron leak at complex III under hypoxic conditions is the underlying mechanism of HPV, the upstream inhibitors should inhibit HPV without being hypoxia mimics, whereas the downstream inhibitors should mimic HPV under normoxic ventilation and also block HPV. These theoretical requirements are fully supported by the effects of rotenone and HQNO in our investigation, but are also partly contradicted by the effects of MPP^+, antimycin A, DNP and FCCP. Under conditions of pre-blocked lung NO synthesis, the complex I inhibitor rotenone specifically inhibited HPV without being a hypoxia mimic, while the complex III inhibitor antimycin A increased normoxic vascular tone and subsequently blocked HPV. In contrast, the complex I inhibitor MPP^+ increased normoxic vascular tone and inhibited HPV, whereas the complex III inhibitor antimycin A did not alter normoxic vascular tone, but did inhibit HPV (Weissmann et al 2003). The uncouplers DNP and FCCP mimicked the hypoxic response and inhibited HPV. Aside from the uncertainties of the effect of the mitochondrial inhibitors and uncouplers on mitochondrial ROS release, these experiments, even if they do not fully support the hypothesis of increased superoxide release from mitochondria at complex III, at least demonstrate that mitochondria are specifically involved in the regulation of HPV.

Measurement of ROS release

In an approach to access lung ROS release from the isolated perfused rabbit lung we established a technique for quantification of exhaled H_2O_2 in the breath condensate of the isolated perfused lungs (Weissmann et al 2004). In these experiments we found a tendency towards a decreased H_2O_2 release during hypoxic ventilation when compared to normoxia. These data correlate well with previous investigations that quantified intravascular release of lung ROS (Archer et al 1993). In contrast, when rabbit isolated pulmonary artery smooth muscle cells where exposed for one hour to normobaric hypoxia (1% O_2), we detected an increased fluorescence signal in comparison to cells exposed to normoxic conditions when applying the 2′,7′-dichlorofluorescein method (Paddenberg et al 2003b). These experiments indicate increased ROS production in pulmonary artery smooth muscle cells maintained under hypoxic conditions.

Concluding remarks

In summary, our data from experiments in isolated perfused rabbit lungs favour the conclusion that during short-term alveolar hypoxia (10 min) an increase in superoxide followed by hydrogen peroxide occurs, and triggers HPV. As stated in the introduction there is currently a contradiction between investigations favouring the concept of increased ROS release in hypoxia as a trigger for HPV and decreased ROS release in hypoxia as a trigger for HPV. What may be the cause of this apparent paradox? (1) Using an exclusively pharmacological approach (e.g. inhibitors), one has to consider the non-specific activity in terms of unknown or known side-effects of such agents. However, repeated investigations with a variety of inhibitors directly interfering with ROS metabolism in our experimental approach reduce the probability of a misleading interpretation of the inhibitor effects. (2) For the interpretation of our data and analogous experiments in general with pharmacological interventions in intact lung systems, their effect on normoxic vascular tone is most important. This aspect thus must be evaluated in detail. In principle, the possibility remains that the effect of an intervention on normoxic vascular tone may be masked by vasodilators present in the experimental setting. At least in our investigations, by blockade of the major known vasodilatory pathways (NO and prostacyclin), we can also exclude such effects with high probability. (3) Concerning direct measurement of ROS generation, there is ongoing discussion as to the reliability of the techniques employed. When we compare whole lung H_2O_2 release in the breath condensate in our experiments and intravascular ROS release measured by others with the ROS generation in isolated smooth muscle cells from pulmonary arteries, it turned out that a decrease occurred in the intact organ, in contrast to an increase observed at the cellular level. Thus, we speculate that the opposing views concerning the question of whether ROS are up- or down-regulated under hypoxic conditions are a question of cellular and subcellular distribution of ROS generation. To trigger HPV, only a subcellular increase in ROS in or in the vicinity of the oxygen sensing system may be sufficient, whereas lung overall ROS release is decreased during hypoxia.

Acknowledgements

This work was supported by the Deutsche Forschungsgemeinschaft, SFB547, project B7. The authors thank Dr Rory Morty for linguistic editing of the manuscript.

References

Archer SL, Huang J, Henry T, Peterson D, Weir EK 1993 A redox-based O2 sensor in rat pulmonary vasculature. Circ Res 73:1100–1112

Archer SL, Hampl V, Nelson DP, Sidney E, Peterson DA, Weir EK 1995 Dithionite increases radical formation and decreases vasoconstriction in the lung. Evidence that dithionite does not mimic alveolar hypoxia. Circ Res 77:174–181

Archer SL, Reeve HL, Michelakis E et al 1999 O2 sensing is preserved in mice lacking the gp91 phox subunit of NADPH oxidase. Proc Natl Acad Sci USA 96:7944–7949

Bokoch GM, Knaus UG 2003 NADPH oxidases: not just for leukocytes anymore! Trends Biochem Sci 28:502–508

Fishman AP 1976 Hypoxia on the pulmonary circulation. How and where it acts. Circ Res 38:221–231

Grimminger F, Weissmann N, Spriestersbach R, Becker E, Rosseau S, Seeger W 1995a Effects of NADPH oxidase inhibitors on hypoxic vasoconstriction in buffer-perfused rabbit lungs. Am J Physiol 268:L747–L752

Grimminger F, Spriestersbach R, Weissmann N, Walmrath D, Seeger W 1995b Nitric oxide generation and hypoxic vasoconstriction in buffer-perfused rabbit lungs. J Appl Physiol 78:1509–1515

Grimminger F, Weissmann N, Spriestersbach R, Becker E, Rosseau S, Seeger W 1995c Effects of NADPH oxidase inhibitors on hypoxic vasoconstriction in buffer-perfused rabbit lungs. Am J Physiol 268:L747–L752

Hillier SC, Graham JA, Hanger CC, Godbey PS, Glenny RW, Wagner WW Jr 1997 Hypoxic vasoconstriction in pulmonary arterioles and venules. J Appl Physiol 82:1084–1090

Jensen KS, Micco AJ, Czartolomna J, Latham L, Voelkel NF 1992 Rapid onset of hypoxic vasoconstriction in isolated lungs. J Appl Physiol 72:2018–2023

Leach RM, Hill HM, Snetkov VA, Robertson TP, Ward JP 2001 Divergent roles of glycolysis and the mitochondrial electron transport chain in hypoxic pulmonary vasoconstriction of the rat: identity of the hypoxic sensor. J Physiol 536:211–224

Liu JQ, Sham JS, Shimoda LA, Kuppusamy P, Sylvester JT 2003 Hypoxic constriction and reactive oxygen species in porcine distal pulmonary arteries. Am J Physiol Lung Cell Mol Physiol 285:L322–L333

Marshall C, Mamary AJ, Verhoeven AJ, Marshall BE 1996 Pulmonary artery NADPH-oxidase is activated in hypoxic pulmonary vasoconstriction. Am J Respir Cell Mol Biol 15: 633–644

Michelakis ED, Hampl V, Nsair A et al 2002 Diversity in mitochondrial function explains differences in vascular oxygen sensing. Circ Res 90:1307–1315

Mohazzab KM, Fayngersh RP, Kaminski PM, Wolin MS 1995 Potential role of NADH oxidoreductase-derived reactive O2 species in calf pulmonary arterial PO2-elicited responses. Am J Physiol 269:L637–L644

Monaco JA, Burke-Wolin T 1995 NO and H2O2 mechanisms of guanylate cyclase activation in oxygen-dependent responses of rat pulmonary circulation. Am J Physiol 268: L546–L550

Paddenberg R, Goldenberg A, Faulhammer P, Braun-Dullaeus RC, Kummer W 2003a Mitochondrial complex II is essential for hypoxia-induced ROS generation and vasoconstriction in the pulmonary vasculature. Adv Exp Med Biol 536:163–169

Paddenberg R, Ishaq B, Goldenberg A et al 2003b Essential role of complex II of the respiratory chain in hypoxia-induced ROS generation in the pulmonary vasculature. Am J Physiol Lung Cell Mol Physiol 284:L710–L719

Rounds S, McMurtry IF 1981 Inhibitors of oxidative ATP production cause transient vasoconstriction and block subsequent pressor responses in rat lungs. Circ Res 48:393–400

Staub NC 1985 Site of hypoxic pulmonary vasoconstriction. Chest 88:240S–245S

Thomas HM, III, Carson RC, Fried ED, Novitch RS 1991 Inhibition of hypoxic pulmonary vasoconstriction by diphenyleneiodonium. Biochem Pharmacol 42:R9–12

Von Euler USLG 1946 Observations on the pulmonary arterial blood pressure in the cat. Acta Physiol Scandinav 12:301–320

Waypa GB, Chandel NS, Schumacker PT 2001 Model for hypoxic pulmonary vasoconstriction involving mitochondrial oxygen sensing. Circ Res 88:1259–1266

Waypa GB, Marks JD, Mack MM, Boriboun C, Mungai PT, Schumacker PT 2002 Mitochondrial reactive oxygen species trigger calcium increases during hypoxia in pulmonary arterial myocytes. Circ Res 91:719–726

Weissmann N, Grimminger F, Walmrath D, Seeger W 1995 Hypoxic vasoconstriction in buffer-perfused rabbit lungs. Respir Physiol 100:159–169

Weissmann N, Grimminger F, Voswinckel R, Conzen J, Seeger W 1998a Nitro blue tetrazolium inhibits but does not mimic hypoxic vasoconstriction in isolated rabbit lungs. Am J Physiol 274:L721–L727

Weissmann N, Seeger W, Conzen J, Kiss L, Grimminger F 1998b Effects of arachidonic acid metabolism on hypoxic vasoconstriction in rabbit lungs. Eur J Pharmacol 356:231–237

Weissmann N, Voswinckel R, Hardebusch T et al 1999 Evidence for a role of protein kinase C in hypoxic pulmonary vasoconstriction. Am J Physiol 276:L90–L95

Weissmann N, Tadic A, Hanze J et al 2000 Hypoxic vasoconstriction in intact lungs: a role for NADPH oxidase-derived H(2)O(2)? Am J Physiol Lung Cell Mol Physiol 279:L683–L690

Weissmann N, Winterhalder S, Nollen M et al 2001a NO and reactive oxygen species are involved in biphasic hypoxic vasoconstriction of isolated rabbit lungs. Am J Physiol Lung Cell Mol Physiol 280:L638–L645

Weissmann N, Grimminger F, Olschewski A, Seeger W 2001b Hypoxic pulmonary vasoconstriction: a multifactorial response? Am J Physiol Lung Cell Mol Physiol 281:L314–L317

Weissmann N, Ebert N, Ahrens M et al 2003 Effects of mitochondrial inhibitors and uncouplers on hypoxic vasoconstriction in rabbit lungs. Am J Respir Cell Mol Biol 29:721–732

Weissmann N, Vogels H, Schermuly RT et al 2004 Measurement of exhaled hydrogen peroxide from rabbit lungs. Biol Chem 385:259–264

Wilson HL, Dipp M, Thomas JM, Lad C, Galione A, Evans AM 2001 Adp-ribosyl cyclase and cyclic ADP-ribose hydrolase act as a redox sensor. a primary role for cyclic ADP-ribose in hypoxic pulmonary vasoconstriction. J Biol Chem 276:11180–11188

DISCUSSION

Ward: I was pleased to see that none of the inhibitors that you used mimic hypoxia in hypoxic pulmonary vasoconstriction (HPV), whereas they ablate it in most cases. We see exactly the same thing, which is different from what Steven Archer and Ken Weir have reported.

Schumacker: With the exception of antimycin A and cyanide, the inhibitors all fail to mimic the hypoxic response in our hands. They just block subsequent responses to hypoxia.

Ward: The only thing we have seen that might possibly give constriction in the isolated pulmonary artery is a low concentration of cyanide (10 μM). Surprisingly however, this did not affect Ca^{2+}. If we used 1 mM cyanide, which is the only level sufficient to cause complete reduction of the NADH to the same level as zero oxygen, then HPV is completely ablated. This sort of fits in with what Paul Schumacker was saying, because if electron transport is completely blocked through complex IV there is nothing that can happen.

Weissmann: There are some differences in our work compared to Paul's. In our hands, rotenone did not induce vasoconstriction, whereas in your experiments it (transiently) did. Antimycin in your experiments induced vasoconstriction and inhibited HPV, whereas in our experiments it only inhibited HPV. In contrast HQNO, a second complex III inhibitor, in our experiments mimicked hypoxia as it induced vasoconstriction in normoxia and inhibited HPV at the same time. The same effect we observed for cyanide. Perhaps species differences play a role here.

Archer: One thing I find hard to explain (unless the mitochondrial theory of O_2 sensing is true) is why rotenone and antimycin dilate the ductus, and systemic arteries whilst causing pulmonary vasoconstriction-a perfect parallel with hypoxia. Moreover, they ablate the subsequent hypoxic response—surely this is more than a coincidence or a function of 'non-specificity' of the inhibitors. If we are trying to find a universal (or at least a widely conserved) sensor, it should work in many of the specialized O_2-sensing tissues (e.g. the ductus, pulmonary artery and possibly the carotid body, not just the pulmonary artery). From my biased perspective, it is interesting that in many of your experiments you found radicals going down but chose to ignore it, but then when they went up you chose to accept it!

Weissmann: What I said is that I don't think it makes much sense to measure overall H_2O_2 or superoxide release detectable in the intravascular space or in the alveolar space. This represents total O_2 radical release. There is no doubt that the total O_2 release is going down during hypoxia and the cells are shifted to a more reduced state. But this does not exclude that HPV is triggered by a localized and compartimentalized increase in reactive oxygen species (ROS) in a distinct compartment of the cell. There may be NADPH oxidases or mitochondria as possible O_2 sensors located in direct neighbourhood and association to Ca^{2+} or K^+ channels. An increase in superoxide and H_2O_2 can then trigger HPV by directly and only locally affecting these channels. But if you measure total cell ROS release it is going down.

Archer: I hope that whichever theory is proven correct (by history) it would not be a proof based on a 'secret compartment'. Like closed hearings, secret compartments are almost always a cover for theories that are wrong. You may need a more precise technique for measurement. If it is true and relevant it should ultimately be measurable.

Ward: It doesn't have to be hidden. Subcellular compartmentalization is known to be a very important issue for cell function. Look at where NADPH oxidase and mitochondria are in some cells. The subcellular organelles or processes responsible for signalling might be very close to the sarcolemma. You could have localized signalling. The signal you pick up with indicators within the cytosol might be quite different from what is happening locally. We know that this happens to Ca^{2+}, with complex interactions between the peripheral sarcoplasmic reticulum, mitochondria and sarcolemma that may be independent of what is happening in the deeper cytosol. We can also see a rise in Ca^{2+} that is not coupled to tension.

Archer: That flies in the face of seminal findings regarding the properties of HPV *in vivo* and *ex vivo* by many labs. Both our group and McMurtry's groups in Colorado find that 80% of acute HPV is obliterated by specific inhibitors of L-type Calcium channels (e.g. verapamil, nisoldipine, etc.). This is also true in humans. Yes, there are compartments within the cells, and this is undisputed, but we can still measure those compartments.

Ward: I'm not sure that we can: this is the problem. The time and spatial resolution of current techniques do not allow us to visualize sub-sarcolemmal microdomains in any meaningful way.

Duchen: In all these preparations, compartments are populations of cells. Your data with PKC inhibitors would suggest that the hypoxic activation of the oxidase is going through some hypoxic sensitivity of PKC. Is that right?

Weissmann: This is one explanation. This would mean that the NADPH oxidase isn't the oxygen sensor, but instead it is PKC. On the other hand, perhaps the assembly of the NADPH oxidase triggered by PKC is a prerequisite for this complex to function as an oxygen sensor. First there has to be an assembly, and then the regulation may be in terms of changes of conformation of the NADPH oxidase.

Chandel: You could imagine that mitochondria that could release limited amounts of oxidants that could activate PKCs and Rac which would then use the NADPH oxidase to amplify and sustain that signal. I am attracted to this because it unifies the two.

Weissmann: Perhaps mitochondria are required for a functioning NADPH oxidase system.

Chandel: We need to look at Rac translocation or PKC activation in the presence of rotenone or other mitochondrial inhibitors to see whether this is ablated. There is precedent for this connection between mitochondrial ROS funnelling the NADPH oxidase, not in hypoxia but with angiotensin II.

Weir: I have a comment on the NADPH oxidase participation. As Steven mentioned this morning we used the GP91 knockout and didn't see any change in HPV. We have subsequently used the p47 knockout; again, HPV was not reduced. Way back in the 1970s Babior showed that at least the neutrophil GP91 NADPH oxidase reduced production of superoxide during hypoxia (Gabig et al 1979). It behaved as if ROS goes down with hypoxia.

Archer: You showed that at 2.5 and 1% oxygen it was inhibited.

Weissmann: NADPH oxidases are still candidates for the oxygen sensor. As you said, there are many different subunits: the gp91, Nox1, Nox4, p22, NoxO1, NoxA1, p47 and p67. It is known that many of these isoforms can substitute the others. They still could be oxygen sensors and some specific NADPH oxidases might be oxygen sensors. The arguments we have for this are only the inhibitor studies, though.

Chandel: With these knockouts you should be able to show that it either has a functional oxidase or not from the biochemical properties.

Weissmann: You are right. We have now changed our model to the isolated mouse lung model. We are generating Nox1 knockouts, and other knockouts concerning the NADPH oxidase, to see whether these contribute to the regulation of HPV.

Chandel: The interpretation is that there are all these isoforms out there. All you have to do is look and see whether they still have oxidase activity.

Aaronson: I haven't really used the isolated lung, but a lot of people who do seem to use angiotensin to gee up the response. Do you?

Weissmann: No, there was no priming by vasoconstrictors.

Aaronson: Angiotensin does itself produce ROS. Also, flow across the endothelium will cause it to produce ROS. There is a basal level of ROS production so it is possible that this contributes to the response, or setting up the response. It is possible antioxidants could act to reduce this basal level of ROS production.

Weissmann: On the other hand, we also have a vasoconstrictor response if we do the preparation in a non-recirculating manner. The strength of HPV is quite the same as when the buffer is going back to the lung. For the detection of specific mechanisms of HPV the system should be as simple as possible.

Schumacker: Regarding the measurement of either exhaled or circulating levels of H_2O_2, I congratulate you for trying, but I am not surprised that you didn't find anything. The levels in the cell are signalling levels and you are looking for signals that would be significantly diluted from those levels. Moreover, Claire Doerschuk has studied the isolation of neutrophils in the pulmonary circulation. Her studies have shown that neutrophils accumulate in pulmonary microvessels and spend a long time transiting through the lung. When a lung is excised and stimulated with things like phorbol, you may be stimulating the NADPH oxidase of these marginated neutrophils, which then release superoxide into the extracellular space. This is going to generate noise and confuse the signal. I congratulate your measurements on intracellular ROS production because I think this is probably what is most important for signalling.

Weissmann: You are absolutely right. We have quantified this for the isolated rabbit lung with a blood-free perfusion. There is an amount of neutrophils sticking inside the vessels that surpasses the amount of circulating neutrophil in rabbits by one to three times even in well-flushed isolated lungs. If you look at our data from PMA stimulation, we certainly also stimulate the intravascular neutrophils. It is interesting to me that we see an increase at 5% of O_2. This is a hint that an NADPH oxidase can produce an increase in the amount of superoxide in hypoxia. Further experiments are needed to clarify this.

Schumacker: You showed PKC acting on the NADPH oxidase. As you know, there are at least a dozen isoforms of PKC. Is it possible that the isoforms being activated are acting on multiple targets, including possibly downstream from the mitochondrial target, other than NADPH oxidase. Could the ryanodine receptor be one of those alternate targets, for example?

Weissmann: The PKC data I presented were from quite old experiments. As we have no good inhibitors for NADPH oxidase we looked at how we could affect the system. In accompanying cell culture experiments we showed that neutrophils from the rabbit pulmonary circulation were activated by PKC activators. I fully agree that PKC has a wide range of effects in terms of phosphorylation and perhaps Ca^{2+} sensitization.

Chandel: p47 does have a phosphorylation site for PKC. It has been shown that this is important for its regulation.

López-Barneo: For all these drugs that inhibit HPV, it would be nice to have an idea of how clean they are. Do they block the L-type Ca^{2+} channels, for example?

Weir: We were saying earlier that rotenone and antimycin would block K^+ channels. Do they also block L-type Ca^{2+} channels? For instance, we know that diphenylene iodonium, an inhibitor of NADPH oxidase, can block both potassium and L-type Ca^{2+} channels (Weir et al 1994).

Kemp: I have no idea. It would be nice to know.

Ward: We have used 100 nM rotenone. It has a very small effect on 80 mM K^+ depolarization and the associated rise in Ca^{2+}, suggesting that it does not affect L-type channels much. We haven't looked at antimycin.

López-Barneo: Antimycin and rotenone don't do much to Ca^{2+} channels, at least in carotid body. But what about the other drugs, such as inhibitors of PKC. We need to check what happens with the L-type Ca^{2+} channels.

Peers: This should be done with caution. Neuronal Ca^{2+} channels can be modulated by phorbol esters via PKC-dependent and -independent mechanisms.

Archer: PKC activators and endothelin also block K_v channels in the pulmonary artery smooth muscle cells (PASMCs).

Weissmann: I am not aware that these mitochondrial inhibitors which interfere with ROS directly interfere with the Ca^{2+} channels. Moreover, I never would have relied on one, two or three inhibitors in our model to draw the conclusion that ROS are going up. But the redundancy in the application of a large variety of inhibitors that all affect ROS and gave consistent results from my point of view are convincing

López-Barneo: From my experience ion channels are very promiscuous.

Weissmann: But you wouldn't expect all these inhibitors to have an effect on Ca^{2+} current?

Buckler: Is it true that FCCP abolishes HPV?

Archer: I am not sure. Jason Yuan does have a paper showing that FCCP mimics the electrophysiological effects of hypoxia on PASMCs (Yuan et al 1996).

Ward: It abolishes it in our hands if we give sufficient concentrations. I wouldn't say it 'mimics' HPV because it dumps Ca^{2+} and gets rid of the mitochondrial Ca^{2+}. It will also remove any mitochondrial involvement in Ca^{2+} homeostasis, so could potentially increase cytosolic Ca^{2+} by removing a buffer.

Buckler: But it abolishes HPV.

Ward: There is a constriction.

Buckler: Do you get an additional constriction when you introduce hypoxia in the presence of FCCP?

Ward: No, but you may have already done some serious damage. High concentrations of FCCP cause complete mitochondrial depolarization, of course.

Weissmann: In our hands it also produces vasoconstriction. It inhibits HPV but also produces severe oedema. It is not a good tool to use in the isolated lung. After 30 min the lung has severe oedema.

Buckler: How long does it take to get an HPV?

Weissmann: HPV starts within seconds, and it reaches a maximum within 4 to 5 minutes.

Buckler: So you could do a short experiment.

Weissmann: But the oedema starts to develop early also, which makes the experiment difficult to interpret.

Buckler: What about oligomycin?

Ward: It blocks HPV in our hands.

References

Gabig T, Bearman S, Babior B 1979 Effects of oxygen tension and pH on the respiratory burst of human neutrophils. Blood 53:1133–1139

Weir EK, Wyatt CN, Reeve HL, Huang J, Archer SL, Peers C 1994 Diphenylene iodonium blocks both potassium and calcium currents in isolated pulmonary artery smooth muscle cells. J Appl Physiol 76:2611–2615

Yuan XJ, Sugiyama T, Goldman WF et al 1996 A mitochondrial uncoupler increases K_{Ca} currents but decreases K_v currents in pulmonary artery myocytes. Am J Physiol 270:C321–331

General discussion III

Kemp: I have a comment and a question for the hypoxic pulmonary vasoconstriction (HPV) crowd. It seems to me that there are many ways to look at HPV. You can take a reductionist approach or look at the whole tissue, for example. Each is good for answering specific questions. I am thinking of the business of being able to detect rises in ROS, which is mostly done in isolated cells. There is clearly a technical limitation here: we should be able to put Norbert Weissmann's system together with Paul Schumacker's system, and measure reactive oxygen species (ROS) online in an isolated perfused lung, or in an animal. Working in the alveolar field with a different hat on, I remember a brace of papers by Bhattacharya, where he was able to load up pulmonary arterial smooth muscle cells (SMCs), endothelial cells and alveolar type 2 cells (Wang et al 2001) in an isolated perfused lung and visualize changes of these fluorochromes. You might be able to answer these questions quite simply by using the separate techniques together. It is technically challenging, but doable.

Archer: That's what we do with the chemiluminescence technique where we have simultaneous pressure and ROS measurement. There is a reversible signal that goes up and down in seconds of a change in inspired O_2 concentration in proportion to pO2 (Archer et al 1989). This is a technique that may be controversial in this room, but is not controversial among people studying ischaemia reperfusion in heart, for example.

Schumacker: In this system are you putting a photomultiplier tube on the surface of the lung and measuring the effluence from the lung?

Archer: We measure from the surface of the lung. The other challenge for all of us is not to be complacent, and in the same lab study pulmonary artery, cultured cells and freshly isolated cells. It is important to look at all levels. It is intellectually lazy to say that our lab is an isolated perfused lung lab and not a cell lab, for instance.

Gurney: There are differences between the lung and the heart. In heart you can put a photomultiplier right on the myocytes. Vascular myocytes lie below the surface of the lung and they are only one of several cell types there.

Archer: It depends on how much effort one wants to make. If you look at Wiltz Wagner's work, the vessels near the surface of the lung are in fact small resistance arteries (Wagner 1993, Hanger et al 1995). So when we do windows on the lung we are looking at vessels that are relevant to HPV. As you get nearer the hilum you are getting to large arteries that are not so relevant for HPV.

Duchen: It would be nice to go to the next step and look with a multiphoton imaging system.

Kemp: I don't think you even need to be using multiphoton. You can do it with standard microscopy. You are only going to get 300–400 microns in the confocal.

Gurney: You don't get anything like that with confocal. To get that deep you need to use two photon. Even with multiphoton it is quite hard to image the vessels under the surface.

Duchen: What probe do you use, then? You can't use luminescent probes. You would have difficulty transfecting your hinge, and we don't trust any of the other dyes.

Kemp: My remembering of Bhattachayra's work is that he was looking at alveolar type 2 cells in a lung.

Kummer: He is looking at the alveolus on the surface of the lung, and this lung is not ventilated. If the lung is ventilated there is so much movement that it isn't possible to see anything. It might not be the best system for this purpose.

Duchen: If I was an undergraduate student reading this book, I would be deeply confused. For the sake of the book, is there any way we can resolve the differences that people see with these different preparations? How is it possible that we can take similar approaches and similar drugs, and see exactly opposite responses? How can one explain these differences that seem so fundamental?

Chandel: The problem has been that most of the work has been done with drugs.

Duchen: We are using the same drugs.

Gonzalez: The problem is even more fundamental. In Chicago it has been shown that HPV is due to an increase in ROS and in Edmonton it has been shown that HPV is produced by a decrease in ROS. That is, American and Canadian hypoxias do the opposite regarding ROS production

Archer: There are two ways to look at life: glass half empty and glass half full. The reality is that there is a lot of agreement. We could turn the light on the carotid body field and if you were critical you would find just as much discord. Most labs agree that there are oxygen-sensitive K^+ channels in the pulmonary circulation that are involved in HPV. Many people who don't agree with this have never looked. Of the labs that have looked there are very few negative reports saying there are no oxygen-sensitive channels. Many of these labs carry their theories at least as far as accepting that there is hypoxia-induced pulmonary artery smooth muscle cell (PASMC) depolarization. There are quite a few labs who will say there is a redox sensor: it is either an oxidase or it is the mitochondria. We probably don't have a consensus, but it is not chaos. The dissonance is because the relevant variables (ROS, IK, Em, mitochondrial function, etc.) are hard to measure and perhaps because there is an excessive devotion to reductionist models and the use of culture cells without adequate reference to the properties of HPV in humans and in freshly isolated resistance pulmonary arteries.

Gonzalez: Perhaps it is because we are looking in the wrong place. We are pushing too far with ROS. Perhaps ROS are driving us off the track for the real thing.

Archer: That is possible. ROS may reflect mitochondrial function and redox coupling, for example.

Chandel: My comment about the drugs is that we have learned how rotenone can partition depending on how it is added, for example. Students or postdocs read the papers and follow the methods, but if they don't know about the problem with these drugs it could lead to variable results.

Kumar: When they are used in the carotid body every lab reports a stimulation. There isn't a lab in the world that has failed to get stimulation with a blocker! In the pulmonary artery there appears to be a variation between labs. There doesn't seem to be any obvious commonality between the carotid body world and the pulmonary one, except perhaps for oxygen-sensitive K^+ channels and elevated intracellular Ca^{2+} which seems to be a consistent light in this darkness. I agree with Professor Gonzalez: perhaps ROS is a parallel system and we are missing something.

Sylvester: I have been teasing Jeremy that compartmentation is the last refuge of scoundrels, and so is redundancy. Doubtlessly, responses to hypoxia are important for survival in many different ways. Perhaps God thought it would be stupid to make these important responses dependent on a single sensing-transduction system! Perhaps there are multiple systems that are expressed differently in various cell types or in the same cell under different conditions. Such redundancy doesn't help our confusion, but it may exist.

Chandel: HIF-1α is not redundant.

Sylvester: Not yet.

Aaronson: It seems like one important thing in the pulmonary circulation is the endothelium. In cells we don't have the endothelium. In some cases there is flow in experimental set-ups. We don't: we use isometric recording. We find there is a lot of influence of NO, which will also interact with superoxide. This could be a reason for the variability between the different preparations.

Archer: If we tried to reproduce each other's work, this would be useful. When you say that there isn't a consistent response to mitochondrial inhibitors, I don't see many papers where people even give them. I am equally guilty of not studying, say, endothelial function. One thing I observe with the carotid field is that everyone gives the same stimuli over and over again. I mean this in a flattering way: this type of repetition does help the field advance. In the pulmonary circulation we have many groups, working with different models, who for whatever reason don't directly test competing theories.

Duchen: It is always difficult to persuade journals to publish something that someone else has done already, which is probably one of the reasons we don't do this.

Kumar: We could start with things like working at the right temperature and with bicarbonate buffer, and with time courses appropriate for physiological responses.

Ward: Using the same degree of hypoxia would be good. Some studies use rather severe hypoxia that does not mimic the physiological situation.

Kumar: The thing about oxygen is that two levels don't give a hypoxic response curve. We quite often only see normoxia versus a single level of hypoxia.

Duchen: One of the things I've missed in many studies is a clear oxygen response curve. We often see response to hypoxia in the presence or absence of a drug, but levels of oxygen defined as hypoxia vary.

Rich: It may be useful if you got together and decided the window of possible concentrations that is reasonable for each of these inhibitors that you use. For example, someone talked about using 1 μM cyanide, but this does not inhibit cytochrome oxidase in a turning-over mitochondrion. If you agree to the window of conditions it would make comparisons of results from different labs much easier. Furthermore, it is not just the concentration that matters but also the delivery time, the difficulty of getting these chemicals back out, and even whether ethanol is present in stock solutions. Using plastic tubing is another variable: antimycin A sticks to plastic, for example, as does rotenone. You should perhaps agree on a set of common protocols.

Kumar: Some people work with superfused preparations, some with single cells and others with perfused organs. There are always access problems. The 'same' concentration will also have to be specified with regard to the preparation.

Nurse: There is also the issue of reversibility and reaching a steady state. It is unsatisfying if we can't reverse the response because we don't know whether we are looking at a real phenomenon.

Duchen: One drug that will never reverse is oligomycin.

López-Barneo: Antimycin A is almost irreversible.

References

Archer SL, Nelson DP, Weir EK 1989 Simultaneous measurement of oxygen radicals and pulmonary vascular reactivity in the isolated rat lung. J Appl Physiol 67:1903–1911

Hanger CC, Hillier SC, Presson RG Jr et al 1995 Measuring pulmonary microvessel diameters using video image analysis. J Appl Physiol 79:526–532

Wagner WW Jr 1993 Invited editorial on 'demonstration of pulmonary vascular perfusion by electron and light microscopy'. J Appl Physiol 75:1437–1438

Wang PM, Ashino Y, Ichimura H, Bhattacharya J 2001 Rapid alveolar liquid removal by a novel convective mechanism. Am J Physiol Lung Cell Mol Physiol 281:L1327–L1334

The role of twin pore domain and other K^+ channels in hypoxic pulmonary vasoconstriction

Alison M. Gurney[1] and Shreena Joshi

Department of Physiology and Pharmacology, Strathclyde Institute for Biomedical Sciences, University of Strathclyde, 27 Taylor Street, Glasgow G4 0NR, UK

Abstract. Hypoxic pulmonary vasoconstriction (HPV) describes the vasoconstrictor response of pulmonary arteries to hypoxia, which directs blood flow towards better ventilated areas of the lung. Exactly how pulmonary arteries sense oxygen and mediate this response is widely debated and several hypotheses have emerged. One has smooth muscle K^+ channels as the primary O_2 sensor, hypoxia causing K^+ channel inhibition, membrane depolarization and voltage-dependent Ca^{2+} influx. Even if this mechanism is not the primary response of pulmonary arteries to hypoxia, inhibition of K^+ channel activity probably plays a role in HPV, due to enhanced membrane excitability and Ca^{2+} influx. Hypoxia inhibits several different K^+ channels expressed in pulmonary artery smooth muscle, most from the K_v class of voltage-gated K^+ channels, but the properties of many K_v channels are incompatible with a role in initiating HPV. Twin-pore domain K^+ channels have emerged as prime candidates for controlling the resting membrane potential of cells. The identification of the twin-pore channel, TASK, in pulmonary artery smooth muscle, along with reports that it is inhibited by hypoxia, raises the possibility that a member of this family of channels acts as an O_2 sensor in pulmonary artery. An unidentified low-threshold, voltage-dependent K^+ channel might also contribute.

2005 Signalling pathways in acute oxygen sensing. Wiley, Chichester (Novartis Foundation Symposium 272) p 218–233

Pulmonary arteries from many species remain relaxed after isolation from the lung, contracting only when exposed to vasoactive substances, depolarizing stimuli or hypoxia (e.g. Casteels et al 1977a, Boe & Simonsson 1980, Madden et al 1985, Bonnet et al 2001). This lack of mechanical activity is reflected in electrical quiescence in the smooth muscle cells. The resting membrane potential of pulmonary artery smooth muscle (PASM) is maintained at a steady level, in the range −50 to −60 mV in the intact vessels of several species (Casteels et al 1977b, Hara et al 1980, Suzuki & Twarog 1982, Haeusler 1983, Madden et al 1985). Contraction can be activated without a change in membrane potential, but is often associated with graded

[1]Current address Faculty of Life Sciences, The University of Manchester, Manchester M13 9PT.

membrane depolarization and occasionally action potential generation (Casteels et al 1977a, Hara et al 1980, Suzuki & Twarog 1982, Haeusler 1985, Madden et al 1985). This opens L-type Ca^{2+} channels when the membrane potential reaches their activation threshold, above −40 mV (Clapp & Gurney 1991a).

Several factors contribute to the electrical quiescence of pulmonary artery smooth muscle. The high density of outwardly directed, voltage-activated K^+ current relative to inward Ca^{2+} current (Clapp & Gurney 1991a,b) means that any stimulus causing membrane depolarization would preferentially activate K^+ channels, resulting in an opposing efflux of K^+ and repolarization. Electrical quiescence is also promoted by a resting membrane potential below the voltage threshold for activating L-type Ca^{2+} channels. This is determined by the steep trans-membrane K^+ concentration gradient and relatively high membrane permeability to K^+ (Casteels et al 1977b), which results in K^+ efflux from the cell through K^+-selective channels, driving the membrane towards the K^+ equilibrium potential (E_K) at around −80 mV (Casteels et al 1977b, Haeusler 1983). The importance of K^+ channels for regulating membrane potential and averting spontaneous action potential firing is illustrated by the need for K^+ channel blockade in order for action potentials to be generated in PASM (Casteels et al 1977b, Haeusler & Thorens 1980, Hara et al 1980).

Hypoxic pulmonary vasoconstriction (HPV) involves mechanisms intrinsic to the pulmonary vasculature, since the response is retained in pulmonary arteries dissected free from the lung (e.g. Madden et al 1985, Ward & Aaronson 1999, Gurney 2002) and therefore independent of circulating humoral substances or neuronal inputs. A role for changes in the muscle membrane ionic conductance was first indicated in 1985, when hypoxia was found to induce depolarization and action potential generation concurrently with contraction in cat small intrapulmonary arteries (Harder et al 1985, Madden et al 1985). It is now clear that PASM depolarization is characteristic of HPV in several species and that the depolarization is associated with inhibition of K^+ conductance. A possible role for K^+ channels in HPV was first suggested in 1982 when it was noted that K^+ channel inhibitors facilitated HPV in isolated rat pulmonary arteries (McMurtry et al 1982). Subsequent studies confirmed that 4-aminopyridine (4-AP), a blocker of voltage-gated K channels, enhances HPV in perfused lungs and isolated arteries (Hasunuma et al 1991, Barman 1998, Archer et al 1996). This suggests either that activation of 4-AP-sensitive K^+ channels during HPV normally limits the amplitude of the contractile response or that membrane depolarization due to K^+ channel inhibition by 4-AP provides a priming depolarization that facilitates the primary constrictor action of hypoxia (Turner & Kozlowski 1997). Thus K^+ channels may be modulators rather than mediators of HPV. There is no doubt however that hypoxia inhibits K^+ channels in PASM and that this leads to membrane depolarization (McCulloch et al 1999, Gurney 2002). The molecular nature of the

K^+ channels underlying the depolarization is less clear. Below we outline recent work that seeks to clarify not only which K^+ channels in pulmonary artery are O_2 sensitive, but also which are open and contribute to K^+ efflux at the resting membrane potential and could therefore give rise to depolarization when inhibited by hypoxia.

Methods

Most experiments were performed on smooth muscle cells isolated from rabbit pulmonary arteries (Clapp & Gurney 1991b). The identification of K^+ channels expressed in PASM and their roles in O_2 sensing were studied using a range of techniques. Whole-cell, patch clamp recording was used to record K^+ currents under voltage clamp and membrane potential under current clamp conditions (Evans et al 1996, Osipenko et al 1997). Experiments were all carried out in the presence of 10 μM glibenclamide and 10 mM tetraethylammonium (TEA) chloride to block K_{ATP} and BK_{Ca} channels, respectively. These channels are not major contributors to the resting membrane potential (Clapp & Gurney 1992, Osipenko et al 1997), although their activation will lead to membrane hyperpolarization and vasodilation. Standard reverse transcriptase polymerase chain reaction (RT-PCR) methods, applied to total RNA extracted from isolated pulmonary arteries, were used to investigate K^+ channel expression, using sequence specific primer pairs (Gurney et al 2003). Single-cell RT-PCR used similar methods applied to individual smooth muscle cells harvested from preparations of isolated cells (Gurney et al 2003).

Results and discussion

PASM contains many K^+ channel subunits, several of which are inhibited by hypoxia (McCulloch et al 1999, Coppock & Tamkun 2001, Patel & Honoré 2001, Gurney et al 2002). Many of them can be ruled out as mediators of hypoxic depolarization because their voltage dependence means that they would be closed at the resting membrane potential: K^+ channel inhibition can only result in depolarization if the channel is open when the hypoxic insult is delivered. In identifying the K^+ channels involved in HPV, it is essential to demonstrate that they are constitutively active at the resting membrane potential. We therefore began by establishing the nature of the K^+ channels that contribute to the resting, or background K^+ conductance and therefore give rise to the resting membrane potential.

As illustrated in Fig. 1, in the presence of 10 μM glibenclamide and 10 mM TEA, pulmonary artery smooth muscle cells retain a resting membrane potential at around −50 mV, close to that recorded from intact vessels. Under these conditions the voltage-gated $K_v2.1$ and $K_v3.1$ channels, as well as K_{ATP} and BK_{Ca}, would be blocked (Coetzee et al 1999). It is therefore concluded that these channels contribute

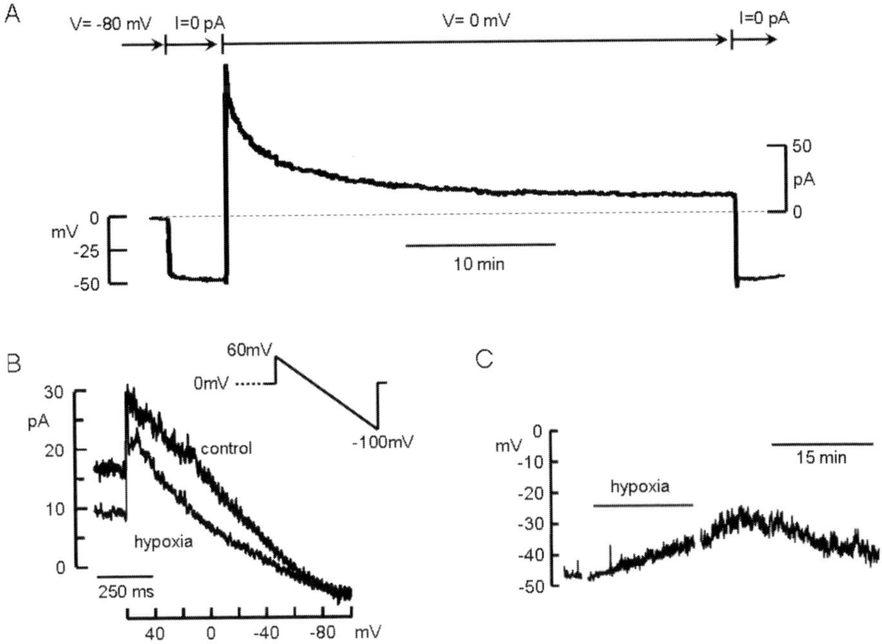

FIG. 1. Properties of the resting K$^+$ conductance. (A) A freshly isolated myocyte was clamped at −80 mV and the current recorded before switching off the voltage clamp and recording the resting membrane potential for 5 min. The cell was then clamped at 0 mV for >40 min before re-recording the resting potential, which had not changed. (B) Current recorded during a 1 s hyperpolarizing ramp from 60 mV to −100 mV, applied after clamping the cell at 0 mV for several minutes. (C) The effect of hypoxia on resting membrane potential. Adapted from Osipenko et al (1997).

insignificantly to the resting membrane potential. Thus although hypoxia inhibits $K_v2.1$ and $K_v3.1$ (Patel et al 1997, Osipenko et al 2000), these actions are unlikely to underlie hypoxia-induced depolarization. Figure 1A shows that the resting potential is also unchanged after clamping cells at 0 mV for more than 30 min, a procedure that inactivates around 90% of the voltage-activated K$^+$ current, including $K_v1.5$ (Snyders et al 1993). This current can therefore also be excluded as a major determinant of the resting potential and its inhibition is unlikely to mediate hypoxia-induced depolarization. As demonstrated by the response to a 1 s hyperpolarizing voltage ramp, the residual, non-inactivating current (I_{KN}) rectifies at negative potentials but remains outward below −60 mV (Fig. 1B), indicating that it is active at and could contribute to the resting potential. Importantly, it is reduced by hypoxia (Fig. 1B) with sufficient inhibition at −50 mV to explain the depolarizing effect of hypoxia, which is apparent under the same conditions (Fig. 1C). Increasing the extracellular K^+ concentration causes the reversal potential of I_{KN} to shift to more positive potentials (Fig. 2A) as predicted for a K^+-selective channel (Evans et al 1996). The current retains voltage-dependent activation/deactivation when the extracellular and intracellular K^+ concentrations are equal (130 mM), indicating that at least part of I_{KN} is voltage gated. It is clear from Fig. 2A that the activation

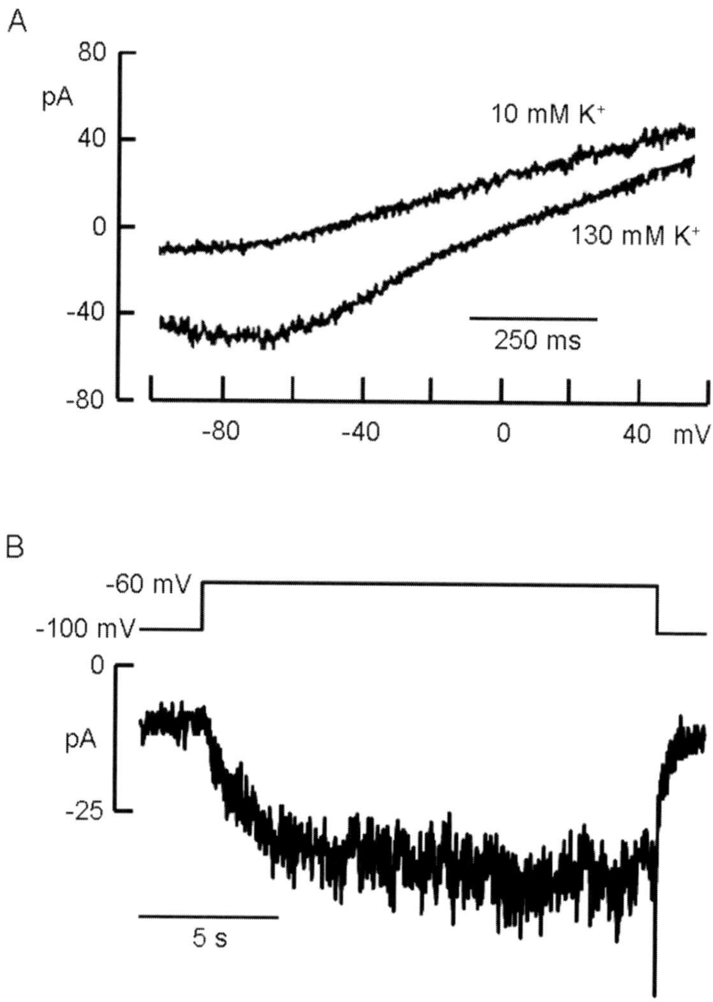

FIG. 2. Voltage-dependent and voltage-independent components of I_{KN}. (A) Influence of extracellular K^+ on the non-inactivating current. The current was recorded as in Fig. 1B, with a 1 s hyperpolarizing ramp from 60 to −100 mV after clamping the cell at 0 mV, but is plotted on a more conventional voltage scale. (B) Low threshold current activated by a voltage step from −100 to −60 mV in the presence of 130 mM extracellular K^+. Adapted from Evans et al (1996).

threshold lies more negative than −60 mV, which is very low for a voltage-gated channel. Applying voltage steps from −100 mV to 60 mV in the presence of 130 mM K^+ reveals a low threshold, voltage-gated current that activates with a slow time constant, on the order of 1 s (Fig. 2B). The large increase in inward current seen below −80 mV when the K^+ concentration is elevated (Fig. 2A) suggests, however, that there is also a voltage-independent component to I_{KN}.

Figure 3 illustrates the pharmacology of the resting membrane potential and I_{KN}. So far, any drug that has been found to modulate the resting potential caused equivalent modulation of I_{KN}, consistent with a causal link between them. Thus, millimolar concentrations of 4-AP cause depolarization and inhibit I_{KN} (Fig. 3A). Quinine has no effect on membrane potential or I_{KN} when tested at 10 μM, but it inhibits both at higher concentrations (Fig. 3B). Both membrane potential and I_{KN} are also inhibited by 100 μM Zn^{2+} (Fig. 3C), 10 μM anandamide and 100 μM bupivicaine (Gurney et al 2003, Gardener et al 2004), but are potentiated by halothane (Fig. 4C). An important finding was that the resting potential and I_{KN} are exquisitely sensitive to the extracellular pH (Fig. 4A,B). Acidosis inhibits I_{KN} and causes depolarization while alkalosis has the opposite effects (Gurney et al 2003, Gardener et al 2004). Moreover, acidosis inhibits the effect of halothane (Fig. 4D), indicating that they act on the same current. These properties are all consistent with the involvement of the two-pore domain channel, TASK, a name derived from TWIK-related acid-sensitive K^+ channel, where TWIK (Two-pore, Weakly Inwardly rectifying K^+) was the first discovered member of the family (Patel & Honoré 2001). Further support for this is provided by RT-PCR experiments employing several primer pairs designed to specifically identify TASK channel sequences. The mRNA for TASK-1 was identified in suspensions of freshly isolated smooth muscle cells (Fig. 5A), demonstrating TASK expression in pulmonary arteries. Its localization to myocytes was confirmed by single-cell RT-PCR (Fig. 4B) and immunostaining of the channel protein (Gurney et al 2003, Gardener et al 2004). Since TASK forms a voltage-independent background K^+ channel it is most likely to be responsible for the voltage-independent component of I_{KN}. Importantly, hypoxia has been reported to inhibit TASK-1 channels (Lewis et al 2001), implicating them as possible mediators of the hypoxia-induced depolarization of PASM.

The nature of the voltage-dependent component of I_{KN} is not yet clear. While $K_v1.5$ has been suggested, the main evidence supporting this candidate is inhibition of K^+ current and membrane depolarization by the K_v1 blockers, 4-AP and correolide, and inhibition of HPV by 4-AP (Archer et al 2004). The latter effect of 4-AP contradicts most studies that found it to enhance HPV, including an earlier one from the same group, which employed the same preparation and concentration of 4-AP (Archer et al 1996). The reason for this discrepancy is unclear. $K_v1.5$ is unlikely to contribute significantly, because millimolar 4-AP is needed to inhibit I_{KN} and

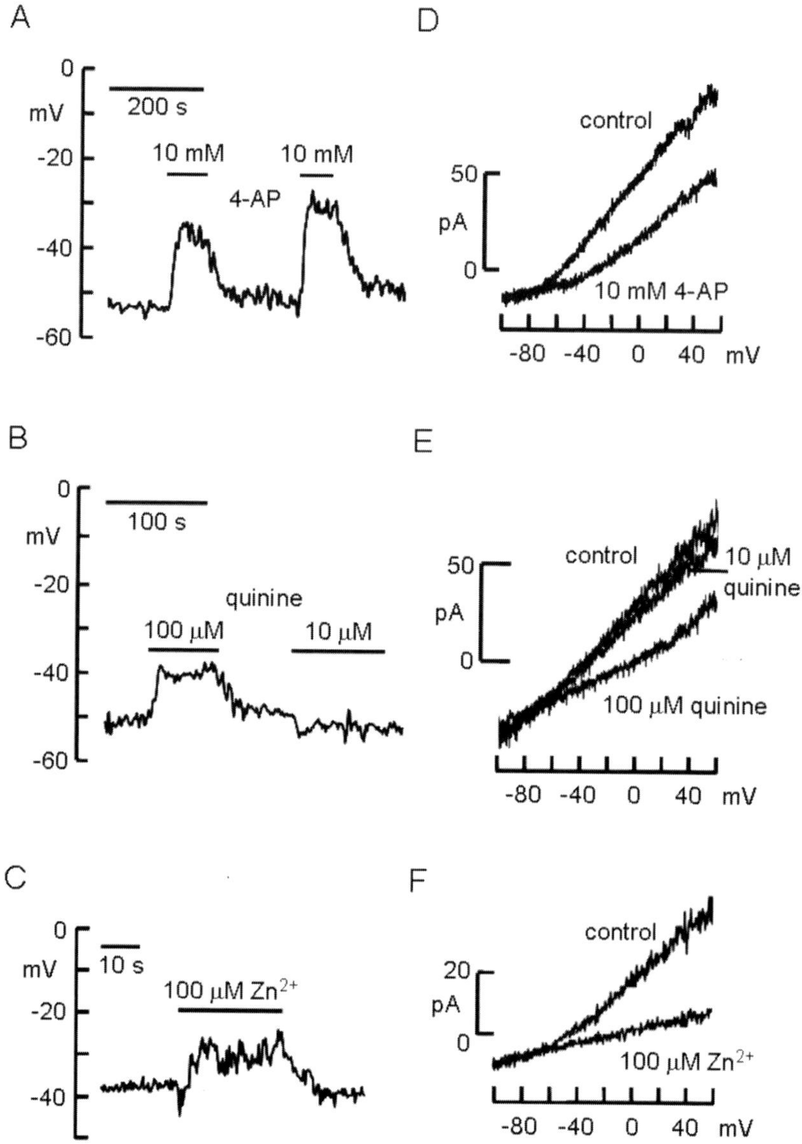

FIG. 3. Pharmacology of the resting K^+ conductance and membrane potential. (A,B,C) Effects of 4-AP (10 mM), quinine (10 and 100 μM) and Zn^{2+} (100 μM) on resting membrane potential. (D,E,F) Effects of the same agents on I_{KN} recorded as in Fig. 1B and plotted as in Fig. 2A. Adapted from Osipenko et al (1997) and Gurney et al (2003).

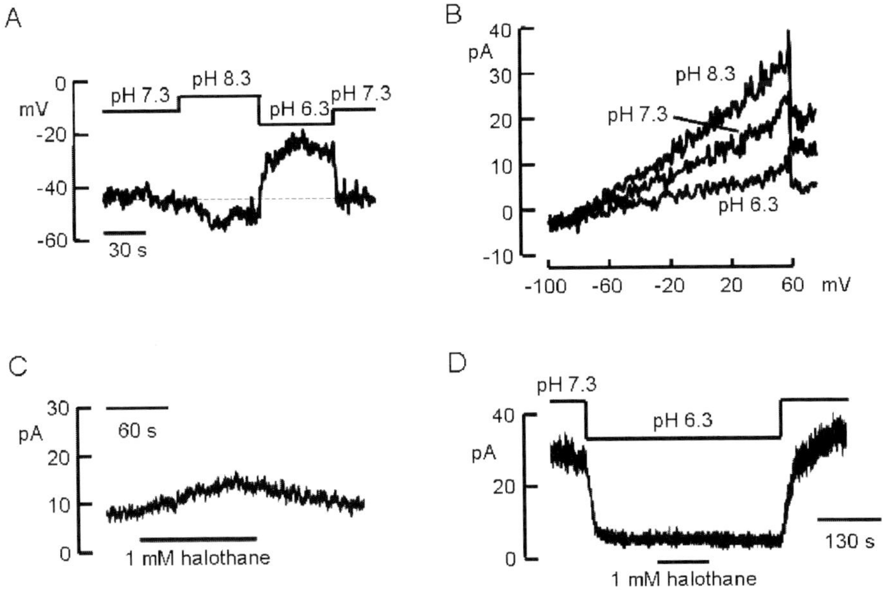

FIG. 4. TASK-like properties of I_{KN}. (A) Influence of extracellular pH on resting potential. (B) Effect of extracellular pH on I_{KN} recorded as in Fig. 1B and plotted as in Fig. 2A. (C) Effect of halothane (1 mM) on I_{KN} recorded continuously at 0 mV and pH 7.3. (D) Effect of halothane when pH is reduced to 6.3. Adapted from Gurney et al (2003).

produce depolarization compared with an EC_{50} <400 μM for block of $K_v1.5$ channels (Coetzee et al 1999). $K_v1.2$ is also a poor candidate due to its block by charybdotoxin, which does not affect I_{KN}, the resting potential or their modulation by hypoxia (Osipenko et al 1997). The voltage dependence and kinetics of I_{KN} are reminiscent of the neuronal M current, which is mediated by members of the KCNQ class of voltage-gated K^+ channels (Robbins 2001). For that reason we have now begun to investigate the expression and potential role of KCNQ channel subunits in the pulmonary circulation. As shown in Fig. 5C, the mouse lung expresses mRNA encoding all of the known KCNQ subunits (KCNQ1–5). All five were also detected in pulmonary blood vessels excised from the lung. Moreover, the pharmacological marker of KCNQ channels, linopirdine (Robbins 2001), is a potent constrictor of pulmonary arteries (P. Balan, A. Gurney, unpublished data). Thus although highly speculative at present, KCNQ channels appear to be promising candidates for a role in regulating the resting membrane potential and possibly mediating the depolarizing action of hypoxia in PASM. The O_2 sensitivity of KCNQ channels has not been investigated, so it remains to be determined if their inhibition could contribute to the depolarizing action of hypoxia.

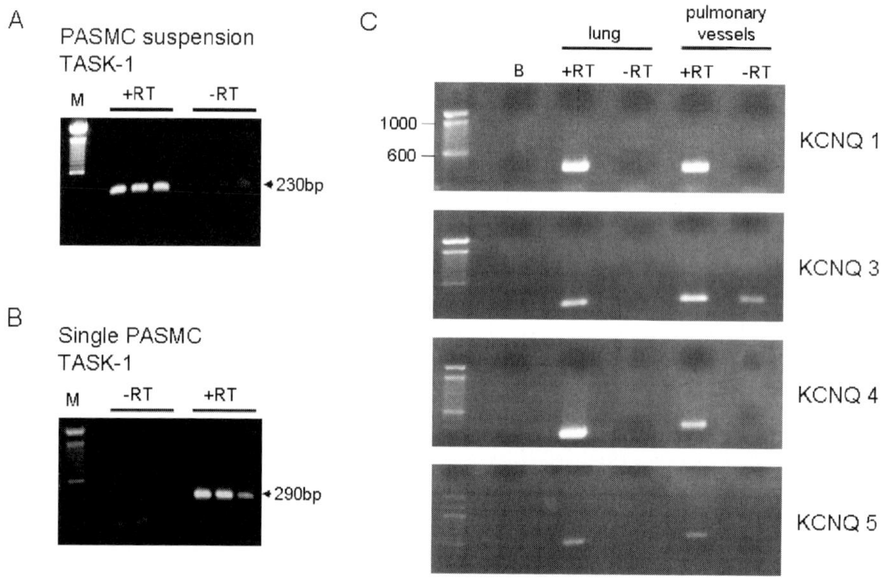

FIG. 5. Products of RT-PCR reactions visualised with ethidium bromide. (A) TASK-1 specific primers used with suspensions of rabbit pulmonary artery muscle cells. Each lane represents a separate reaction. (B) TASK-1 specific primers used in single cell RT-PCR. Each lane represents a single rabbit cell, harvested from a cell suspension. (C) Primer pairs specific for the KCNQ subunits indicated, used with RNA preparations extracted from samples of mouse lung or pulmonary blood vessels. Reactions in the presence (+RT) and absence (−RT) of reverse transcriptase. M, size markers.

Conclusions

HPV is associated with depolarization of the smooth muscle cell membrane in pulmonary arteries, mediated by the inhibition of a K^+ conductance that is comprised of voltage-dependent and voltage-independent components. The latter appears to be mediated by a two-pore domain K^+ channel with properties similar to recombinant TASK channels, possibly TASK-1. It is therefore comparable with the O_2-sensitive K^+ conductance reported in other chemosensory cells, such as those in the carotid body (Buckler et al 2005, this volume) and neuroepithelial bodies of the lung (Hartness et al 2001). Similar mechanisms may therefore operate to sense O_2 in different cell types. The voltage-dependent component of the resting K^+ conductance in PASM cells is mediated by a low threshold, slowly activating channel that fails to inactivate even during prolonged depolarization. Its identity remains to be determined, but channels of the KCNQ family are promising molecular candidates.

Acknowledgements

Grants from the BBSRC, British Heart Foundation, Wellcome Trust, Scottish Hospitals Endowment Research Trust and Tenovus Scotland are gratefully acknowledged.

References

Archer SL, Huang JMC, Reeve HL et al 1996 Differential distribution of electrophysiologically distinct myocytes in conduit and resistance arteries determines their response to nitric oxide and hypoxia. Circ Res 78: 431–442

Archer SL, Wu X-C, Thébaud B et al 2004 Preferential expression and function of voltage-gated, O_2-sensitive K^+ channels in resistance pulmonary arteries explains regional heterogeneity in hypoxic pulmonary vasoconstriction: Ionic diversity in smooth muscle cells. Circ Res 95: 308–319

Barman SA 1998 Potassium channels modulate hypoxic pulmonary vasoconstriction. Am J Physiol 275:L64–L70

Boe J, Simonsson BG 1980 Adrenergic receptors and sympathetic agents in isolated human pulmonary arteries. Eur J Respir Dis 61:195–202

Bonnet S, Hyvelin JM, Bonnet P, Marthan R, Savineau JP 2001 Chronic hypoxia-induced spontaneous and rhythmic contractions in the rat main pulmonary artery. Am J Physiol Lung Cell Mol Physiol 281:L183–L192

Buckler KJ, Williams BA, Varas Orozco R, Wyatt CN 2005 The role of TASK-like potassium channels in oxygen sensing in the carotid body. In: Signalling pathways in acute oxygen sensing (Novartis Found Symp 272). Wiley, Chichester, p 73–84

Casteels R, Kitamura K, Kuriyama H, Suzuki H 1977a Excitation-contraction coupling in the smooth muscle cells of the rabbit main pulmonary artery. J Physiol 271: 63–79

Casteels R, Kitamura K, Kuriyama H, Suzuki H 1977b The membrane properties of the smooth muscle cells of the rabbit main pulmonary artery. J Physiol 271:41–61

Clapp LH, Gurney AM 1991a Modulation of calcium movements by nitroprusside in isolated vascular smooth muscle cells. Pflügers Arch 418:462–470

Clapp LH, Gurney AM 1991b A simple method for cell isolation: characterisation of the major outward currents in rabbit pulmonary artery. Exp Physiol 76:677–693

Clapp LH, Gurney AM 1992 ATP-sensitive potassium channels regulate the resting potential of arterial cells. Am J Physiol 262:H916-H920

Coetzee WA, Amarillo Y, Chiu J et al 1999 Molecular diversity of K^+ channels. Ann NY Acad Sci 868:233–285

Coppock EA, Tamkun MM 2001 Differential expression of K_V channel α- and β-subunits in the bovine pulmonary arterial circulation. Am J Physiol 281:L1350–L1360

Evans AM, Osipenko ON, Gurney AM 1996 Properties of a novel K^+ current that is active at resting potential in rabbit pulmonary artery smooth muscle cells. J Physiol 496:407–420

Gardener MJ, Johnson IT, Burnham MP, Edwards G, Heagerty AM, Weston AH 2004 Functional evidence of a role for two-pore domain potassium channels in rat mesenteric and pulmonary arteries. Br J Pharmacol 142:192–202

Gurney AM 2002 Multiple Sites of oxygen sensing and their contributions to hypoxic pulmonary vasoconstriction. Resp Physiol Neurobiol 132:43–53

Gurney AM, Osipenko ON, MacMillan D, Kempsill EJ 2002 Potassium channels underlying the resting potential of pulmonary artery smooth muscle cells. Clin Exp Pharmacol Physiol 29:330–333

Gurney AM, Osipenko ON, MacMillan D, McFarlane KM, Tate RJ, Kempsill FEJ 2003 The two-pore domain K channel, TASK-1, in pulmonary artery smooth muscle cells. Circ Res 93:957–964

Haeusler G 1983 Contraction, membrane potential, and calcium fluxes in rabbit pulmonary arterial muscle. Fed Proc 42:263–268
Haeusler G 1985 Contraction of vascular muscle as related to membrane potential and calcium fluxes. J Cardiovasc Pharmacol 7(suppl 6):S3–S8
Haeusler G, Thorens S 1980 Effects of tetraethylammonium chloride on contractile, membrane and cable properties of rabbit artery muscle. J Physiol 303:203–224
Hara Y, Kitamura K, Kuriyama H 1980 Actions of 4-aminopyridine on vascular smooth muscle tissues of the guinea-pig. Brit J Pharm 68:99–106
Harder DR, Madden JA, Dawson C 1985 Hypoxic induction of Ca^{2+}-dependent action potentials in small pulmonary arteries of the cat. J Appl Physiol 59:1389–1393
Hartness ME, Lewis A, Searle GJ, O'Kelly I, Peers C, Kemp PJ 2001 Combined antisense and pharmacological approaches implicate hTASK as an airway O_2 sensing K^+ channel. J Biol Chem 276: 26499–26508
Hasunuma K, Rodman DM, McMurtry IF 1991 Effects of K^+ channel blockers on vascular tone in the perfused rat lung. Am Rev Respir Dis 144:884–887
Lewis A, Hartness ME, Chapman CG et al 2001 Recombinant hTASK1 is an O_2-Sensitive K^+ Channel. Biochem Biophys Res Commun 285:1290–1294
Madden JA, Dawson CA, Harder DR 1985 Hypoxia-induced activation in small isolated pulmonary arteries from the cat. J Appl Physiol 59:113–118
McCulloch KM, Osipenko ON, Gurney AM 1999 Oxygen-sensing potassium currents in pulmonary artery. Gen Pharmacol 32: 403–411
McMurtry IF, Rounds S, Stanbrook HS 1982 Studies of the mechanism of hypoxic pulmonary vasoconstriction. Adv Shock Res 8:21–33
Osipenko ON, Evans AM, Gurney AM 1997 Properties of the oxygen-sensing potassium current in rabbit pulmonary artery myocytes. Br J Pharmac 120:1461–1470
Osipenko ON, Tate RJ, Gurney AM 2000 Potential role for Kv3.1b channels as oxygen sensors. Circ Res 86:534–540
Patel AJ, Honoré E 2001 Properties and modulation of mammalian 2P domain K^+ channels. Trends Neurosci 24:339–346
Patel AJ, Lazdunski M, Honoré E 1997 $K_V2.1/K_V9.3$, a novel ATP-dependent delayed-rectifier K^+ channel in oxygen-sensitive pulmonary artery myocytes. EMBO J 22:6615–6625
Robbins J 2001 KCNQ potassium channels: physiology, pathophysiology, and pharmacology. Pharmacol Ther 90:1–19
Snyders DJ, Tamkun MM, Bennett PB 1993 A rapidly activating and slowly inactivating potassium channel cloned from human heart. Functional analysis after stable mammalian cell culture expression. J Gen Physiol 101:513–543
Suzuki H, Twarog BM 1982 Membrane properties of smooth muscle cells in pulmonary hypertensive rats. Am J Physiol 242:H907–H1015
Turner JL, Kozlowski RZ 1997 Relationship between membrane potential, delayed rectifier K^+ currents and hypoxia in rat pulmonary arterial myocytes. Exp Physiol 82:629–645
Ward JP, Aaronson PI 1999 Mechanisms of hypoxic pulmonary vasoconstriction: can anyone be right? Respir Physiol 115:261–271

DISCUSSION

Buckler: Can you estimate the proportion of K^+ current at the resting potential that is carried through TASK-like channels?

Gurney: Not yet.

Buckler: Have you shown that they are both independently oxygen sensitive?

Gurney: I have no idea whether the oxygen sensitivity reflects the voltage-dependent component, the voltage-independent component or both. All I can say is that the composite current is inhibited by hypoxia. We do now have ways of separating the two components so we will go back and look at them individually.

Buckler: What if you look at the voltage dependence of oxygen-sensitive currents by subtracting control minus hypoxia to see what the oxygen-sensitive bit looks like?

Gurney: We have only done this in the presence of normal K^+. We would need to use high K^+ in order to differentiate between the two components.

Buckler: If there is an element of strong voltage activation (to the oxygen sensitive current), this would suggest that KCNQ has to be oxygen sensitive as well as TASK. If it looks just vaguely GHK (i.e. conforms to the Goldman Hodgkin Katz constant field equation for ionic current through a non-voltage gated pore) it suggests that it might just be a weakly outwardly rectifying TASK-like current.

Gurney: The current is very small and I am not sure we would be able to pick this out. In physiological K^+ it would look voltage dependent, whichever component is involved. We need to do the experiment in high K^+, or once we have selectively blocked individual components. We should be able to do this now we have some idea of their pharmacology.

Nurse: If you were to add 4-AP to the mix of drugs you use to block the currents, would this linearize the current–voltage relationship?

Gurney: It is not linearized, but then the experiments were all carried out in normal K^+.

López-Barneo: Does 4-AP block TASK-1 channels?

Gurney: No. The effect of 4-AP we see is probably on the voltage-dependent channel.

López-Barneo: Do quinine and quinidine have similar effects? Is this at the same concentrations?

Gurney: Yes. Both these drugs reduce the non-inactivating current at the same concentrations ($\geq 100\,\mu M$) that act on TASK channels.

López-Barneo: Do you see an increase or decrease in input resistance in your preparation during hypoxia? You mentioned before that you see a non-selective cationic conductance.

Gurney: We have never specifically looked at the effect of hypoxia on cation conductance. When we have looked at its effects on the background current, we have always done it after first blocking or inactivating everything else.

López-Barneo: The amount of current that is blocked by 10 mM 4-AP is quite large: about 50%. Does this mean that 50% of current present after everything is inactivated belongs to the slowly activating current?

Gurney: The amount of block with 4-AP and other drugs varies from cell to cell. With zinc, for example, we saw pronounced inhibition in some cells while in others it was small. With halothane, some cells showed little effect while in others there was clear facilitation. We therefore think that the proportion of the slowly activating (voltage-dependent) current varies from cell to cell.

López-Barneo: Is the effect of halothane and pH related to TASK-1 current?

Gurney: The pH sensitivity might be due to its effect on that current. The pH effect was really robust and reproducible, so I'm guessing it may affect both components. KCNQ channels are pH-sensitive, which is another reason that they seem a good candidate.

Kemp: 4-AP sensitivity is quite problematic.

Gurney: I don't think 4-AP is working on TASK channels.

Kemp: You aren't left with much once you have added 4-AP.

Gurney: It varies from cell to cell, but on average 10 mM 4-AP blocks around 50% of the non-inactivating current at 0 mV.

Kemp: In support of your idea that you may be looking at TASK-3, your data with anandamide would be quite a useful pharmacological tool. This is because using the human recombinant TASK-3, we find that anandamide is a potent blocker.

Gurney: Was it the rat TASK-1 that Eric Honoré looked at? They published anandamide inhibition of TASK-1 (Maingret et al 2001).

Kemp: I'm not convinced by those data.

Gurney: The pharmacology of TASK channels seems confusing. From what Keith said in his talk, it seems that the early studies suggesting Zn^{2+} selectivity for TASK-1 (Leonoudakis et al 1998, Kim et al 2000) may have been misleading, as more recent papers indicate that Zn^{2+} is a selective blocker of TASK-3, with virtually no effect on TASK-1 (Clarke et al 2004). Is it possible that the same is true for anandamide?

Buckler: Two labs got this wrong! One lab said that TASK-1 was zinc sensitive and another lab said that TASK-3 wasn't.

Kemp: If you look at the anandamide data in Eric's paper (Maingret et al 2001) there does seem to be a significant inhibition with anandamide of TASK-3. We have done this with human TASK-3 in HEK 293 cells, and it is a potent blocker.

Gurney: Arthur Weston's group also has evidence from work on intact vessels that anadamide and bupivicaine block the TASK channels in pulmonary artery (Gardener et al 2004).

Kummer: Your siRNA knockdown in isolated vessels is a very promising approach. How long after taking these preparations into culture do they keep this property of constriction in response to hypoxia?

Gurney: We have only just begun to look at this and all I can say is that they retain the ability to respond to agonists after several days.

Kummer: We have been doing this on lung slices. We get good transfection but even after a day in culture there is almost no response of the controls to hypoxia, and after two days it is entirely lost.

Gurney: Elaine Hunter in my group has been looking at hypoxia and I believe she has seen contractions after two days, but I don't know how reproducible it is.

Sylvester: Have you looked at the effect of hypoxia on the non-inactivating current–voltage relationship?

Gurney: We have, using voltage ramps, and hypoxia inhibits it from around −65 mV to positive potentials.

Sylvester: Does the morphology of the currents change? In other words, does it look like the same channels are contributing during hypoxia and normoxia, or could additional channels contribute to the shift? Is it possible that some new type of channel is recruited to determine membrane potential under hypoxic conditions?

Gurney: I don't have any evidence for this. It looks like we inhibit one component, but we don't know which of the two components this is. We originally thought it might be inhibiting TASK, because Paul has shown that recombinant TASK is sensitive to hypoxia, but others have disagreed.

Kumar: If the point of hypoxic pulmonary vasoconstriction (HPV) is to cause blood diversion from areas of hypoxia to more ventilated areas, in these underventilated areas of the lung *in vivo* you may get acidosis as well. Why isn't HPV renamed APV instead? TASK is an acid-sensitive current. There seems to be a lot of chasing around for the hypoxia sensitivity of this current when there may be a perfectly good physiological stimulus in pH.

Gurney: I don't know what acidosis does in HPV. We have looked at the effects of changing the extracellular pH on contraction, and it is not as you would predict from blocking the current. It is not entirely reproducible, either, but acidosis could be doing other things besides inhibiting TASK channels.

Kumar: It is a good physiological stimulus in an underventilated part of the lung and it could cause constriction.

Buckler: Where does the acidosis come from? Is it oxygen deprivation to the pulmonary vascular myocytes causing lactate production?

Kumar: It is an underventilated part of the lung with mixed venous blood returning.

Buckler: So it is the difference between P_{ACO2} in the ventilated lung versus venous P_{CO2}, so it will be about a 5 Torr difference.

Kumar: It is not a lot, but after a while the blood stops flowing.

Archer: I don't agree with the K_v discussion. Many labs find that K_v channels form the mechanism for HPV and are open at resting potential. If they are blocked with antibodies or knocked out, this gets rid of HPV (see Archer SL et al 2004). This correlates with the actual physiology in the lung. In the ductus, when you put it in tissue culture, you remove the oxygen-sensitive channels over 48 h. If these channels are restored it starts constricting again. The physiology of acidosis was well characterized 25 years ago by Marshall & Marshall in the pulmonary circulation (Marshall et al 1984). They found that acidosis didn't have much effect on HPV.

López-Barneo: It shows a very steep pH sensitivity. In your plot at pH 6.7 you got almost 100% of blockade.

Gurney: Yes, the background K^+ current is very pH sensitive.

Archer: There is work by Brimioulle et al (1990) who looked at the effects of acidosis induced by respiratory versus metabolic means. They are dissociated. They found that in intact anaesthetized dogs, metabolic acidosis and alkalosis, respectively, enhance and reverse HPV. Respiratory acidosis did not affect HPV and respiratory alkalosis blunted HPV, which suggests a pH-independent vasodilating effect of CO_2.

Gurney: We didn't find it constricted either.

Archer: Just like with the mitochondria, I think it is important to try to figure out why we find different things. One suggestion would be that we have highly different patch clamping set-ups. Your patch clamp configuration is set up to exclude all the things that we are looking at.

Gurney: If you exclude them all, you still have the resting membrane potential. That is my argument.

Archer: I do think those other K_v channels are oxygen sensitive, whatever their relevant role.

López-Barneo: An important point is to know how much TASK-1, or a mixture of TASK-1 plus the other channel, contribute to HPV. There are two simple experiments that would help answer this. First, the pH experiment: if you go to pH 6.5 and still have HPV, this means that TASK channels are not important. The second experiment is with quinine, which is a potent blocker at 100–200 μM. If this is applied and there is still HPV, it means that besides TASK1 there are several channels contributing to HPV.

Gurney: At those concentrations quinine would already be blocking K_v currents. It is not very selective. Zinc or anandamide may be better alternatives.

López-Barneo: The problem with zinc is that it will block Ca^{2+} channels. Is there a specific TASK-1 inhibitor?

Buckler: No, there isn't a specific one.

Chandel: Is there a mouse knockout of TASK-1?

Buckler: People are working on it.

Kemp: I was intrigued with your siRNA experiments. I suppose HPV not withstanding, you can still test whether TASK is controlling the membrane potential there. You show very clearly that your siRNA knocks TASK-1 completely out. You'd therefore expect those cells without collapsed membrane potentials to be significantly depolarized.

Gurney: Yes, I would also expect the tissue to behave differently. In a few of the contraction experiments we looked to see whether we have altered the response to 4-AP. We showed before that in chronic hypoxic rats the blood vessels are much more sensitive to 4-AP, an effect due to down-regulation of the non-inactivating current (Osipenko et al 1998). If this is due to down-regulation of TASK, we would expect to see the same thing in the siRNA treated vessels. Initial experiments look promising.

References

Archer SL, Wu XC, Thebaud B et al 2004 Preferential expression and function of voltage-gated, O_2-sensitive K^+ channels in resistance pulmonary arteries explains regional heterogeneity in hypoxic pulmonary vasoconstriction: ionic diversity in smooth muscle cells. Circ Res 95: 308–318

Brimioulle S, Lejeune P, Vachiery JL et al 1990 Effects of acidosis and alkalosis on hypoxic pulmonary vasoconstriction in dogs. Am J Physiol 258:H347–353

Gardener MJ, Johnson IT, Burnham MP, Edwards G, Heagerty AM, Weston AH 2004 Functional evidence of a role for two-pore domain potassium channels in rat mesenteric and pulmonary arteries. Br J Pharmacol 142:192–202

Kim Y, Bang H, Kim D 2000 TASK-3, a new member of the tandem pore K^+ channel family. J Biol Chem 275:9340–9347

Leonoudakis D, Gray AT, Winegar BD et al 1998 An open rectifier potassium channel with two pore domains in tandem cloned from rat cerebellum. J Neurosci 18:868–877

Maingret F, Patel AJ, Lazdunski M, Honore E 2001 The endocannabinoid anandamide is a direct and selective blocker of the background K^+ channel TASK-1. EMBO J 20:47–54

Marshall C, Lindgren L, Marshall BE 1984 Metabolic and respiratory hydrogen ion effects on hypoxic pulmonary vasoconstriction. J Appl Physiol 57:545–550

Osipenko ON, Alexander D, MacLean MR, Gurney AM 1998 Influence of chronic hypoxia on the contributions of non-inactivating and delayed rectifier K currents to the resting potential and tone of rat pulmonary artery smooth muscle. Br J Pharmacol 124:1335–1337

AMP-activated protein kinase couples mitochondrial inhibition by hypoxia to cell-specific Ca^{2+} signalling mechanisms in oxygen-sensing cells

A. Mark Evans, D. Grahame Hardie*, Antony Galione†, Chris Peers‡, Prem Kumar§ and Christopher N. Wyatt

*Division of Biomedical Sciences, School of Biology, University of St. Andrews, Fife KY16 9TS, *Division of Molecular Physiology, School of Life Sciences, Wellcome Trust Biocentre, University of Dundee, Dow Street, DD1 5EH, †University Department of Pharmacology, University of Oxford, Oxford OX1 3QT, ‡Institute for Cardiovascular Research, University of Leeds, Leeds LS2 9JT, and §Department of Physiology, The Medical School, University of Birmingham, Birmingham B15 2TT, UK*

Abstract. It is generally accepted that relatively mild hypoxia inhibits mitochondrial oxidative phosphorylation in O_2-sensing cells, and thereby mediates, in part, cell activation. However, the mechanism by which this process is coupled to discrete, cell-specific Ca^{2+} signalling mechanisms remains elusive. We considered the possibility that hypoxia may increase the cellular ATP/AMP ratio, increase the activity of AMP-activated protein kinase (AMPK) and thereby evoke Ca^{2+} signals in O_2-sensing cells. Co-immunoprecipitation identified $\alpha1\beta2\gamma1$ as the primary AMPK isozyme in pulmonary arterial smooth muscle, whilst the tissue-specific distribution of AMPK activities and their activation by hypoxia suggested that the AMPK-$\alpha1$ catalytic subunit isoform is key to the regulation of O_2-sensing cells. Strikingly, 3D reconstruction of immunofluorescence images showed AMPK-$\alpha1$ to be located throughout the cytoplasm of pulmonary arterial smooth muscle cells and, by contrast, targeted to the plasma membrane in carotid body glomus cells. Consistent with these observations Ca^{2+} imaging, tension recording and electrophysiology demonstrated that AMPK, like hypoxia, activates each cell type via discrete Ca^{2+} signalling mechanisms: cyclic ADP-ribose-dependent Ca^{2+} mobilization from the sarcoplasmic reticulum via ryanodine receptors in pulmonary arterial smooth muscle cells and voltage-gated Ca^{2+} influx into carotid body glomus cells. Thus, metabolic-sensing by AMPK underpins the cell-specific response of O_2-sensing cells to hypoxia.

2005 Signalling pathways in acute oxygen sensing. Wiley, Chichester (Novartis Foundation Symposium 272) p 234–258

Specialized O_2-sensing cells within the body have evolved as vital homeostatic mechanisms that monitor O_2 supply and alter respiratory and circulatory function, as well as the capacity of the blood to transport O_2. By this means, arterial pO_2 is maintained within physiological limits. Two key systems involved are the pulmonary arteries and the carotid body (Heymans et al 1930, von Euler & Liljestrand 1946). Constriction of pulmonary arterial smooth muscle by hypoxia optimizes ventilation-perfusion matching in the lung (von Euler & Liljestrand 1946), whilst carotid body excitation by hypoxia initiates corrective changes in breathing patterns via increased sensory afferent discharge to the brain stem (Heymans et al 1930). In both cases it is generally accepted that relatively mild hypoxia inhibits mitochondrial oxidative phosphorylation and that this underpins, at least in part, cell activation (Heymans et al 1930, Gonzalez et al 2002, Mills & Jobsis 1972, Rounds & McMurtry 1981, Duchen & Biscoe 1992, Leach et al 2001, Wyatt & Buckler 2004). However, the mechanism by which inhibition of mitochondrial oxidative phosphorylation couples to discrete, cell-specific Ca^{2+} signalling mechanisms has remained elusive, and extensive investigation of previous hypotheses has delivered conflicting data and failed to unite the field since its inception in 1930 (for review see, Gonzalez et al 2002). Recently, the AMP-activated protein kinase (AMPK) cascade has come to prominence as a sensor of metabolic stress that appears to be ubiquitous throughout eukaryotes (Hardie et al 2003, Hardie 2004). AMPK is activated by many different metabolic stresses, including heat shock and metabolic poisons in hepatocytes (Corton et al 1994), exercise in skeletal muscle (Winder & Hardie 1996) and ischaemia and hypoxia in the heart (Marsin et al 2000). AMPK complexes are heterotrimers comprising a catalytic α subunit and regulatory β and γ subunits which monitor the cellular AMP/ATP ratio as an index of metabolic stress (Hardie et al 2003). Through the action of adenylate kinase, any small increase in the cellular ATP/ADP ratio is converted into a much larger decrease in the ATP/AMP ratio (Hardie & Hawley 2001). Binding of AMP to two sites in the γ subunit triggers activation of the kinase via phosphorylation of the α subunit at Thr172, an effect antagonized by high concentrations of ATP (Hawley et al 1996). This phosphorylation is catalysed by upstream kinases (AMPK kinases) the major form of which is a complex between the tumour suppressor kinase, LKB1, and two accessory subunits, STRAD and MO25 (Hawley et al 2003, Shaw et al 2004, Woods et al 2003). As well as promoting phosphorylation by LKB1, AMP binding to the γ subunits also inhibits dephosphorylation by protein phosphatases and causes allosteric activation of the phosphorylated enzyme: this triple mechanism ensures a very sensitive response to small changes in the cellular AMP/ATP ratio (Hardie et al 1999).

The primary targets for AMPK had previously been presumed to be mainly involved in energy metabolism, but it is now recognized that AMPK can also target non-metabolic processes (Hardie et al 2003). Given that inhibition of mitochondrial oxidative phosphorylation by hypoxia would be expected to promote a rise in the AMP/ATP ratio (Hardie et al 2003) we considered the proposal (Evans 2004) that AMPK activation may mediate pulmonary artery constriction and carotid body excitation by hypoxia.

AMP-activated protein kinase, cyclic ADP-ribose-dependent sarcoplasmic reticulum Ca^{2+} release and hypoxic pulmonary vasoconstriction

In 1946, von Euler and Liljestrand demonstrated that hypoxia without hypercapnia induced constriction within the pulmonary circulation and proposed that hypoxic pulmonary vasoconstriction (HPV) might aid ventilation/perfusion matching at the alveoli, by diverting blood flow away from poorly ventilated areas of the lung (von Euler & Liljestrand 1946). In contrast, systemic arteries dilate in response to tissue hypoxemia, in order to match local perfusion to local metabolism (Roy & Sherrington 1890). Clearly, therefore, one would expect a mediator of HPV to offer in some way the pulmonary selectivity required to elicit this critical and distinguishing characteristic of the pulmonary arterial tree.

Pulmonary selective distribution of a specific AMPK isozyme

Western blot (Durante et al 2002) and co-immunoprecipitation (Cheung et al 2000) analyses of pulmonary arterial smooth muscle lysates identified the presence of the $\alpha1$, $\alpha2$, $\beta2$, $\gamma1$, $\gamma2$ and $\gamma3$ subunits of AMPK. Furthermore, immunoprecipitate kinase assays (Cheung et al 2000) using isoform-specific antibodies revealed that the $\gamma2$ subunit isoform accounted for 40% and the $\gamma1$ isoform for 60% of the total AMPK activity in pulmonary arterial smooth muscle lysates, whilst the $\gamma3$ subunit isoform accounted for an insignificant fraction (Fig. 1A). Most significantly, however, these studies showed that the catalytic $\alpha1$ subunit isoform accounted for 80–90% and the $\alpha2$ isoform for only 10–20% of the total catalytic activity in

FIG. 1. AMPK subunit isoforms, their activation by hypoxia and Ca^{2+} signalling in pulmonary arterial smooth muscle. (A) AMPK activity immunoprecipitated from pulmonary arterial smooth muscle with anti-$\alpha1$, -$\alpha2$, -$\gamma1$, -$\gamma2$ and -$\gamma3$ antibodies. (B) AMPK activity immunoprecipitated from pulmonary and systemic arterial smooth muscle with anti-$\alpha1$ and -$\alpha2$ antibodies. (C) activation of AMPK in pulmonary arterial smooth muscle by AICAR, phenformin (ambient pO_2) and by switching from physiological normoxia (154–160 Torr; 1 h) to hypoxia (16–21 Torr, 1 h). (D) (i), brightfield image of a fixed pulmonary arterial smooth muscle cell; (ii) z-section showing staining by antibodies to AMPK-$\alpha1$ and DAPI staining of the nucleus; (iii) 3D reconstruction. Effect on Fura-2 fluorescence ratio (F340/F380) in isolated pulmonary arterial smooth muscle cells of: (E) phenformin (10 mM) and (F) AICAR (1 mM), with and without extracellular Ca^{2+} (+1 mM EGTA). (G) Relative increase in the Fura-2 fluorescence in response to phenformin in the presence and absence of extracellular Ca^{2+}, and AICAR in the presence and absence of extracellular Ca^{2+}, ryanodine (10 μM) and caffeine (10 mM), and 8-bromo-cADPR (300 μM), respectively.

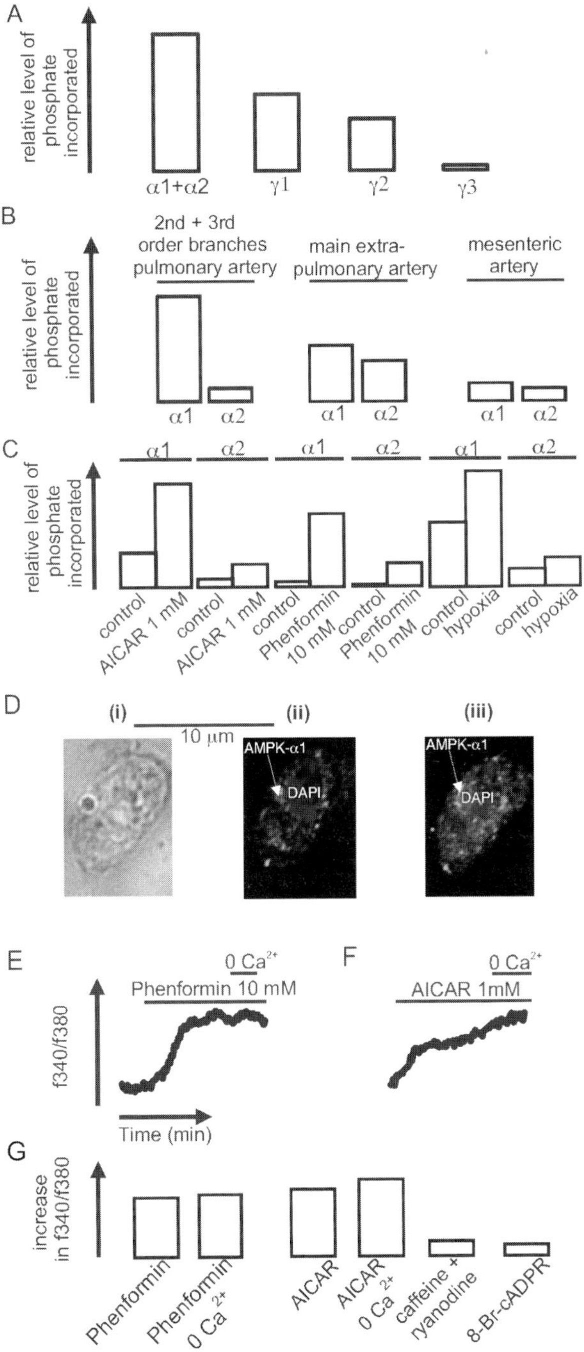
A
relative level of phosphate incorporated
α1+α2
γ1
γ2
γ3
B
2nd + 3rd order branches pulmonary artery
main extra-pulmonary artery
mesenteric artery
α1 α2
C
control
AICAR 1 mM
Phenformin 10 mM
hypoxia
D
(i)
(ii)
(iii)
10 μm
AMPK-α1
DAPI
E
0 Ca2+
Phenformin 10 mM
f340/f380
Time (min)
F
AICAR 1mM
G
increase in f340/f380
Phenformin
Phenformin 0 Ca2+
AICAR
AICAR 0 Ca2+
caffeine + ryanodine
8-Br-cADPR

smooth muscle lysates from 2nd and 3rd order branches of the pulmonary arterial tree (Fig. 1B). The predominant AMPK subunit composition in pulmonary arterial smooth muscle from 2nd and 3rd order branches of the pulmonary arterial tree is likely, therefore, to be $\alpha1\beta2\gamma1$. This contrasted markedly with analyses of the AMPK activities in the main pulmonary artery that feeds the lung and systemic (mesenteric) arteries. Thus, the activity of the $\alpha1$ subunit isoform was lower in smooth muscle from the main pulmonary artery when compared to the smaller branches, and was lower still in the systemic (mesenteric) arteries. By contrast, that of the $\alpha2$ isoform was the same or higher (Fig. 1B). Thus, AMPK-$\alpha1$ activity is inversely related to pulmonary artery diameter, as is the magnitude of pulmonary artery constriction by hypoxia (Kato & Staub 1966). Furthermore, AMPK-$\alpha1$ activity is much higher in 2nd and 3rd order branches of the pulmonary arterial tree when compared with systemic (mesenteric) arteries, which dilate rather than constrict in response to hypoxia (Roy & Sherrington 1890). The differential arterial distribution of the AMPK-$\alpha1$ catalytic activity is consistent, therefore, with this isozyme playing an important role in HPV, the critical and distinguishing characteristic of pulmonary arteries (von Euler & Liljestrand 1946).

Hypoxia triggers a rise in the AMP/ATP ratio and thereby activates AMPK-$\alpha1$ activity in pulmonary arterial smooth muscle

As mentioned previously, cells initially respond to metabolic stress via the conversion by adenylate kinase of any small increase in the cellular AMP/ATP ratio into a much larger increase in the AMP/ATP ratio (Hardie & Hawley 2001). Consequent binding of AMP to two sites in the γ subunits of AMPK triggers activation of the kinase via phosphorylation by upstream AMPK kinases (Hawley et al 1996). If inhibition of mitochondrial oxidative phosphorylation and consequent activation of AMPK were to mediate chemotransduction by hypoxia, therefore, we would expect a pre-requisite increase in the AMP/ATP ratio on exposure of pulmonary arterial smooth muscle to hypoxia. Consistent with this, capillary electrophoresis (Schrauwen et al 2004) showed that the AMP/ATP ratio rose twofold on switching from physiological normoxia (155–160 Torr) to hypoxia (16–21 Torr), and that this was associated with a concomitant, twofold increase in AMPK activity (Fig. 1C).

We therefore studied the effects on AMPK activity in pulmonary arterial smooth muscle of two drugs known to activate AMPK via discrete mechanisms. The first was phenformin, which inhibits Complex I of the mitochondrial respiratory chain (El-Mir et al 2000, Owen et al 2000) and thereby activates AMPK by increasing the cellular AMP/ATP ratio (D.G. Hardie et al, unpublished data). The second was 5-aminoimidazole-4-carboxamide ribonucleoside (AICAR), which by contrast is metabolized to yield the AMP mimetic 5-aminoimidazole-4-carboxamide ribonucleotide (ZMP) and thereby activates AMPK without affecting the cellular AMP/ATP ratio (Corton et al 1995, Owen et al 2000). In common with the effects of hypoxia, both phenformin (10 mM) and AICAR (1 mM), irrespective of their mechanism of action, were found to increase AMPK activity (Fig. 1C).

Most importantly with respect to the present study, all three stimuli were found to increase AMPK-α1 activity to a greater extent than they increased AMPK-α2 activity in pulmonary arterial smooth muscle (Fig. 1C), and in each case AMPK activation was associated with the phosphorylation of a classical AMPK substrate in the smooth muscle, acetyl CoA carboxylase (not shown).

AMPK-α1 is targeted to sites throughout the cytoplasm in isolated pulmonary arterial smooth muscle cells

To determine the cellular distribution of AMPK-α1, isolated pulmonary arterial smooth muscle cells were labelled with sequence specific antibodies to the AMPK-α1 catalytic subunit isoform (Durante et al 2002) and FITC (490 nm excitation, 518 nm emission) conjugated secondary antibodies. Consistent with an intracellular site of action, 3D reconstruction of deconvolved z-sections acquired by fluorescence imaging (Kinnear et al 2004) revealed that AMPK-α1 was targeted at sites located throughout the cytoplasm, but excluded from the nucleus (Fig. 1D).

AMPK activation and hypoxia mobilize ryanodine-sensitive sarcoplasmic reticulum Ca^{2+} stores in pulmonary arterial smooth muscle

Despite their different modes of action, hypoxia (16–21 Torr; Dipp et al 2001), phenformin (10 mM) and AICAR (1 mM) evoked an increase in intracellular Ca^{2+} concentration in isolated pulmonary arterial smooth muscle cells, as reported by the Fura-2 fluorescence ratio (F340/F380; Fig. 1E, Dipp et al 2001). In each case the increase in intracellular Ca^{2+} concentration was resistant to removal of extracellular Ca^{2+} (Figs 1E–F; Dipp et al 2001). However, prior depletion of sarcoplasmic reticulum (SR) stores by pre-incubation of cells with ryanodine (10 μM) and caffeine (10 mM) abolished the increase in intracellular Ca^{2+} induced by hypoxia (Dipp et al 2001, Dipp & Evans 2001) and AICAR (1 mM; Fig. 1G).

Most significantly, SR Ca^{2+} release by AMPK activation was also abolished by blocking the Ca^{2+} mobilizing messenger cyclic adenosine diphosphate-ribose, an endogenous regulator of ryanodine receptors (cADPR; Lee et al 1989, Galione et al 1991), with a selective cADPR antagonist (Dipp & Evans 2001) 8-bromo-cADPR (100 μM). Thus, AMPK activation triggers cADPR-dependent SR Ca^{2+} release via ryanodine receptors in isolated pulmonary arterial smooth muscle, as does hypoxia (Salvaterra & Goldman 1993, Dipp et al 2001, Wilson et al 2001, Dipp & Evans 2001).

This latter finding is significant, because we have previously shown that the enzyme activities for the synthesis and metabolism of cADPR confer a degree of pulmonary selectivity, that cADPR accumulation in pulmonary arterial smooth muscle is augmented by hypoxia and that cADPR-dependent SR Ca^{2+} release is a pre-requisite for maintained HPV (Wilson et al 2001, Dipp & Evans 2002).

ADP-ribosyl cyclase and cADPR hydrolase activities are differentially distributed in pulmonary versus systemic artery smooth muscle

As with the distribution of AMPK-α1 activity, our findings with respect to the tissue-specific distribution of the enzyme activities for cADPR synthesis and metabolism were striking. These were at least an order of magnitude higher in homogenates of pulmonary arterial smooth muscle than in those of aortic or mesenteric arterial smooth muscle (Fig. 2A–B; Wilson et al 2001). Thus, the differential distribution of these enzyme activities afford a further degree of pulmonary selectivity over and above that offered by AMPK-α1. Of further significance was the finding that the level of these enzyme activities was also inversely related to pulmonary artery diameter (Fig. 2A–B; Wilson et al 2001). The ADP-ribosyl cyclase/cADPR hydrolase activities afford, therefore, a further degree of selectivity for pulmonary versus systemic arteries over and above that offered by AMPK-α1 activity, as would be expected of a mediator of HPV. Furthermore, the relative capacity of the smooth muscle from the main pulmonary artery and the 2nd and 3rd order branches to synthesize cADPR may also determine the inverse relationship between the diameter of pulmonary arteries and the magnitude of constriction by hypoxia (Kato & Staub 1966).

Hypoxia increases cADPR content in pulmonary arterial smooth muscle

The possibility that hypoxia may promote HPV, in part, by increasing cADPR accumulation in pulmonary arterial smooth muscle gained further support from direct measurements of cADPR content using a [^{32}P]cADPR binding assay. Hypoxia (16–21 Torr) increased cADPR levels twofold in 2nd order branches of the pulmonary artery, and 10-fold in 3rd order branches (Fig. 2C–D, Wilson et al 2001). Consistent with constriction by hypoxia, the distribution of AMPK-α1 activity and the distribution of the enzyme activities for cADPR synthesis and metabolism, respectively, the increase in cADPR accumulation induced by hypoxia was inversely related to pulmonary artery diameter (Wilson et al 2001).

β-NADH augments cADPR accumulation from β-NAD$^+$

β-NADH levels in pulmonary arterial smooth muscle increase in response to hypoxia (Shigemori et al 1996, Leach et al 2002), which may result from both the

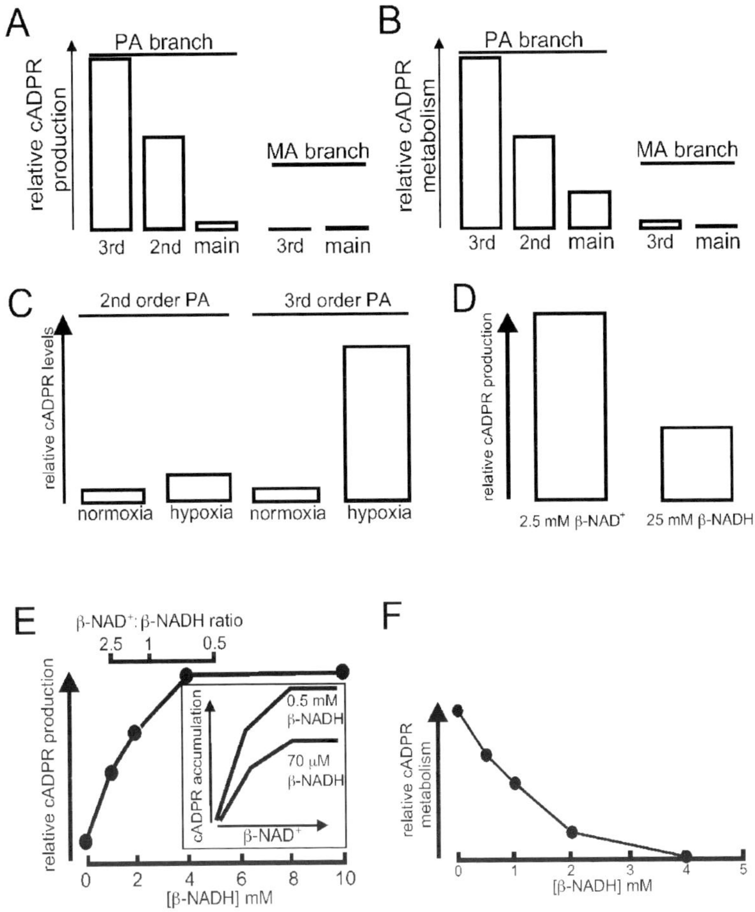

FIG. 2. Synthesis and metabolism of cADPR in pulmonary arterial smooth muscle, and the effect of hypoxia on cADPR levels. (A) synthesis of cADPR from β-NAD$^+$ in smooth muscle homogenates from a series of pulmonary and systemic arteries. (B) Metabolism of cADPR in smooth muscle homogenates from a series of pulmonary and systemic arteries. (C) Relative cADPR levels in 2nd and 3rd order branches of the pulmonary artery in the presence of normoxia (154–160 Torr) and hypoxia (16–21 Torr), respectively. PA, pulmonary artery; MA, mesenteric artery. (D) Relative cADPR accumulation from 2.5 mM β-NAD$^+$ and 25 mM β-NADH, respectively. (E) Effect of increasing β-NADH concentration on cADPR accumulation from 2.5 mM β-NAD$^+$. Inset, predicted effect of increased β-NADH by hypoxia on curve of cADPR accumulation versus β-NAD$^+$ concentration. (F) Concentration-dependent effect of β-NADH on cADPR metabolism.

inhibition of mitochondrial oxidative phosphorylation by hypoxia and the subsequent acceleration of anaerobic glycolysis by AMPK (Hardie et al 2003). We therefore considered the possible impact of β-NADH on cADPR accumulation in pulmonary arterial smooth muscle. Consistent with β-NAD$^+$ being the primary substrate for cADPR synthesis (Lee et al 1989) we found β-NADH to be a poor substrate for cADPR synthesis. In fact, when applied to pulmonary artery smooth muscle homogenates, 25 mM β-NADH produced a total cADPR yield that amounted to no more than 30% of that derived from 2.5 mM β-NAD$^+$ (Fig. 2E; Wilson et al 2001). Thus, we considered the possibility that the β-NADH/β-NAD$^+$ ratio may be a significant factor, rather than the absolute level of either nucleotide. Surprisingly, β-NADH induced a concentration-dependent and synergistic increase in cADPR accumulation from a fixed concentration of β-NAD$^+$ (Fig. 2F; Wilson et al 2001). Moreover, the range over which a change in the β-NADH/β-NAD$^+$ ratio augmented cADPR accumulation was equivalent to the ratio change predicted from direct measurement of β-NAD$^+$ and β-NADH in extracts from pulmonary arteries during normoxia and moderate hypoxia (Shigemori et al 1996, Wilson et al 2001). In short, β-NADH may shift the curve for cADPR production from β-NAD$^+$ to the left and raise maximal cADPR accumulation (Fig. 2F, inset). Furthermore, β-NADH inhibited cADPR metabolism in pulmonary arterial smooth muscle homogenates in a concentration-dependent manner, equivalent in range to that by which β-NADH augmented cADPR synthesis. Thus, β-NADH may both augment the synthesis of cADPR from β-NAD$^+$ and inhibit cADPR metabolism. However, these effects alone may be insufficient to account for the degree of cADPR accumulation observed in response to hypoxia (Evans & Dipp 2002, Evans 2004). AMPK, in addition to promoting β-NADH formation by accelerating anaerobic glycolysis in response to hypoxia (Hardie et al 2003), may therefore act in synergy with β-NADH to promote cADPR accumulation via ADP-ribosyl cyclase.

cADPR-independent and cADPR-dependent phases of smooth muscle SR Ca^{2+} release contribute to pulmonary artery smooth muscle constriction by hypoxia

HPV is biphasic when induced by switching from a normoxic (155–160 Torr) to a hypoxic gas mixture (16–21 Torr; Fig. 3A). Thus, hypoxia induces an initial transient constriction (phase 1), followed by a slow tonic constriction (phase 2; Dipp & Evans 2001, Dipp et al 2001, Wilson et al 2001). The initial transient constriction peaks within 5–10 min of the hypoxic challenge, whilst the secondary, tonic constriction peaks after 30–40 min (Dipp & Evans 2001, Dipp et al 2001, Wilson et al 2001). In arteries without endothelium, we have shown that the initial transient constriction is followed by a maintained plateau constriction, which persists for the duration of exposure to hypoxia and reverses rapidly on return to normoxia (Fig. 3B; Dipp & Evans 2001, Dipp et al 2001). Thus, the development of the slow tonic

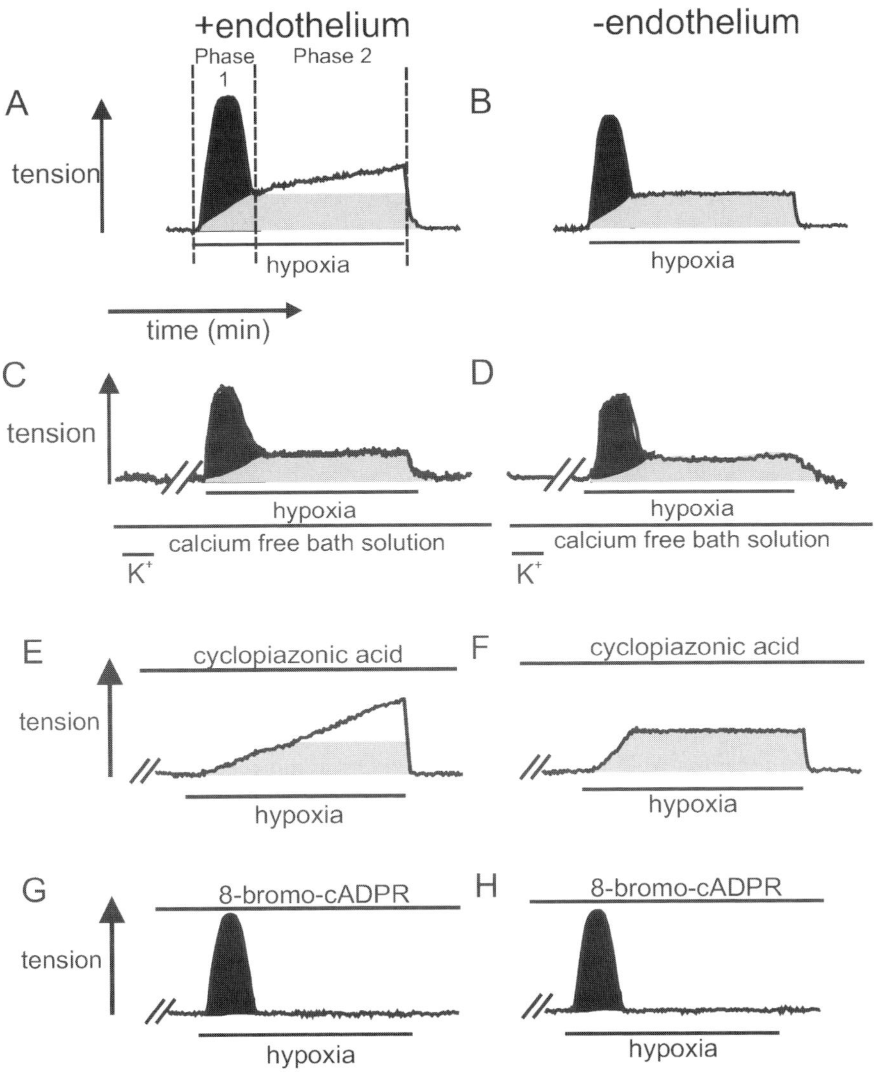

FIG. 3. Mobilization of sarcoplasmic reticulum Ca^{2+} stores via cADPR-independent and cADPR-dependent mechanisms underpins HPV. (A) Record indicating phase 1 and phase 2 of the response of an intact pulmonary artery ring to hypoxia (16–21 Torr). (B) Constriction by hypoxia (16–21 Torr) of a pulmonary artery ring without endothelium. (C) Constriction by hypoxia of an intact pulmonary artery ring in the absence of extracellular Ca^{2+}. (D) Constriction by hypoxia of a pulmonary artery ring without endothelium and in the absence of extracellular Ca^{2+}. (E) Constriction by hypoxia of an artery ring with endothelium in the presence 8-bromo-cADPR. (F) Constriction by hypoxia of an artery ring without endothelium in the presence of 8-bromo-cADPR. (G) Constriction by hypoxia of an artery ring with endothelium in the presence of cyclopiazonic acid. (H) Constriction by hypoxia of an artery ring without endothelium in the presence of cyclopiazonic acid. Black indicates phase 1 constriction of smooth muscle, grey indicates phase 2 constriction of smooth muscle, and white indicates endothelium-dependent component of phase 2 constriction.

constriction is endothelium-dependent (Dipp et al 2001), whilst the initial transient constriction and maintained plateau are mediated by mechanisms intrinsic to the smooth muscle (Dipp & Evans 2001, Dipp et al 2001, Wilson et al 2001). Surprisingly, considering the proposal that inhibition of O_2-sensitive K^+ channels and subsequent voltage-gated Ca^{2+} influx triggers HPV (Post et al 1992, Olschewski et al 2002, Osipenko et al 1997), we found that maintained HPV in pulmonary arteries without endothelium could be evoked even after removal of extracellular Ca^{2+} (Fig. 3D; Dipp et al 2001). Importantly, and in marked contrast, constriction by K^+ (i.e. depolarization and voltage-gated Ca^{2+} influx) was abolished (Fig. 3C–D, Dipp et al 2001). Of further significance was our finding that removal of extracellular Ca^{2+} abolished the endothelium-dependent constriction by hypoxia (Fig. 3C). Thus, contrary to our findings on the smooth muscle, we find that transmembrane Ca^{2+} influx into the pulmonary artery endothelium in response to hypoxia is required to initiate vasoconstrictor release (Dipp et al 2001).

Given the above, we investigated the possibility that the component of HPV intrinsic to the smooth muscle was triggered by the mobilization of intracellular Ca^{2+} stores. Depletion of ryanodine-sensitive SR Ca^{2+} stores by pre-incubation with ryanodine (10 μM) and caffeine (10 mM) abolished constriction of pulmonary arteries by hypoxia (not shown), whilst constriction by K^+ remained unaffected in the presence of ryanodine, caffeine and hypoxia (Dipp et al 2001, Dipp & Evans 2001). Clearly, therefore, smooth muscle SR Ca^{2+} release via ryanodine receptors is a prerequisite for HPV, whilst voltage-gated Ca^{2+} influx is not.

Strikingly, however, in arteries with and without the endothelium, 8-bromo-cADPR, a selective cADPR antagonist, had no effect on phase 1 of HPV, but abolished phase 2 in the presence of the endothelium, and blocked the maintained plateau constriction observed in arteries without endothelium (Fig. 3E–F; Dipp & Evans 2001, Wilson et al 2001). Thus, while cADPR-dependent SR Ca^{2+} release may maintain acute HPV in isolated pulmonary artery rings, it does not mediate the phase 1 constriction (Dipp & Evans 2001, Wilson et al 2001). In marked contrast, we found the sarco/endoplasmic reticulum Ca^{2+} ATPase (SERCA) antagonist cyclopiazonic acid to block only phase 1 of HPV, without any perceptible effect on the maintained phase 2 constriction of pulmonary arteries with or without endothelium (Fig. 3G–H; Dipp et al 2001). This is despite our finding that Ca^{2+} release from ryanodine-sensitive SR stores in the smooth muscle underpins both phases of HPV. We propose, therefore, that phase 1 may be mediated, in part, by the mobilization of Ca^{2+} from a peripheral SR compartment that is served by a cyclopiazonic acid-sensitive SERCA. This may be driven by a fall in ATP supply to this cellular compartment that is only sufficient for consequent inhibition of Ca^{2+} sequestration by this SERCA pump to its associated SR store (Dipp & Evans 2001, Evans & Dipp 2002). However, to achieve this whilst allowing for a second phase of maintained cADPR-dependent SR Ca^{2+} release, one would require the presence of a second, spatially segregated SR Ca^{2+} store that is served by a discrete SERCA subtype insensitive to cyclopiazonic acid (Boittin et al 2003, Dipp & Evans 2001, Evans & Dipp 2002).

If the aforementioned proposals are confirmed, a secondary action of hypoxia may be to deplete peripheral SR compartments that are in close apposition to the plasma membrane, served by a cyclopiazonic acid-sensitive SERCA pump, and which normally mediate vasodilation (Boittin et al 2003, Evans & Dipp 2004). This could explain, at least in part, why β-adrenoceptor induced pulmonary artery dilation is attenuated by hypoxic pulmonary hypertension (Boittin et al 2003, McIntyre et al 1994).

Maintained cADPR-dependent constriction of pulmonary arteries by AMPK

Before we could be certain of a role for AMPK in HPV a detailed comparison of the effects of hypoxia and AMPK activation by AICAR on isolated pulmonary arteries was required. Consistent with the time course of the 2nd maintained phase of pulmonary artery constriction by hypoxia (Dipp et al 2001, Dipp & Evans 2001), AICAR induced a slow, sustained constriction of pulmonary artery rings (Fig. 4A). This was reversed rapidly on washing, consistent with the active metabolite of AICAR, ZMP, being metabolised at 37 °C (Corton et al 1995). There was little evidence, however, of any component of constriction equivalent to the transient phase 1 constriction of pulmonary arteries in response to hypoxia (see above). However, it may be premature to rule out any contribution of AMPK to this event, as this may be obscured by the pharmacokinetics of AICAR.

The precise characteristics of the response to AMPK activation by AICAR also compared well with the phase 2 constriction by hypoxia. Removal of the pulmonary artery endothelium reduced the constriction in response to hypoxia and AICAR by approximately 28% and 29%, respectively ($n = 4$; Fig. 4A–E). Furthermore, the endothelium-dependent component of constriction by AICAR (Fig. 4A–E) and hypoxia (Fig. 4E, Dipp et al 2001) was abolished upon removal of extracellular Ca^{2+}. In contrast, constriction mediated by mechanisms intrinsic to the smooth muscle was not, although it is notable that in rat pulmonary arteries this component of constriction was attenuated in the absence of extracellular Ca^{2+} (Fig. 4E). Thus, maintained constriction of rat pulmonary artery smooth muscle by AICAR and hypoxia, respectively, exhibits a marginal dependence on transmembrane Ca^{2+} influx. In this respect it is of major significance that depletion of SR Ca^{2+} stores with caffeine (10 mM) and ryanodine (10 μM) or blockade of cADPR with 8-bromo-cADPR completely abolished the constriction of pulmonary arteries, with or without endothelium, by both AICAR and hypoxia (Fig. 4C–E). Thus, the partial dependence of smooth muscle constriction on extracellular Ca^{2+} must be dependent in some way upon SR store depletion-activated Ca^{2+} influx. Clearly, however,

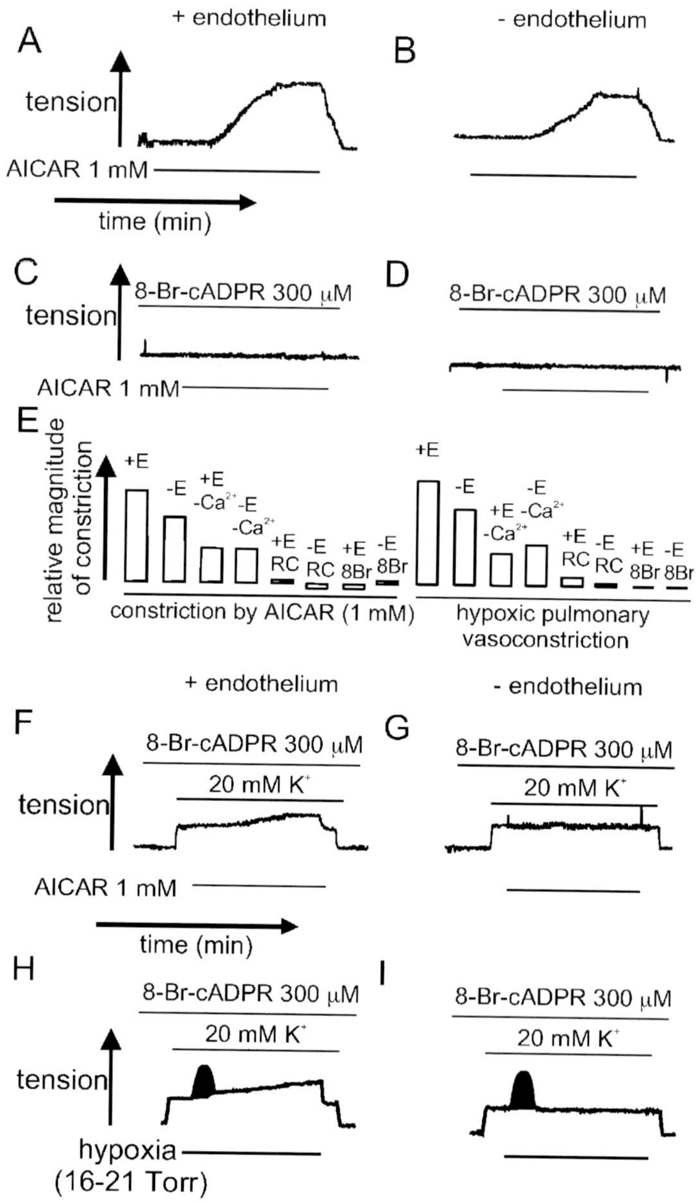

FIG. 4. AMPK activation by AICAR replicates maintained hypoxic pulmonary vasoconstriction. Constriction of isolated pulmonary artery ring in response to AMPK activation by AICAR (1 mM) with (A) and without (B) the endothelium, and the effect (C and D) of pre-incubation with 8-bromo-cADPR (300 μM). (E) Mean relative constriction of isolated pulmonary artery rings by AICAR (1 mM; left panel) and hypoxia (right panel) under the conditions studied; E, endothelium; RC, ryanodine and caffeine; 8Br, 8-bromo-cADPR. Constriction of a pulmonary artery ring with (F) and without (G) endothelium in response to AICAR (1 mM) after pre-incubation with 8-bromo-cADPR (300 μM) and submaximal pre-constriction by K^+ (20 mM). Constriction of a pulmonary artery ring with (H) and without (I) endothelium in response to hypoxia (16–21 Torr) after pre-incubation with 8-bromo-cADPR (300 μM) and submaximal pre-constriction by K^+ (20 mM). Note: black indicates cADPR-independent phase 1 constriction.

this process is not activated by either hypoxia (Kang et al 2003) or AMPK activation by AICAR *per se*. *Note:* constriction by hypoxia of rabbit pulmonary arteries without endothelium, unlike those from the rat, appears insensitive to removal of extracellular Ca^{2+} for the maximum duration studied (40 min; Dipp et al 2001).

Importantly, pre-constriction of pulmonary arteries by K^+-induced depolarization restored the endothelium-dependent component of constriction by AICAR in the continued presence of 8-bromo-cADPR (Fig. 4F), but not that mediated by mechanisms intrinsic to the smooth muscle (Fig. 4G). This is consistent with the effects of 8-bromo-cADPR on HPV (Fig. 4H–I, Dipp et al 2001, Dipp & Evans 2001, Wilson et al 2001), and reflects the fact that the endothelium-derived vasoconstrictor released by hypoxia is unable to constrict pulmonary arterial smooth muscle in the absence of SR Ca^{2+} release by cADPR (Dipp et al 2001, Dipp & Evans 2001, Wilson et al 2001), because it promotes constriction by sensitizing the contractile apparatus to Ca^{2+} (Robertson et al 2001).

AMPK, voltage-gated Ca^{2+} influx and carotid body excitation

The aforementioned findings strongly support our assertion that inhibition of mitochondrial oxidative phosphorylation by hypoxia leads to a rise in the cellular AMP/ATP ratio and consequent activation of AMPK, and that this in combination with an increase in β-NADH formation mediates HPV, in part, by cADPR-dependent SR Ca^{2+} release in pulmonary arterial smooth muscle. However, it is clear that cADPR-dependent SR Ca^{2+} release is not the primary mechanism of chemotransduction by hypoxia in other O_2-sensing cell types. We therefore sought to determine whether or not metabolic-sensing by AMPK mediates Ca^{2+} signalling in other O_2-sensing cells, and focused our attention on the carotid body glomus cell. These cells are also stimulated by hypoxia and, at least in part, via inhibition of mitochondrial oxidative phosphorylation (Heymans et al 1930, Mills & Jobsis 1972, Duchen & Biscoe 1992, Wyatt & Buckler 2004). In this case, however, excitation is primarily mediated by voltage-gated Ca^{2+} influx, rather than by Ca^{2+} release from intracellular stores, which leads ultimately to neurosecretion (Buckler & Vaughan-Jones 1994, Pepper et al 1995, Wyatt et al 1995, Wyatt & Buckler 2004). Immunofluorescence imaging of AMPK-α1 and 3D reconstruction of acquired z sections (Kinnear et al 2004) revealed that, in marked contrast to its cytoplasmic distribution in pulmonary arterial smooth muscle cells (PASMCs), the AMPK-α1 catalytic subunit isoform was almost entirely restricted to the plasma membrane of carotid body glomus cells (Fig. 5A). Thus, the spatial localization of AMPK in carotid body glomus cells is consistent with it targeting plasma membrane delimited processes such as voltage-gated Ca^{2+} influx. Consistent with this proposal and with the effects of hypoxia, AMPK activation by AICAR induced an increase in intracellular Ca^{2+} concentration in acutely isolated carotid body glomus cells. This increase was abol-

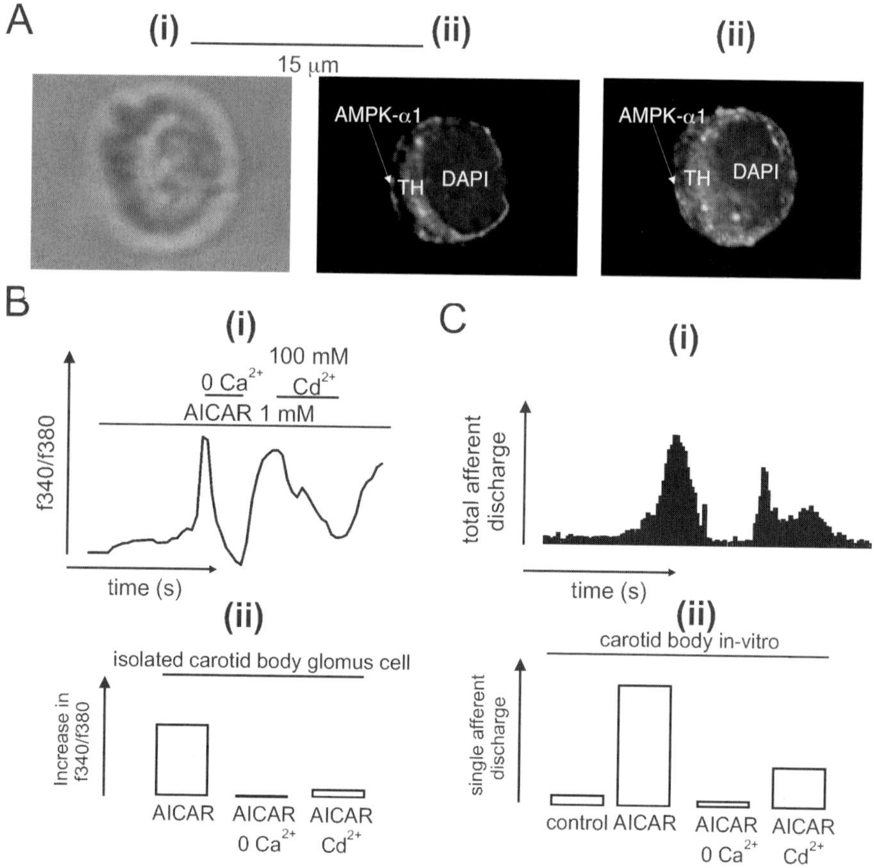

FIG. 5. AMPK activation by AICAR activates carotid body glomus cells by voltage-gated Ca^{2+} influx as does hypoxia. (A)(i) brightfield image of a fixed carotid body glomus cell, (ii) z-section showing staining by antibodies to AMPK-α1, tyrosine hydroxylase (TH) and DAPI staining of the nucleus, (iii) 3D reconstruction. (B)(i) Effect of AMPK activation by AICAR on Fura-2 fluorescence ratio (F340/F380) in an isolated carotid body glomus cell with and without extracellular Ca^{2+} (+ 1 mM EGTA) and Cd^{2+} (100 μM). (B)(ii) mean relative change for experiments in B(i). (C)(i) Frequency histogram of action potential spikes shows the effect of AICAR on multiple-fibre afferent discharge from the carotid body *in vitro* with and without extracellular Ca^{2+} (+1 mM EGTA) and Cd^{2+} (100 μM). (C)(ii), mean relative change in total afferent fibre discharge in response to AICAR in the absence and presence of extracellular Ca^{2+} and Cd^{2+} (100 μM).

ished by removal of extracellular Ca^{2+} and attenuated by blockade of voltage-gated Ca^{2+} influx with Cd^{2+} (Fig. 5B). Furthermore, AICAR, like hypoxia (Pepper et al 1995), induced an increase in single fibre sensory afferent discharge from the isolated carotid body. This too was abolished by removal of extracellular Ca^{2+} and was attenuated by blockade of voltage-gated Ca^{2+} influx with Cd^{2+} (Fig. 5C). Thus,

AMPK activation reproduced the precise excitatory effects of hypoxia on isolated carotid body glomus cells and on the carotid body *in vitro*.

In summary, previous studies have established that hypoxia promotes pulmonary artery constriction and carotid body sensory afferent discharge in part by inhibiting oxidative phosphorylation by mitochondria (Heymans et al 1930, Mills & Jobsis 1972, Rounds & McMurtry 1981, Duchen & Biscoe 1992, Leach et al 2001, Wyatt & Buckler 2004). Our findings now suggest that inhibition of mitochondrial oxidative phosphorylation by hypoxia leads to a rise in the cellular AMP/ATP ratio, consequent AMPK activation and the initiation of cell-specific Ca^{2+} signalling mechanisms in pulmonary arterial smooth muscle and the carotid body (Fig. 4E). Thus, the characteristic response of each tissue type to AMPK activation precisely mirrors that seen in response to hypoxia: (A) constriction of pulmonary arteries by cADPR-dependent SR Ca^{2+} release in the smooth muscle cells, with a secondary component of constriction driven by the pulmonary arterial endothelium (Dipp et al 2001, Wilson et al 2001, Dipp & Evans 2001); (B) voltage-gated Ca^{2+} influx into carotid body glomus cells and a consequent increase in afferent fibre discharge (Buckler & Vaughan-Jones 1994, Pepper et al 1995, Wyatt et al 1995, Wyatt & Buckler 2004). We propose, therefore, that AMPK acts as the primary metabolic sensor and effector in O_2-sensing cells. This view unites for the first time the mitochondrial and Ca^{2+} signalling hypotheses for chemotransduction by hypoxia (Gonzalez et al 2002, Wyatt & Buckler 2004). Our data also suggest that high levels of AMPK-α1 expression, together with a reliance on mitochondrial oxidative phosphorylation for ATP production, may explain why O_2-sensing cells are particularly sensitive to hypoxia. In common with its effects on mammalian pulmonary arteries, hypoxia also constricts the arteries that feed the gills of the Pacific hagfish and the skin of the amphibian *Xenopus laevis* (Malvin & Walker 2001, Olson et al 2001). Thus, the ability to target acute O_2-sensitive Ca^{2+} signalling mechanisms to specialized cells must have been a relatively early evolutionary development. It seems likely, therefore, that AMPK-dependent Ca^{2+} signalling by hypoxia may contribute to O_2-sensitive Ca^{2+} signalling cascades in a number of other systems, such as neonatal adrenomedullary chromaffin cells where hypoxia initiates catecholamine release required to prepare the newborn lung for breathing (Thompson & Nurse 1998), hypoxia-induced neuro-excitability that contributes to excitotoxicity and neuronal apoptosis (Jiang & Haddad 1994), and ischaemic damage to cardiac muscle (Marsin et al 2000).

References

Boittin F-X, Dipp M, Kinnear NP, Galione A, Evans AM 2003 Vasodilation by the calcium mobilising messenger cyclic ADP-ribose. J Biol Chem 278:9602–9608

Buckler KJ, Vaughan-Jones RD 1994 Effects of hypoxia on membrane potential and intracellular calcium in rat neonatal carotid body type I cells. J Physiol 476:423–428

Cheung PC, Salt IP, Davies SP, Hardie DG, Carling D 2000 Characterization of AMP-activated protein kinase gamma-subunit isoforms and their role in AMP binding. Biochem J 346:659–669

Corton JM, Gillespie JG, Hardie DG 1994 Role of the AMP-activated protein kinase in the cellular stress response. Curr Biol 4:315–324

Corton JM, Gillespie JG, Hawley SA, Hardie DG 1995 5-Aminoimidazole-4-carboxamide ribonucleoside: a specific method for activating AMP-activated protein kinase in intact cells? Eur J Biochem 229:558–565

Dipp M, Evans AM 2001 cADPR is the primary trigger for hypoxic pulmonary vasoconstriction in the rat lung *in-situ*. Circ Res 89:77–83

Dipp M, Nye PCG, Evans AM 2001 Hypoxic release of calcium from the sarcoplasmic reticulum of pulmonary artery smooth muscle. Am J Physiol 281:L318–L325

Duchen MR, Biscoe TJ 1992 Relative mitochondrial membrane potential and $[Ca^{2+}]_i$ in type I cells isolated from the rabbit carotid body. J Physiol 450:33–61

Durante PE, Mustard KJ, Park SH, Winder WW, Hardie DG 2002 Effects of endurance training on activity and expression of AMP-activated protein kinase isoforms in rat muscles. Am J Physiol Endocrinol Metab 283: E178–186

El-Mir MY, Nogueira V, Fontaine E, Averet N, Rigoulet M, Leverve X 2000 Dimethylbiguanide inhibits cell respiration via an indirect effect targeted on the respiratory chain complex I. J Biol Chem 275:223–228

Evans AM 2004 Hypoxia cell metabolism and cADPR accumulation. In: Yuan X-J (ed) Hypoxic pulmonary vasoconstriction: cellular and molecular mechanisms. Kluwer Academic Publications, p 313–338

Evans AM, Dipp M 2002 Hypoxic pulmonary vasoconstriction: cyclic adenosine diphosphate-ribose, smooth muscle calcium stores and the endothelium. Resp Physiol Neurobiol 132:3–15

Evans AM, Dipp M 2004 Calcium mobilisation by hypoxia in pulmonary artery smooth muscle. In: X-J Yuan (ed) Hypoxic pulmonary vasoconstriction: cellular and molecular mechanisms, Kluwer Academic Publications, Dordrecht p 81–101

Galione A, Lee HC, Busa WB 1991 Ca^{2+}-induced Ca^{2+} release in sea urchin egg homogenates: modulation by cyclic ADP-ribose. Science 253:1143–1146

Gonzalez C, Sanz-Alfayate G, Agapito T, Gomez-Nino A, Rocher A, Obeso A 2002 Significance of ROS in oxygen sensing in cell systems with sensitivity to physiological hypoxia. Respiratory Physiol Neurobiol 132:17–41

Hardie DG 2004 The AMP-activated protein kinase pathway—new players upstream and downstream. J Cell Sci 117:5479–5487

Hardie DG, Salt IP, Hawley SA, Davies SP 1999 AMP-activated protein kinase: an ultrasensitive system for monitoring cellular energy charge. Biochem J 338:717–722

Hardie DG, Hawley SA 2001 AMP-activated protein kinase: the energy charge hypothesis revisited. BioEssays 23:1112–1119

Hardie DG, Scott JW, Pan DA, Hudson ER 2003 Management of cellular energy by the AMP-activated protein kinase system. FEBS Lett 546:113–120

Hawley SA, Davison M, Woods A et al 1996 Characterization of the AMP-activated protein kinase kinase from rat liver and identification of threonine 172 as the major site at which it phosphorylates AMP-activated protein kinase. J Biol Chem 271:27879–27887

Hawley SA, Boudeau J, Reid JL et al 2003 Complexes between the LKB1 tumor suppressor, STRAD$\alpha\beta$ and MO25α/β are upstream kinases in the AMP-activated protein kinase cascade. J Biol 2:28

Heymans C, Bouckaert JJ, Dautrebande L 1930 Sinus carotidien et reflexes respiratoires. II. Influences respiratoires reflexes de l'acidose, de l'alcolose, de l'anhydride carbonique, de l'ion hydrogene et de l'anoxemie: sinus carotidiens et echanges respiratoires dans les poumons et au dela de poumons. Arch Int Pharmacodyn Ther 39:400–408

Jiang C, Haddad GG 1994 A direct mechanism for sensing low oxygen levels by central neurons. Proc Natl Acad Sci USA 91:7198–7201

Kang TM, Park MK, Uhm DY 2003 Effects of hypoxia and mitochondrial inhibition on the capacitative calcium entry in rabbit pulmonary arterial smooth muscle cells. Life Sci 72:1467–1479
Kato M, Staub NC 1966 Response of small pulmonary arteries to unilobar hypoxia and hypercapnia. Circ Res 19:426–440
Kinnear NP, Boittin F-X, Thomas JM, Galione A, Evans AM 2004 Lysosome clusters in arterial smooth muscle: a trigger zone for Ca^{2+} signalling by NAADP and endothelin-1. J Biol Chem 279:54319–54326
Leach RM, Hill HS, Snetkov VA, Robertson TP, Ward JPT 2001 Divergent roles of glycolysis and the mitochondrial electron transport chain in hypoxic pulmonary vasoconstriction of the rat. Identity of the hypoxic sensor. J Physiol 536:211–224
Leach RM, Hill HS, Snetkov VA, Ward JP 2002 Hypoxia, energy state and pulmonary vasomotor tone. Respir Physiol Neurobiol 132:55–67
Lee HC, Walseth TF, Bratt GT, Hayes RN, Clapper DL 1989 Structural determination of a cyclic metabolite of NAD with intracellular calcium-mobilizing activity. J Biol Chem 264:1608–1615
Malvin GM, Walker BR 2001 Sites and ionic mechanisms of hypoxic pulmonary vasoconstriction in frog skin. Am J Physiol Regul Integr Comp Physiol 280:R1308–R1314
Marsin AS, Bertrand L, Rider MH et al 2000 Phosphorylation and activation of heart PFK-2 by AMPK has a role in the stimulation of glycolysis during ischaemia. Curr Biol 10:1247–1255
McIntyre RC, Banerjee A, Bensard DD, Brew EC, Hahn AR, Fullerton DA 1994 Adenosine A1-receptor mechanisms antagonize beta-adrenergic pulmonary vasodilation in hypoxia. Am J Physiol 267:H2179–H2185
Mills E, Jobsis FF 1972 Mitochondrial respiratory chain of carotid body and chemoreceptor response to changes in oxygen tension. J Neurophysiol 35:405–428
Olschewski A, Hong Z, Nelson D, Weir EK 2002 Graded response of K^+ current, membrane, and $[Ca^{2+}]$ to hypoxia in pulmonary arterial smooth muscle. Am J Physiol 283:L1143–L1150
Olson KR, Russell MJ, Forster ME 2001 Hypoxic pulmonary vasoconstriction of cyclostome systemic vessels: the antecedent of hypoxic pulmonary vasoconstriction? Am J Physiol Regul Integr Comp Physiol 280:R198–R206
Osipenko ON, Evans AM, Gurney AM 1997 A novel O_2-sensing potassium channel may mediate hypoxic pulmonary vasoconstriction. Br J Pharmacol 120:1461–1470
Owen MR, Doran E, Halestrap AP 2000 Evidence that metformin exerts its anti-diabetic effects through inhibition of complex 1 of the mitochondrial respiratory chain. Biochem J 348: 607–614
Pepper DR, Landauer RC, Kumar P 1995 Post-natal development of CO_2-O_2 interaction in the rat carotid body *in-vitro*. J Physiol 485:531–541
Post J, Hume J, Archer S, Weir E 1992 Direct role for potassium channel inhibition in hypoxic pulmonary vasoconstriction. Am J Physiol 262:C882–C890
Robertson TP, Ward JPT, Aaronson PI 2001 Hypoxia induces the release of a pulmonary-selective, Ca^{2+}-sensitising, vasoconstrictor from the perfused rat lung. Cardiovasc Res 50:145–150
Rounds S, McMurtry IF 1981 Inhibitors of oxidative ATP production cause transient vasoconstriction and block subsequent pressor responses in rat lungs. Circ Res 48:393–400
Roy CS, Sherrington CS 1890 The regulation of the blood supply of the brain. J Physiol 11:85
Salvaterra CG, Goldman WF 1993 Acute hypoxia increases cytosolic calcium in cultured pulmonary arterial myocytes. Am J Physiol 264:L323–L328
Schrauwen P, Hardie DG, Roorda B et al 2004 Improved glucose homeostasis in mice overexpressing human UCP3: a role for AMP-kinase? Int J Obes Relat Metab Disord 28:824–828
Shaw RJ, Bardeesy N, Manning BD et al 2004 The LKB1 tumor suppressor negatively regulates mTOR signaling. Cancer Cell 6:91–99
Shigemori K, Ishizaki T, Matsukawa S, Sakai A, Nakai T, Miyabo S 1996 Adenine nucleotides via activation of ATP-sensitive K^+ channels modulate hypoxic response in rat pulmonary artery. Am J Physiol 270:L803–L809

Thompson RJ, Nurse CA 1998 Anoxia differentially modulates multiple K^+ currents and depolarizes neonatal rat adrenal chromaffin cells. J Physiol 512:421–434
von Euler US, Liljestrand G 1946 Observations on the pulmonary arterial blood pressure in the cat. Acta Physiol Scand 12:301–320
Wilson HL, Dipp M, Thomas JM, Lad C, Galione A, Evans AM 2001 ADP-ribosyl cyclase and cyclic ADP-ribose hydrolase act as a redox sensor: a primary role for cyclic ADP-ribose in hypoxic pulmonary vasoconstriction. J Biol Chem 276:11180–11188
Winder WW, Hardie DG 1996 Inactivation of acetyl-CoA carboxylase and activation of AMP-activated protein kinase in muscle during exercise. Am J Physiol 270:E299–E304
Woods A, Johnstone SR, Dickerson K et al 2003 LKB1 is the upstream kinase in the AMP-activated protein kinase cascade. Curr Biol 13:2004–2008
Wyatt CN, Buckler KJ 2004 The effect of mitochondrial inhibitors on membrane currents in isolated neonatal rat carotid body type I cells. J Physiol 556:175–191
Wyatt CN, Wright C, Bee D, Peers C 1995 O_2-sensitive K^+ currents in carotid body chemoreceptor cells from normoxic and chronically hypoxic rats and their roles in chemotransduction. Proc Natl Acad Sci USA 92:295–299

DISCUSSION

Duchen: You said that the NADH autofluorescence measurements were done with confocal imaging. Can you identify the site of change? Do you see mitochondrial signals or cytosolic signals change independently?

Evans: There were some punctate regions. However, I can't say I really looked in sufficient detail to comment on the site of β-NADH production.

Ward: We have done imaging in rat pulmonary artery cells, with co-localization. The rise in NADH fluorescence in response to hypoxia is almost entirely mitochondrial.

Duchen: That doesn't surprise me, but I was asking about these agents AICAR and phenformin. From your model, what you would expect is a rise in cytosolic NADH.

Evans: You are correct in the case of phenformin. This compound is a weak inhibitor of complex I and you would therefore expect it to inhibit mitochondrial β-NADH consumption and thereby raise cytoplasmic β-NADH levels. In contrast, AICAR is without direct effect on mitochondrial metabolism, and would not be expected to have marked effects on mitochondrial β-NADH consumption nor cytoplasmic β-NADH. However, we haven't looked carefully enough to determine the site of β-NADH accumulation by phenformin.

Duchen: The time course of that response is slow.

Evans: These experiments were carried out at room temperature, and will be slowed because of this fact. Furthermore, the time course by which phenformin enters the cell is known to be very slow. AICAR also mediates its effects relatively slowly because it first needs to enter the cell, following which adenosine kinase acts to yield the active metabolite of AICAR, ZMP. It is notable that the action of these compounds on intracellular Ca^{2+} concentration is also slow at room temperature, although the increase in β-NADH levels in response to phenformin is faster than the rate at which phenformin elicits an increase in intracellular Ca^{2+} concentration. Thus mitochondrial inhibition by phenformin likely precedes, as our hypothesis predicts, the increase in intracellular Ca^{2+} concentration. It is important to consider, however, the fact that phenformin will not completely block mitochondrial function. It is a very weak inhibitor of complex I compared with many of the classical blockers.

Duchen: How do those changes in NADH fluorescence that you saw compare in scale with the response to hypoxia or metabolic inhibition with cyanide?

Evans: Good question. I don't have an answer.

Nurse: How do you see AMP kinase regulating the Ca^{2+} channels in the glomus cell membranes? Is it a case of regulating the voltage dependence of activation of the channels?

Evans: If the effects are, as we suspect, consistent with the effects of hypoxia, then I would expect AMP-activated protein kinase (AMPK) to induce membrane depolarization by inhibiting the cellular potassium conductance, which would in turn lead to voltage-gated Ca^{2+} influx. We haven't carried out the investigations thoroughly enough, in terms of the electrophysiology, to comment at this stage. However, there is clear evidence of AMPK-dependent phosphorylation of the CFTR Cl^- channel (Hallows 2005). There is also regulation due to phosphorylation by AMPK of the cardiac Na^+ channel and of direct phosphorylation of voltage-gated Ca^{2+} channels. There is also evidence to suggest that there is a consensus sequence for AMPK phosphorylation within Ca^{2+} channel constructs. In short, there are lots of different targets for AMPK, of which ion channels are one. And it seems likely that AMPK-dependent phosphorylation could either inhibit or facilitate ion channel activity in a manner dependent on the nature of the ion channel under investigation.

Chandel: Central to this hypothesis is that hypoxia causes metabolic stress in the mitochondria. At the pO2s that you are using there is no evidence that hypoxia should do that. Mitochondria, in most systems, aren't compromised until you go to 3 Torr or less. It could be that specialized cells, such as those in the carotid body or pulmonary circulation, might have a different kind of cytochrome oxidase and can experience metabolic stress at 15 Torr. This is a speculation.

Evans: That is why we had to carry out the experiments on the effect of hypoxia on the AMP/ATP ratio and AMPK activation in pulmonary artery smooth muscle (PASM). Our data show that hypoxia elicits a clear and consistent increase in the AMP/ATP ratio and consequent AMPK activation in PASM. Oxygen-sensitive tissues could indeed be more susceptible to AMPK activation by hypoxia. This is a central part of our original hypothesis (Evans 2004).

Chandel: I was surprised this happened, because you would not predict it on the basis of the mitochondrial biology. As a control experiment, take another tissue instead of these specialized cells, and see whether it is the same. A lot of people have been looking in hypoxia for AMP kinase activation and they don't see it because hypoxia doesn't alter the ATP/ADP/AMP ratios. Again, there could be something special about these cells bioenergetically.

Sylvester: Regardless of those considerations, pO2s in the 20s cause marked diminution of the phosphorylation potential in isolated systemic vessels but not in pulmonary vessels (Leach et al 1998).

Chandel: That might be true, but in this case we want it the other way round.

Sylvester: I'm just addressing the issue of hypoxia and metabolic stress. Perfusate pO2s in the 20s caused a big decrease of phosphorylation potential in resting systemic arterial smooth muscle (Leach et al 1998) Presumably, the mitochondria were involved in this effect. Of course, pO2 could be lower at the mitochondria because of diffusion gradients.

Duchen: Jeremy Ward, you said that hypoxia causes a big increase in mitochondrial NADPH autofluorescence. At what level of oxygen tension does this occur?

Ward: We have done the dose–response to it. It is at an approximate K_m of 20 mmHg. It is over the sort of range we see physiological responses to hypoxia. We would interpret this as showing that hypoxia over the range we are looking at is actually slowing electron transport, or at least NADH consumption.

Duchen: Is this specifically in PASM?

Ward: Yes. If we look at mesenteric artery, it is difficult to do the stats, but the two points down at the levels we are talking about are pushed to the right in mesenteric artery. In other words, the pulmonary artery seems to be a lot more sensitive to hypoxia in terms of this NADH measure.

Evans: Further to the discussion on the likely impact of hypoxia on the cellular redox state and ROS generation, AMP kinase might have an effect. This might take longer to occur than the time course of events that we are discussing here, but AMP kinase activates β oxidation of fatty acids by mitochondria. What is the impact of this on mitochondrial redox potential?

Rich: With β oxidation a lot of NADH is produced. Hence, I imagine that, rather than have TCA cycle feeding in at several places, you would have primarily NADH donating into the chain. I guess this would result in increased levels of NADH and CO_2 production, together with a high proton motive force.

González: The ATP : AMP ratios you got in pulmonary artery were quite low.

Evans: This is something that we have noticed ourselves. The adenine nucleotides of interest are susceptible to breakdown if the samples are allowed to partially thaw at any stage prior to extraction, and this could result in a marked fall in ATP levels and possibly an increase in AMP levels. We have now carried out further assays. These assays now provide AMP/ATP ratios that are closer to those which one might expect. Under normoxic conditions the AMP/ATP ratio is 0.04, and this increases to 0.083 under hypoxia. This suggests that the adenylate kinase reaction is close to equilibrium, because the AMP/ATP ratio varies approximately as the square of the ADP/ATP ratio.

Gonzalez: It could be important to have the real absolute values. We are linking K+ channels, NADH and ATP, and this link might very well represent some kind of unifying mechanism

Evans: It would be great if it was, but I wouldn't like to claim this now. We can't determine accurately the absolute values for ATP/ADP/AMP using the techniques currently available. Only the nucleotide ratios can be measured with accuracy and reproducibility.

Archer: There are probably a dozen studies looking at L-type Ca^{2+} channel blockers and agonists. These are fairly specific blockers. In the ductus, pulmonary artery and carotid body in our hands L-type Ca^{2+} channel blockers blockers get rid of about 80% of hypoxic pulmonary vasoconstriction (HPV). They do this in the isolated lung and in people. For example, a dose of 20 mg nifedipine eliminates 50% of HPV in humans without lowering systemic blood pressure (Burghuber 1987). This was a dose that did not affect systemic vascular tone. We published a series of papers where we did this in dogs and inhibited HPV (Archer et al 1985). Likewise, Ivan McMurtry showed that verapamil eliminates all HPV in an isolated rat lung (McMurtry et al 1976).

Evans: I agree with you. I initially entered the field of HPV by considering the role of the inhibition of the cellular K^+ conductance by hypoxia and voltage-gated Ca^{2+} influx. We obtained evidence in support of the view that the K^+ conductance in pulmonary arterial smooth muscle was indeed inhibited by hypoxia. However, the depolarization we measured was less than 10 mV and may not be sufficient to promote a marked increase in intracellular Ca^{2+} concentration in PASM. The data we have obtained since are consistent with sarcoplasmic recticulum (SR) Ca^{2+} release triggering HPV, and we have little evidence to support a primary role for voltage-gated Ca^{2+} influx. We can only interpret the data we have. We find that the first transient phase of constriction, which lasts 5–10 minutes, is inhibited by block of SR Ca^{2+} release and that this phase of the response is resistant to the removal of extracellular Ca^{2+}. However, we also find that the phase 1 constriction is abolished by both nifedipine and lanthanum.

Archer: One should be cautious about the meaning of 'no Ca^{2+}'. We have examined the effects of thapsigargin, ryanodine and caffeine on HPV. We could never find a dose of these drugs that inhibited HPV without simultaneously inhibiting constriction to all other stimuli. If we give large doses of these drugs we do block HPV, but we also impair constriction to phenylephrine in KCl. We couldn't find a selective dose. We promise to do more of that if you promise to do some Bay K 8644 experiments.

Peers: You have to be careful with Bay K because it shifts the voltage dependence of L-type Ca^{2+} channels significantly.

Duchen: Can you not reconcile some of this with Ca^{2+}-induced Ca^{2+} release (CICR)?

Evans: Our studies suggest that CICR does not contribute to constriction mediated by voltage-gated Ca^{2+} influx in PASM (Dipp et al 2001, Dipp & Evans 2001). Thus, voltage-gated Ca^{2+} channels may not couple to ryanodine receptors on the SR in pulmonary arterial smooth muscle cells (PASMCs). Indirect support for this view may be taken from our finding that CICR via ryanodine receptors may be initiated via other mechanisms (see Boittin et al 2003, Kinnear et al 2004).

López-Barneo: 10 mM caffeine and 10 μM ryanodine will block HPV because they increase cytosolic Ca^{2+}, which activates maxi-K channels and hyperpolarizes the cells. In this condition, hypoxia-induced depolarization could not reach threshold.

Evans: Over the time course of the pre-incubation used in our experiments we would expect that in isolated PASMCs and isolated pulmonary arteries any Ca^{2+} transient induced by ryanodine and caffeine application would have returned to baseline and/or reached a steady-state before we exposed the preparation to hypoxia. Whatever the case may be the resting level of tone is not significantly different from control prior to exposure to hypoxia (see Dipp et al 2001). Furthermore, neither ryanodine + caffeine, nor blocking of Ca^{2+} mobilization by cADPR affects constriction to depolarization induced by either 20 mM, or 80 mM K^+ (Dipp et al 2001).

López-Barneo: I think you have shown very nicely the distribution of this AMP kinase in different tissues and cell types, but the model looked to me to be a little bit contradictory. You are stressing the cell, ATP goes down, AMP goes up and you are activating a kinase that will spend more ATP.

Evans: This is the point I was trying to emphasize. I knew nothing about AMP kinase a couple of years ago, and our understanding of the cell-specific regulation of this enzyme by cell metabolism is in its infancy. However, we do know that AMP kinase activates catabolic pathways in order to maintain cellular ATP levels. In addition AMPK inhibits non-essential ATP consuming processes. There is also evidence to suggest that the precise effects of AMPK may be cell specific. Of course, if ATP levels drop below a certain point, there will be insufficient ATP to support phosphorylation by AMPK, and this could explain some of the contrary data in the field. Providing there is sufficient ATP, however, AMPK will phosphorylate its targets and there will be a pO2 window within which this can happen.

Gurney: In the experiments you showed where the Ca^{2+} concentration increased, it seemed to stay up.

Evans: These experiments are very expensive and because of this we use a microsuperfusion pipette positioned near to the cell to allow for the use of small volumes of solution. We can't control the temperature of these solutions, and because of this we carry out the experiments at room temperature. The metabolite of AICAR is not rapidly metabolized at these temperatures and the response is, therefore, maintained.

Gurney: The reason I asked is because if you put on anything that evokes Ca^{2+} release, the signal goes up and comes back down as the Ca^{2+} is removed from the cell. In your experiments it seemed to go up and stay up.

Evans: The increase in intracellular Ca^{2+} concentration that we observe is very small. The Fura-2 fluorescence ratio increases by approximately 0.1 ratio units on average. In marked contrast, intracellular dialysis (from a patch pipette) of potent Ca^{2+} mobilizing messengers such as inositol-1,4,5-triphosphate (IP_3) and NAADP, respectively, elicits a marked but transient increase in Fura-2 fluorescence ratio, which is followed by regenerative Ca^{2+} waves or oscillations. These Ca^{2+} mobilizing messengers increase the Fura-2 fluorescence ratio from a basal level of 0.5 to 2. Our findings with the Ca^{2+} mobilizing messenger cADPR, which underpins maintained HPV, were quite different. Intracellular dialysis of cADPR induces a small but maintained increase in Fura-2 fluorescence ratio at room temperature (unpublished observation). When considering the effects of hypoxia and AMPK activation, however, we must also consider the possibility that calcium sequestration mechanisms may also be regulated. If they are inhibited then the increase in intracellular Ca^{2+} concentration triggered by these mechanisms may also be maintained for long periods, even in the absence of extracellular Ca^{2+}.

Buckler: What other products does AICAR get converted into? Its effects are clearly reversible: you alluded to the fact that it is metabolized. Is it possible that it might produce some other compound that could have a direct effect on ryanodine receptors, for example?

Evans: ZMP is metabolized by the same pathway as AMP and it is removed from the cell via this process. Briefly, by metabolism to IMP. According to my collaborator (Grahame Hardie) there is no evidence of any confounding effects of ZMP metabolites. In fact, ZMP is a naturally occurring, but minor intermediate in the purine synthesis cycle. It is usually metabolised to inosine monophosphate (IMP).

Buckler: Does that mean that it could be phosphorylated further into an analogue of ATP?

Evans: There is no evidence that I am aware of to support your proposal in terms of functional significance. However, when very high concentrations of AICAR/ZMP are present for prolonged periods of time, ZTP may be formed. This is not a naturally occurring analogue and it could mimic the effects of ATP and may even inhibit AMPK (Corton et al 1995). But levels are normally immeasurable because such insignificant amounts of ZTP are produced from ZMP.

References

Archer S, Yankovich R, Chesler E et al 1985 Comparative effects of nisoldipine, nifedipine and bepridil on experimental pulmonary hypertension. J Pharmacol Exp Ther 233:12–17

Boittin F-X, Dipp M, Kinnear NP, Galione A, Evans AM 2003 Vasodilation by the calcium mobilising messenger cyclic ADP-ribose. J Biol Chem 278:9602–9608

Burghuber OC 1987 Nifedipine attenuates acute hypoxic pulmonary vasoconstriction in patients with chronic obstructive pulmonary disease. Respiration 52:86–93

Dipp M, Evans AM 2001 cADPR is the primary trigger for hypoxic pulmonary vasoconstriction in the rat lung *in situ*. Circ Res 89:77–83

Dipp M, Nye PCG, Evans AM 2001 Hypoxic release of calcium from the sarcoplasmic reticulum of pulmonary artery smooth muscle. Am J Physiol 281:L318–L325

Evans AM 2004 Hypoxia cell metabolism and cADPR accumulation. In: Yuan X-J (ed) Hypoxic pulmonary vasoconstriction: cellular and molecular mechanisms. Kluwer Academic Publications, p 313–338

Hallows KR 2005 Emerging role of AMP-activated protein kinase in coupling membrane transport to cellular metabolism. Curr Opin Nephrol Hypertens 14:464–471

Kinnear NP, Boittin FX, Thomas JM, Galione A, Evans AM 2004 Lysosome-sarcoplasmic reticulum junctions. A trigger zone for calcium signaling by nicotinic acid adenine dinucleotide phosphate and endothelin-1. J Biol Chem 279:54319–54326

Leach RM, Sheehan DW, Chacko VP, Sylvester JT 2000 Energy state, pH, and vasomotor tone during hypoxia in precontracted pulmonary and femoral arteries. Am J Physiol Lung Cell Mol Physiol 278:L294–L304

McMurtry I, Davidson B, Reeves J et al 1976 Inhibition of hypoxic pulmonary vasoconstriction by calcium antagonists in isolated rat lungs. Circ Res 38:99–104

Role of capacitative Ca^{2+} entry but not Na^+/Ca^{2+} exchange in hypoxic pulmonary vasoconstriction in rat intrapulmonary arteries

Silke Becker, Gregory A. Knock, Vladimir Snetkov, Jeremy P. T. Ward and Philip I. Aaronson[1]

Division of Asthma, Allergy and Lung Biology, Room 2.20, New Hunt's House, King's College London, London SE1 1UL, UK

Abstract. The identities of the Ca^{2+} influx mechanisms underlying hypoxic pulmonary vasoconstriction (HPV), the contractile response of pulmonary arteries (PA) to hypoxia, remain controversial. We investigated the roles of Na^+/Ca^{2+} exchange (NCX) and capacitative Ca^{2+} entry (CCE) in HPV by measuring isometric tension in rat intrapulmonary arteries (IPA) during hypoxia. It has been shown in PA cells that hypoxia raised $[Ca^{2+}]_i$ by inhibiting NCX. We found that removal of Na^+ caused a response resembling HPV, suggesting that HPV could be due to inhibition of NCX. However, prior inhibition of NCX using Na^+ free solution or 3 μM KB-R7943 enhanced rather than prevented HPV, indicating that HPV is not caused by an inhibition of NCX. The CCE blocker 2-APB inhibited both phases of HPV, and the ryanodine receptor blocker dantrolene inhibited the 2nd phase of HPV. Taken together with previous observations, these results suggest a role for CCE in HPV, but suggest that different CCE pathways may be involved for each phase.

2005 Signalling pathways in acute oxygen sensing. Wiley, Chichester (Novartis Foundation Symposium 271) p 259–273

Hypoxic pulmonary vasoconstriction (HPV), the unique contractile response of the pulmonary vasculature to hypoxia, is associated with a rise in $[Ca^{2+}]_i$ in pulmonary artery smooth muscle cells (PASMCs)(Cornfield et al 1993) and isolated arteries (Robertson et al 2000).

---

[1]This paper was presented at the symposium by Philip I. Aaronson to whom correspondence should be addressed.

The mechanisms underlying this rise in $[Ca^{2+}]_i$ are the subject of intense controversy. Following the initial observation by Post et al (1992) that hypoxia caused an inhibition of the K^+ current in canine PASMCs, Weir, Archer and colleagues have presented extensive evidence that HPV is caused by an inhibition of voltage-gated K^+ channels, causing depolarization and the opening of voltage-gated Ca^{2+} channels (Weir & Archer 1995, see also Archer et al 2005, this volume).

However, other mechanisms can also operate to raise $[Ca^{2+}]_i$ in PASMCs during hypoxia. For example, the release of Ca^{2+} from the sarcoplasmic reticulum/endoplasmic reticulum (SR/ER) makes an important and perhaps predominant contribution to HPV when this is elicited in dog (Jabr et al 1997) and rat (Robertson et al 2000, see also Evans et al 2005, this volume) and rabbit (Dipp et al 2001) pulmonary resistance arteries.

There are two potential mechanisms by which intracellular Ca^{2+} release could contribute to the rise in $[Ca^{2+}]_i$. Firstly, the Ca^{2+} released from stores will increase $[Ca^{2+}]_i$, most likely in a transient manner since the limited amount of Ca^{2+} released will quickly be extruded from the cells. Secondly, the depletion of SR/ER $[Ca^{2+}]$ consequent on release may activate capacitative Ca^{2+} entry (CCE), leading to a Ca^{2+} influx which would be sustained as long as the store remained empty. Pulmonary arteries demonstrate a prominent CCE (Gonzalez De La Fuente et al 1995, Ng & Gurney 2001), which may make more of a contribution to contraction than in systemic arteries (Snetkov et al 2003). Moreover, multiple isoforms of TRP proteins, which probably form the channels responsible for CCE, are expressed in PASMCs (e.g. Ng & Gurney 2001).

We previously showed that La^{3+}, which is a potent inhibitor of CCE in many types of cell, almost completely blocked the initial transient phase of HPV (phase 1) in rat isolated pulmonary artery at a concentration of 1 μM. Cyclopiazonic acid (CPA), which depletes the SR/ER of Ca^{2+}, had a similar effect. These observations led us to propose that Ca^{2+} release and CCE underlie phase 1 HPV. La^{3+} at this concentration had no effect on the subsequent sustained phase of HPV (phase 2), whereas CPA caused a partial inhibition, suggesting that phase 2 HPV either did not involve CCE, or was mediated by a separate CCE pathway with different properties. Recently, Ng et al (2005) and Wang et al (2005) have provided direct evidence that CCE makes a major contribution to the hypoxia-mediated rise in $[Ca^{2+}]_i$ in dog and rat PASMCs, respectively. For example, Wang et al (2005) showed that hypoxia caused an increase in $[Ca^{2+}]_i$ which was blocked by the CCE antagonists La^{3+}, Ni^{2+}, and SKF 96365, at concentrations which did not affect the rise in $[Ca^{2+}]_i$ caused by high K^+ depolarization (i.e. by the opening of voltage-gated Ca^{2+} channels). Hypoxia was also associated with an increase in Mn^{2+} quenching of the Fura-2 signal, considered a hallmark of CCE. The rise in $[Ca^{2+}]_i$ was also partially inhibited by the voltage-gated Ca^{2+} channel antagonist nifedipine, although the contribution of these channels to the rise in $[Ca^{2+}]_i$ was unclear, since the CCE antagonists completely inhibited the rise in $[Ca^{2+}]_i$. It is noteworthy that the CCE pathway demonstrated in dog PASMCs was La^{3+} insensitive, while that in rat PASMCs was La^{3+} sensitive, underlining the heterogeneity of CCE pathways that may be activated by hypoxia in smooth muscle cells.

Wang et al (2000) demonstrated that moderate hypoxia inhibits Na^+/Ca^{2+} exchange (NCX) in rat PASMCs. By inhibiting Ca^{2+} extrusion, this would be expected to potentiate the rises in $[Ca^{2+}]_i$ resulting from any type of Ca^{2+} influx or release mechanisms activated by hypoxia. It is also possible that rise in $[Na^+]_i$, which would be expected to occur due to the activation of the CCE pathway which also conducts Na^+ in pulmonary artery cells (Snetkov et al 2003), could inhibit Ca^{2+} extrusion via NCX and contribute to any rise in $[Ca^{2+}]_i$ associated with hypoxic stimulation of CCE. However, whether NCX inhibition plays any role in HPV in intact PA is unknown.

In this paper, we present recent experiments we have performed to further explore the role of both NCX and CCE in HPV in isolated rat intrapulmonary artery (IPA). The results provide additional support for the involvement of CCE in HPV, but are not consistent with a role for NCX in HPV in this preparation.

Methods

HPV was studied by recording isometric force development in rings of rat IPA (i.d. 200–600 microns) mounted in a Mulvany-Halpern type myograph. Normoxia of the Krebs solution was maintained by gassing the recording chamber with 21% O_2/74% N_2/5% CO_2, and hypoxia was imposed by changing the gas to 95%N_2/5% CO_2. Before making IPA hypoxic, they were preconstricted to a level of ~15% of the contraction observed in 80 mM K^+ solution; this 'pretone' greatly potentiates HPV. Time controls, in which HPV was elicited twice under control conditions, were run in parallel with experiments in which a control HPV was followed by one during which a drug was added or the solution changed, in order to correct the response for time dependent changes. These methods and the solutions used have been described in detail previously (Robertson et al 2000, Snetkov et al 2003).

Results and discussion

Is inhibition of NCX involved in HPV?

Wang et al (2000) showed that hypoxia diminished Ca^{2+} extrusion in PASMCs by inhibiting NCX. If this were also to occur in isolated arteries, and to contribute to HPV, inhibition of NCX should mimic the effect of hypoxia on tension development. In addition, once NCX had been inhibited, hypoxia should cause no further effect on tension.

Figure 1 illustrates that the first prediction was fulfilled, since inhibition of NCX caused a contraction which closely resembled HPV (compare with the normal HPV illustrated on the left of Fig. 2). Substitution of the Na^+ in the Krebs solution with Li^+ and choline (see legend), which will inhibit Ca^{2+} extrusion by NCX, led to a small initial relaxation of the prostaglandin F2 alpha ($PGF_{2\alpha}$) pretone, followed by a progressive vasoconstriction (Fig. 1a). Since hypoxia is known to suppress nitric oxide (NO) release, we repeated these experiments in the presence of 1 mM L-NAME to suppress NO production. In this case, the initial relaxation to Na^+ removal was replaced by an initial transient contraction, whereas the subsequent rise in tension was not affected (Fig. 1b). Na^+ removal, and by inference inhibition of NCX, therefore caused a response which closely resembled HPV. Similar results were observed in five experiments.

However, Fig. 2 shows that the second prediction, that prior Na^+ substitution should prevent HPV, was not fulfilled. For these experiments, a control HPV was first recorded. Na^+ was then replaced by Li^+, and then $PGF_{2\alpha}$ was applied to cause pretone. In the absence of Na^+, it was necessary to reduce the $PGF_{2\alpha}$ concentration to maintain pretone at the control level. Under these conditions, phase 1 HPV

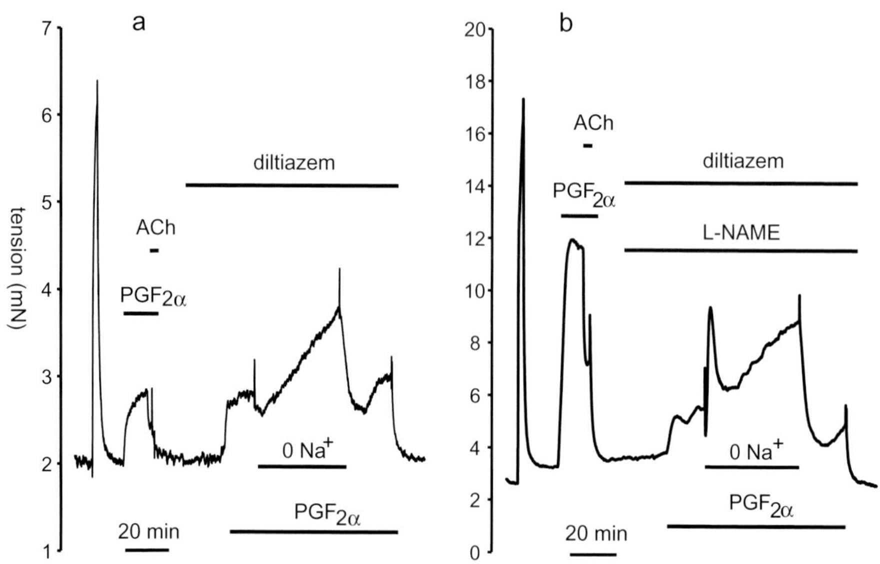

FIG. 1. (a) Substitution of Na^+ (NaCl replaced by LiCl, $NaHCO_3$ replaced by choline HCO_3), in the presence of $PGF_{2\alpha}$ caused a transient relaxation followed by a slow contraction. (b) In the presence of L-NAME (1 mM), Na^+ substitution caused a transient contraction followed by a slow contraction; this resembled the effect of hypoxia shown in Fig. 2.

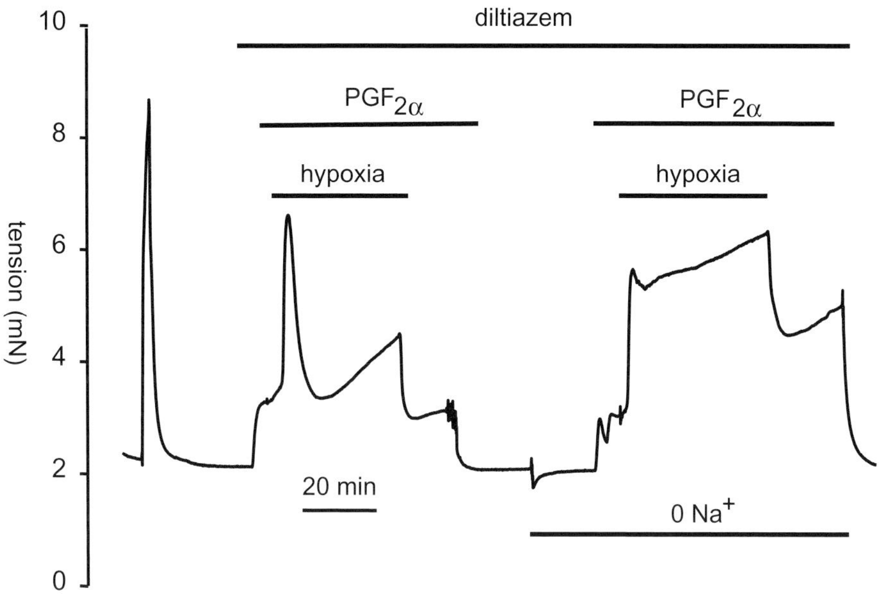

FIG. 2. HPV in rat IPA consists of a transient contraction (phase 1) followed by a smaller slowly developing sustained contraction (phase 2). Substitution of Na^+ 20 min prior to hypoxia significantly enhanced phase 2 HPV. $PGF_{2\alpha}$ was used to set a level of pretone equivalent to 10–15% of the maximal 80 mM K^+ response, which is shown on the extreme left of the trace.

was not significantly different from control. However, phase 2 of hypoxia was markedly (by 110%) and significantly ($P < 0.05$) enhanced in the absence of Na^+. Similar results were obtained when NCX was inhibited using KB-R7943 (3 μM), and also when the ability of NCX to extrude Ca^{2+} was reduced by pretreating IPA with 100 μM ouabain, which causes an increase in $[Na^+]_i$ by inhibiting the Na^+ pump (not shown).

Thus, although inhibition of NCX can cause a contraction similar to that induced by hypoxia, at least in the presence of L-NAME, it does not prevent HPV. Rather, it strongly enhances phase 2 HPV. These results therefore do not support the concept that HPV is caused by, or is significantly enhanced by, hypoxia-mediated inhibition of NCX. Instead, they suggest that NCX is important in limiting the rise in $[Ca^{2+}]$ during contractions to both $PGF_{2\alpha}$ and hypoxia. This is not surprising, since the normal role of NCX is to mediate Ca^{2+} extrusion in smooth muscle cells, unless $[Na^+]_i$ is elevated.

Is CCE involved in HPV?

There are no available agents which selectively inhibit CCE, although the polyvalent cations La^{3+}, Gd^{3+} and Ni^{2+} and the drugs 2-APB and SKF96365, have been used as CCE blockers. Another method of inhibiting CCE is to inhibit Ca^{2+} release from intracellular stores. Dantrolene, a ryanodine receptor antagonist, has been shown to inhibit Ca^{2+} release and CCE in pulmonary arteries (Savineau et al 1996). We examined the effect of 2-APB and dantrolene on HPV in IPA.

Figure 3a shows that 2-APB (75 μM) reduced both the first and second phases of HPV, with a greater effect on the phase 2 (35% inhibition of phase 1, 79% inhibition of phase 2; $n = 16$ for time controls, $n = 7$ for 2-APB; $P < 0.05$ in both cases).

Figure 3b illustrates the effect of 2-APB when a raised K^+ concentration (~40 mM) was used to generate pretone. In this case, HPV in the absence of 2-APB did not demonstrate a distinct phase 1, although there was a slow rise in tension resembling the phase 2 response observed when $PGF_{2\alpha}$ was used as the pretone agent. 2-APB strongly inhibited this slow rise in tension by 82% $n = 7$; $P < 0.05$, similar to its effect on phase 2 when $PGF_{2\alpha}$ was used as the pretone agent.

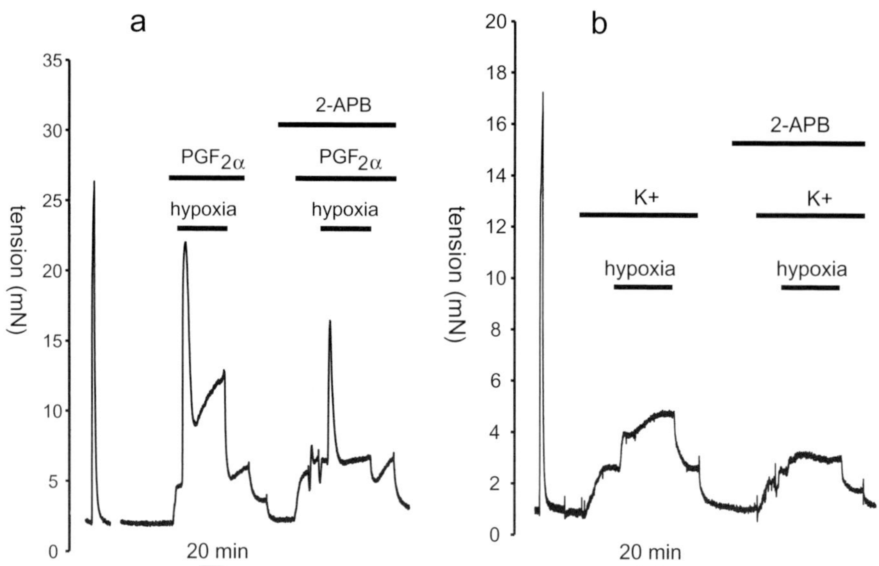

FIG. 3. (a) Application of 2-APB (75 μM) 20 min prior to hypoxia inhibited both phases of HPV elicited with $PGF_{2\alpha}$ as the pretone agent, with the 2nd phase reduced more than the 1st. (b) When a moderately elevated [K^+] (40 mM) was used to set the pretone, hypoxia caused only a small contraction similar to phase 2. This was strongly inhibited by 2-APB.

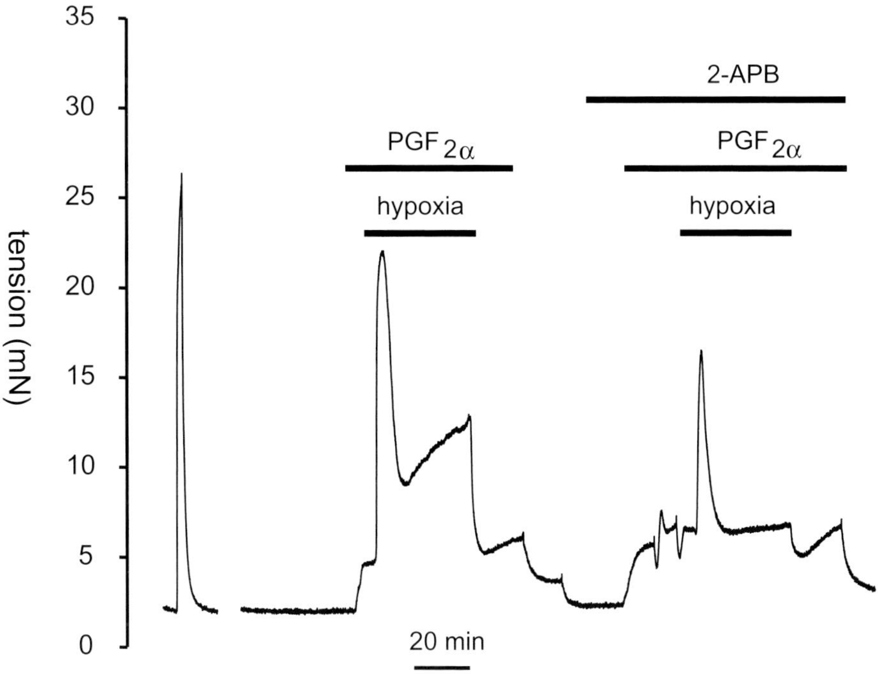

FIG. 4. Dantrolene (50 μM) applied 20 min prior to hypoxia inhibited phase 2 HPV, with a variable effect on phase 1.

As illustrated in Fig. 4, dantrolene (50 μM) also significantly inhibited phase 2 HPV (by 96%, $n = 16$ for time controls, $n = 8$ for dantrolene; $P < 0.05$). It also reduced the mean phase 1 by 36%, although this effect was variable, and did not reach statistical significance.

These results are consistent with a role for CCE in HPV, particularly with regard to phase 2. It is noteworthy that dantrolene, which inhibits Ca^{2+} release via the ryanodine receptor, and 2-APB, which is thought to directly block the CCE pathway, had quantitatively similar effects on both phases. This suggests that it is CCE caused by Ca^{2+} release, rather than Ca^{2+} release *per se*, which contributes to phase 2.

In contrast to these results, we previously showed that CPA, which depletes the SR of Ca^{2+}, had more of an effect on phase 1 than on phase 2 HPV. Similarly, phase 1 HPV was much more potently blocked by La^{3+} than was phase 2 HPV (Robertson et al 2000). Both observations suggested that it was phase 1 rather than phase 2 which was crucially dependent on CCE. However, it is possible to reconcile these findings by considering that the various agents used to inhibit Ca^{2+} release and CCE would also be expected to modify the response to

$PGF_{2\alpha}$, the drug used to generate pretone, and that this could affect the subsequent effect of hypoxia. Work by our laboratory shows that $PGF_{2\alpha}$ causes Ca^{2+} release *via* an inositol-1,4,5-trisphosphate (IP_3)-dependent pathway, and that this activates a La^{3+} and 2-APB-sensitive CCE. The observation that phase 1 HPV occurred when $PGF_{2\alpha}$ but not elevated K^+ was used to provide pretone (Fig. 3) suggests that phase 1 may require previous stimulation of Ca^{2+} release and/or CCE.

Figure 5 presents a speculative model which incorporates these various findings; this takes as its starting point the premise that there exist functionally separate intracellular Ca^{2+} stores in PASMCs, one of which is released by IP_3, and the other by opening of ryanodine receptor-channel (Janiak et al 2001). We propose that $PGF_{2\alpha}$ and hypoxia act to release Ca^{2+} from the IP_3 and ryanodine-sensitive stores, termed stores SR1 and SR2, respectively. Preconstriction with $PGF_{2\alpha}$ results (1) in the release of Ca^{2+} from SR1, leading to the activation of a La^{3+}-sensitive CCE (2). This results in an increase in the Ca^{2+} content of SR2 (3). Hypoxia then initiates phase 1 HPV by releasing Ca^{2+} from SR2 (4). This leads to further stimulation of the CCE pathway (5), in such a way that it becomes less sensitive to La^{3+}. As Ca^{2+} continues to enter via this pathway, it contributes to phase 2 HPV (6). This model would explain why $PGF_{2\alpha}$, but not a moderate elevation of K^+, causes the large initial phase 1 contraction; the latter stimulus, which does not activate CCE and therefore preferentially refill intracellular Ca^{2+} stores, would be less likely to load SR2 with Ca^{2+},

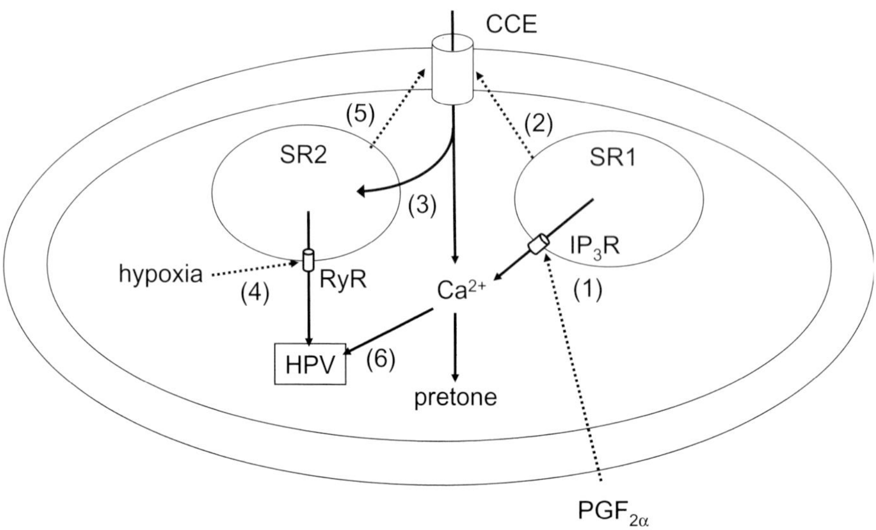

FIG. 5. A model to explain the role of Ca^{2+} release and CCE in HPV with PGF2α or high K^+ solution as pretone agents. See text for explanation.

therefore resulting in a smaller initial hypoxia induced rise in $[Ca^{2+}]_i$. An analogous proposal involving the differential effect of hypoxia on two intracellular Ca^{2+} stores has previously been described by Evans & Dipp (2002), although this differs in its detail from the present scheme.

In summary, our results do not support the concept that NCX contributes an active role to the rise in $[Ca^{2+}]_i$ associated with HPV, either *via* its inhibition or its 'reversal'. Conversely, the effects of dantrolene and 2-APB on HPV in IPA support the growing consensus that hypoxia activates CCE in PASMCs, and that this pathway does play an important role in causing the rise in $[Ca^{2+}]_i$ which underlies HPV. Whether or not the relative contribution to HPV of CCE is greater than that of depolarisation and the opening of voltage-gated Ca^{2+} channels remains to be established.

Acknowledgements

We are grateful to the Wellcome Trust (Grants 062554 and 059564) and the British Heart Foundation (Grant FS/03/022/15153) for their support.

References

Archer AL, Michelakis ED, Thébaud B et al 2005 A central role for O_2-sensitive K^+ channels and mitochondria in the specialized O_2-sensing system. In: Signalling pathways in acute oxygen sensing (Novartis Found Symp 272). Wiley, Chichester, p 157–171

Cornfield DN, Stevens T, McMurtry IF, Abman SH, Rodman DM 1993 Acute hypoxia increases cytosolic calcium in fetal pulmonary artery smooth muscle cells. Am J Physiol 265:L53–56

Dipp M, Nye PC, Evans AM 2001 Hypoxic release of calcium from the sarcoplasmic reticulum of pulmonary artery smooth muscle. Am J Physiol Lung Cell Mol Physiol 281:L318–325

Evans AM, Dipp M 2002 Hypoxic pulmonary vasoconstriction: cyclic adenosine diphosphate-ribose, smooth muscle Ca^{2+} stores and the endothelium. Resp Physiol Neurobiol 132:3–15

Evans AM, Hardie DG, Galione A, Peers C, Kumar P, Wyatt CN 2005 AMP-activated protein kinase couples mitochondrial inhibition by hypoxia to cADPR-dependent calcium mobilisation from the sarcoplasmic reticulum and / or transmembrane calcium influx in oxygen-sensing cells. In: Signalling pathways in acute oxygen sensing (Novartis Found Symp 272). Wiley, Chichester, p 234–252

Gonzalez De La Fuente P, Savineau J-P, Marthan R 1995 Control of pulmonary vascular smooth muscle tone by sarcoplasmic reticulum Ca^{2+} pump blockers: thapsigargin and cyclopiazonic acid. Pflügers Arch – Eur J Physiol 429:617–624

Jabr RI, Toland H, Gelband CH et al 1997 Prominent role of intracellular Ca^{2+} release in hypoxic vasoconstriction of canine pulmonary artery. Brit J Pharmacol 122:21–30

Janiak R, Wilson SM, Montague S, Hume JR 2001 Heterogeneity of calcium stores and elementary release events in canine pulmonary arterial smooth muscle cells. Am J Physiol Cell Physiol 280:C22–33

Ng LC, Gurney AM 2001 Store-operated channels mediate Ca^{2+} influx and contraction in rat pulmonary artery. Circ Res 89:923–929

Ng LC, Wilson SM, Hume JR 2005 Mobilization of SR stores by hypoxia leads to consequent activation of capacitative Ca^{2+} entry in isolated canine pulmonary arterial smooth muscle cells. J Physiol 563(Pt2):409–419

Post JM, Hume JR, Archer SL, Weir EK 1992 Direct role for potassium channel inhibition in hypoxic pulmonary vasoconstriction. Am J Physiol 262:C882–890
Robertson TP, Hague D, Aaronson PI, Ward JPT 2000 Voltage-independent calcium entry in hypoxic pulmonary vasoconstriction of intrapulmonary arteries of the rat. J Physiol 525.3:669–680
Savineau J-P, Gonzalez De La Fuente P, Marthan R 1996 Effect of modulators of tyrosine kinase activity on agonist-induced contraction in the rat pulmonary vascular smooth muscle. Pulm Pharmacol 9:189–195
Snetkov VA, Aaronson PI, Ward JPT, Knock GA, Robertson TP 2003 Capacitative calcium entry as a pulmonary specific vasoconstrictor mechanism in small muscular arteries of the rat. Br J Pharmacol 140:97–106
Wang J, Shimoda LA, Weigand L, Wang W, Sun D, Sylvester JT 2005 Acute hypoxia increases intracellular [Ca^{2+}] in pulmonary arterial smooth muscle by enhancing capacitative Ca^{2+} entry. Am J Physiol Lung Cell Mol Physiol 288:L1059–1069
Wang Y-X, Dhulipala PK, Kotlikoff MI 2000 Hypoxia inhibits the Na^+/Ca^{2+} exchanger in pulmonary artery smooth muscle cells. FASEB J 14:1731–1740
Weir EK, Archer S 1995 The mechanism of acute hypoxic pulmonary vasoconstriction: the tale of two channels. FASEB J 9:183–189

DISCUSSION

Duchen: In your [Ca^{2+}] measurements you are soaking the whole pulmonary artery in Fura-2 and then you are measuring the signal with a photomultiplier. It seems important to know where your signal is coming from, especially if you have dissociation between [Ca^{2+}] and the contraction.

Aaronson: Yes. It is important, but we are not set up to do this. However, we are not the only people who have seen this. Dave Beech at the University of Leeds has also shown that it is possible to get a substantial rise in Ca^{2+} through capacitative Ca^{2+} entry (CCE) with no tension. This is in small cerebral arteries. He uses a preparation of resistance arteriole, which is really just smooth muscle cells, partially dissociated. This rise in Ca^{2+} is not occurring in the endothelium. If the endothelium is taken off we get the same effect. It may not be the localization so much; it may be the temporal effect. Some people feel that contraction requires Ca^{2+} waves, and the frequency of the waves is important. We may not be getting waves in this case.

Archer: I am surprised that you decided that the La^{3+} dose that actually blocked both phases blocks all Ca^{2+} entry. This is not a CCE entry blocker, but a generalized Ca^{2+} entry blocker, full stop.

Aaronson: Not at 1 μM.

Archer: At the highest dose lanthanum inhibited both phases of hypoxic pulmonary vasoconstriction (HPV). One could look at this result as supporting Ca^{2+} entry as being very important to both phases of HPV. Moreover, priming with F2 alpha is not required in intact animals for HPV.

Aaronson: There may be priming. There is flow and agonists are present.

Archer: We don't find that priming is required in isolated resistance pulmonary vessels (Archer et al 2004).

Aaronson: You may have endogenous priming. You can't say that you don't have priming.

Archer: Let me ask my question a different way then. By some cruel twist of fate would it be possible that hypoxia affects the sensitivity of arteries to prostaglandin F2 alpha ($PGF_{2\alpha}$)? When people intentionally use priming, for example by putting angiotensin 2 in the isolated perfused lung model, the AII-induced constriction is long gone by the time hypoxia is given. You are operating off an unchanged baseline. Here, though, you are clearly operating off an elevated baseline tension due to persistent $PGF_{2\alpha}$, and you are assuming that what you are seeing is HPV. How do you know for sure that this is not hypoxia enhanced by $PGF_{2\alpha}$-induced constriction?

Aaronson: I have shown you that we get the same thing with K^+, and we also get the same thing with another agonist called SPC. It gives a very similar effect. We don't see anything when we use angiotensin; nor do we see any effect with endothelin. I can't imagine why it would affect certain agents and not others.

Evans: We found out by accident that we can get HPV in isolated pulmonary arteries without preconstriction. We were following up on Jeremy Ward and Phil Aaronson's work. Whether we use K^+ or $PGF_{2\alpha}$ preconstriction we still get a response equivalent to that observed without pretone. The characteristics of HPV do not appear to be altered dramatically by pre-constriction with these agents.

Aaronson: In terms of the Ca^{2+} channel blockers, our argument would be that they are inhibiting an endogenous vascular pretone. Blood vessels have tone. Pulmonary arteries may have less tone, but they still probably have some.

Archer: In normal humans the pulmonary circulation has very little active tone.

Aaronson: But they may have a level of pre-excitation. We get rises in Ca^{2+} that don't cause tone. I think *in vivo* you will have tone.

Archer: There is not much tone in normal pulmonary circulation.

Westwick: You were alluding to TRP channel involvement, and you have shown the various TRPCs are present by PCR. Have you looked at whether there is any different distribution in mesenteric arteries compared with pulmonary?

Aaronson: We have not.

Westwick: Have you looked for TRP channels other than TRPCs, such as TRPM2?

Aaronson: We have really only just started studying this.

Sylvester: Phil Aaronson very kindly referred to some of our results, perhaps because they are so supportive of his own! In our studies, we examined Ca^{2+} responses of smooth muscle cells from rat distal pulmonary arteries using Fura-2 fluorescence (Wang et al 2005). Hypoxia consistently increased intracellular Ca^{2+} concentration, and this increase was completely dependent on influx of extracellular Ca^{2+}. If the cells were exposed to nifedipine at concentrations that completely blocked the Ca^{2+} response to KCl, the Ca^{2+} response to hypoxia was reduced but not eliminated. Thus, voltage-dependent Ca^{2+} channels might be playing some role, but they are not the whole story. We then assessed the possibility that CCE contributed to the Ca^{2+} response to hypoxia. In the presence of nifedipine under normoxic conditions, there was essentially no Ca^{2+} entry, as measured by Mn^{2+} quenching. During hypoxia, basal nifedipine-insensitive Ca^{2+} entry was increased and could be blocked by agents such as SKF96365, $LaCl_3$, and $NiCl_2$ at concentrations shown previously to block CCE in these cells (Wang et al 2004). We also showed that CCE elicited by cyclopiazonic acid (CPA) was enhanced by hypoxia. The hypoxic Ca^{2+} response could be inhibited by the various CCE antagonists at concentrations which inhibited CCE but not voltage-dependent Ca^{2+} exchange (VDCE). We then examined the isolated rat lung (Weigand et al 2005), where zero extracellular Ca^{2+} completely blocked HPV. Interestingly, both nifedipine and CCE antagonists completely blocked HPV, but only nifedipine blocked the pressor response to high K^+. I think that both voltage-dependent Ca^{2+} channels (VDCCs) and store-operated channels (SOCs) are involved in the hypoxic response. One possible explanation is that primary activation of SOCs, which are thought to be non-specific cation channels, caused depolarization and secondary activation of VDCCs.

Gurney: In the records I showed, there wasn't any evidence for activation of inward current by hypoxia. The reversal potential of the background current did not seem to change. However, we have not looked closely at it.

Sylvester: Another possibility is that local increases in Ca^{2+} concentration due to CCE altered activity of Cl^- or K^+ channels, leading to depolarization and secondary Ca^{2+} influx through VDCCs. Whatever the mechanism, I think both channels are involved. Perhaps hypoxic contraction cannot occur unless both SOCs and VDCCs are activated. This may be consistent with Phil Aaronson's interesting observations that increases in Ca^{2+} can occur without contraction, perhaps due to compartmentation. In any case, we feel strongly that CCE is involved in the response of pulmonary arterial smooth muscle (PASM) to hypoxia. I was just made aware of a study from Joe Hume's group (Ng et al 2005), which makes essentially the same conclusion.

Duchen: It is quite plausible that CCE channels are being activated by pathways other than store emptying. There is evidence for activation of CCE pathways independently of endoplasmic reticulum (ER) status, either by redox modulation or arachidonic or some other signalling pathway.

Sylvester: If CPA completely depleted the stores, the enhancing effect of hypoxia on CPA-induced CCE must be explained by something other than enhanced depletion.

Ward: At the moment we don't have the pharmacological tools to look at this. We don't know whether it is potentiation of the same channel, which is possible, or activation of a completely different channel with the same pharmacological profile. We have the relative La^{3+} insensitivity of one pathway, but it is still highly sensitive to APB. This is in the absence of any inositol-1,4,5-triphosphate (IP_3) receptor effect.

Sylvester: I hope I never have to use La^{3+} again! It is a very difficult agent to use and has all sorts of effects. For example, higher concentrations increased the basal level of intracellular Ca^{2+} concentration, perhaps due to inhibition of plasma membrane Ca^{2+} ATPase or some other effect.

Gurney: Are you sure it isn't interfering with the Fura-2?

Sylvester: It could well be.

Ward: High concentrations of La^{3+} cause odd effects. We use 1 and 10 μM, and HEPES-based solutions with no phosphate to prevent chelation and precipitation.

Sylvester: Above 10 μM is where we began to see this effect. Even below this concentration, there are issues of chelation and precipitation.

López-Barneo: You have shown that hypoxia causes HPV on top of 80 mM K^+. You are proposing that this signal comes from Ca^{2+} release from stores induced by hypoxia. Have you tested whether inhibitors of the G protein pathway will block this response? In your images you were proposing that hypoxia releases Ca^{2+} through the ryanodine-sensitive stores.

Aaronson: We haven't done that; it's a good idea.

López-Barneo: We recently published a paper on smooth muscle cells (del Valle-Rodriguez et al 2003) in which we show that the change of conformation of the Ca^{2+} channel by itself, in the absence of any extracellular Ca^{2+} influx, can produce Ca^{2+} release from the sarcoplasmic reticulum (SR). The Ca^{2+} channel is coupled to the G protein signalling cascade. The Ca^{2+} channels produce downstream activation of a G protein, IP_3 synthesis and Ca^{2+} release. In high K^+, could hypoxia induce a shift to the left in the conductance-voltage curve of the L-type Ca^{2+} channel, which would lead to Ca^{2+} release from the stores? In the pulmonary circulation we have data showing that hypoxia potentiates Ca^{2+} channel activity. I would suggest you check whether the Ca^{2+} release induced by hypoxia in both conditions disappears after blockade of the G protein pathway. My point is, in high K^+, hypoxia produces release from the stores. How is hypoxia doing this?

Aaronson: It also does it in normal K^+.

Ward: We would say that it is probably through Mark Evans's cADP ribose mechanism. This might be consistent with an insensitive store.

Evans: I don't know why there are these subtle differences between our findings and yours. However, it was a surprise to me when I analysed Michelle's (Dipp) data on HPV from isolated rat arteries. The data had been sitting around for ages, and it looked qualitatively similar to data we obtained from isolated rabbit pulmonary arteries. It was only because we switched species that I actually analysed the measures of the maintained phase of constriction by hypoxia that we observe in the absence of the pulmonary artery endothelium. About 50% of this maintained constriction is dependent on transmembrane Ca^{2+} influx in rat PASM; in marked contrast, constriction by hypoxia of rabbit PASM did not exhibit any dependence on extracellular calcium over the time course of our experiments (40 min). As I said, if we block SR Ca^{2+} release by hypoxia in rat PASM then we do not see any constriction. Thus, it is possible that Ca^{2+} mobilization from SR stores leads to the activation of a store refilling current, and that this process is required for full and maintained constriction of pulmonary arteries by hypoxia. However, our data suggest that Ca^{2+} influx via a store-refilling current is not activated by hypoxia *per se*, but as a consequence of SR Ca^{2+} release. Nonetheless, there would appear to be some coupling between these two processes. It seems likely that the Ca^{2+} provided by the store-refilling current is not going directly into cytoplasmic sites to support constriction, but is being delivered to the SR i.e. Ca^{2+} entry via the store refilling current does not bring about constriction in its own right.

When you applied dantrolene to block SR Ca^{2+} release, my interpretation of what you saw was that somehow tone had dropped in the arteries and below the basal level of pre-constriction provided by $PGF_{2\alpha}$. Superimposed on this drop in tone it would appear that you record the endothelium-dependent component of constriction by hypoxia. This would suggest that there may be some processes dependent on SR Ca^{2+} release that may be contributing to your overall response, but for some reason you are not quite resolving this in terms of constriction?

Aaronson: We need to measure Ca^{2+}. Most smooth muscle relaxes to hypoxia. Pulmonary artery is unique. I believe that there is also a component of relaxation to hypoxia in the pulmonary artery, and this is confounding things.

Archer: A point of clarification. Jim Sylvester, you said that your work was inconsistent with Phil's. Did I hear you say that most or all of the HPV went away with a Ca^{2+} channel blocker, and you are postulating that both CCE and the L-type Ca^{2+} channel are important? This seems to be in opposition to the recent presentations that indicate that HPV occurs without any contribution from extracellular Ca^{2+} in one model or, in another lecture, results solely from CCE.

Sylvester: Yes.

Aaronson: L-type Ca^{2+} channel blockers have some effect on the first phase, so there may be some role for these channels.

Evans: It is the same in our hands. Phase 1 of HPV is blocked by nifedipine at $10\,\mu M$, even in the absence of extracellular Ca^{2+}.

Sylvester: We didn't study vessels, so I can't talk about phases. In the isolated lung model (Weigand et al 2005), both types of inhibitors were highly effective at blocking HPV. But the store-operated entry inhibitors were not effective at blocking pressor responses to KCl, whereas nifedipine blocked KCl responses completely.

Aaronson: This fits better with my own biased view of the world!

Archer: This fits better with my own biased view of the world. I have never had any problem with the notion that CCE contributes something to the net hypoxic pressor response, but there are just too many papers by too many groups that have found an obligator role for Ca^{2+} entry via the L-type Ca^{2+} channel to discount this observation.

Sylvester: I believe that VDCCs play a role. One might be able to eliminate the influence of these channels under certain experimental conditions and still have

HPV. This doesn't mean that they are not active in the normal situation. I think SOCs and VDCCs might act in series rather than in parallel, with the SOC first and the VDCC second. They are linked together such that activation of one leads somehow to activation of the other. In a parallel arrangement, they might both be active and independent, but neither alone would be sufficient to raise intracellular Ca^{2+} concentration high enough to trigger contraction, perhaps due to compartmentation. I like Philip's data; however, in PASMCs we still get 40–50% of the Ca^{2+} response to hypoxia after nifedipine treatment.

Chandel: Are there any other signalling pathways that link hypoxia to the Ca^{2+} changes you see?

Aaronson: We have some evidence that tyrosine kinase inhibitors have a very strong effect on phase 2. The rho kinase blocker abolishes phase 2.

Chandel: Do MAP kinases do anything?

Aaronson: MAP kinase blockers don't seem to have any effect. Interestingly there is literature suggesting that p38 is involved, but we have looked at this and think it is probably an artefact. The p38 blocker used releases NO.

Chandel: Did you use the SB compound?

Aaronson: Yes, we used SB-203580. We get a nice block of HPV, but in the presence of the eNOS blocker L-NAME, we don't get any block. We have additional evidence that SB-203580 causes endothelium-dependent relaxation even at low concentrations, which is blocked by L-NAME and endothelial denudation.

References

Archer SL, Wu XC, Thebaud B et al 2004 Preferential expression and function of voltage-gated, O_2-sensitive K^+ channels in resistance pulmonary arteries explains regional heterogeneity in hypoxic pulmonary vasoconstriction. Ionic diversity in smooth muscle cells. Circ Res 95:308–318

del Valle-Rodriguez A, Lopez-Barneo J, Urena J 2003 Ca^{2+} channel-sarcoplasmic reticulum coupling: a mechanism of arterial myocyte contraction without Ca^{2+} influx. EMBO J 22:4337–4345

Ng LC, Wilson SM, Hume JR 2005 Mobilization of sarcoplasmic reticulum stores by hypoxia leads to consequent activation of capacitative Ca^{2+} entry in isolated canine pulmonary arterial smooth muscle cells. J Physiol 563:409–419

Wang J, Shimoda LA, Sylvester JT 2004 Capacitative calcium entry and TRPC channel proteins are expressed in rat distal pulmonary arterial smooth muscle. Am J Physiol Lung Cell Mol Physiol 286:L848–858

Wang J, Shimoda LA, Weigand L, Wang W, Sun D, Sylvester JT 2005 Acute hypoxia increases intracellular $[Ca^{2+}]$ in pulmonary arterial smooth muscle by enhancing capacitative Ca^{2+} entry. Am J Physiol Lung Cell Mol Physiol 288:L1059–1069

Weigand L, Foxson J, Wang J, Shimoda LA, Sylvester JT 2005 Inhibition of hypoxic pulmonary vasoconstriction by antagonists of store-operated Ca^{2+} and nonselective cation channels. Am J Physiol Lung Cell Mol Physiol 289:L5–13

Final general discussion

Duchen: In this final discussion, I'd like us to think once again of someone reading this book. This should be a fabulous resource, because there isn't anywhere else where all this material is put together in one place. What has struck me as a relative outsider to the field is that everyone gives a wonderful presentation with an internally consistent story, but it is completely different to everyone else's story. It would be useful to understand how everyone sees their work fitting in with everyone else's. In both hypoxic pulmonary vasoconstriction (HPV) and carotid body the story is different, but we have roles for TASK channels, K_v channels, possibly KNCQ channels, BK channels, haem oxygenase, NADPH oxidase, mitochondria, reactive oxygen species (ROS), mechanisms of release of Ca^{2+} from internal stores or bursts of Ca^{2+} influx, and the role of AMP-dependent kinase. How do these fit together? I'm sure everyone's work is fine; but how do we make a coherent model from all of this?

Gurney: Perhaps I should start because I have no vested interest in my channels being the mediators. My feeling is that there are multiple signals that are modulated by hypoxia. There may be several pathways, all of which are either additive or modulate each other. From the point of view of K^+ channel involvement, the depolarization produced by hypoxia is not that large. It may take the membrane towards the threshold for Ca^{2+} channel activation, but I suspect it may not be enough by itself to take it all the way there, at least in the pulmonary circulation. By being closer to threshold, however, the membrane would be more easily excitable and the chance of activating Ca^{2+} channels when another signal comes along would be increased. I think it is unlikely that K^+ channel inhibition is the mediator of HPV; it could be a modulator though. I don't feel that K^+ channels are the vital thing that triggers it off.

Nurse: Another point worthy of consideration can be seen from the comparative approach. Even the carotid bodies of rats and rabbits show quite different properties. It is easier to find spontaneous activity in the chemoreceptor cells of the rabbit than it is of the rat. Hence examination of parallel systems could be beneficial. In a cell that is spontaneously active you could argue that you don't really need background K^+ current modulation in order to obtain a response to hypoxia. But in the case where a cell is quiescent it becomes important to depolarize the membrane, for example by inhibiting a background K^+ current. I see the idea of having both systems working in parallel as a way of increasing the safety factor, thus ensuring that there is a signal produced at the critical time when the stimulus is applied.

Duchen: It's an interesting point: if you compare it to the eye, there is a light sensor. Perhaps here we don't need to look for a specific oxygen sensor.

Buckler: We wouldn't try to explain the cardiac action potential on the basis of the activity of one ion channel. There is no reason to presuppose that there has to be only one ion channel involved in mediating responses to hypoxia. If we look at the way neurotransmitters or hormones work in modulating the electrical activity of many other cells, they often work through multiple targets. We shouldn't get too bogged down arguing which is the oxygen-sensitive ion channel. It's a combination. We should be developing a more sophisticated electrophysiological model of what these channels are doing, including the oxygen-insensitive ones whose activity may be modulated by other things as this could also have consequential effects on apparent oxygen sensitivity.

Duchen: This sounds like your next grant application!

López-Barneo: We shouldn't go away with the idea that everything is a mess. There are a number of facts that have been well documented in the last 10–15 years. We have a concept for the carotid body. There are glomus cells, which are excitable and have a family of K^+ channels that are inhibited by hypoxia. This produces cell depolarization, Ca^{2+} influx and secretion. In this framework, which is well established, you can understand new phenomena like long-term facilitation. In my view, whether or not TASK is important in the rat or rabbit is a secondary question. Our major problem is to understand how oxygen is sensed. We are a little lost on this. We have heard some good models, but they need some more experimental support. We are discussing the nature of the oxygen sensor a lot because unfortunately we are using very dirty tools, relying too much on pharmacology.

Chandel: I have a proposal. Why don't we all start investing in mouse models? Forget about the rat, rabbit and dog. The mouse has some problems. For example, it doesn't give great HPV, particularly in the C57 Bl/6 strain. But it gives a nice HIF response. Mice are good for studying this because they are amenable to genetic manipulation.

López-Barneo: I would encourage the use of genetic models too.

Chandel: In order to do this everyone has to get on board and use the mouse as the species of choice.

Duchen: It is a nice idea, but there are obstacles. First, the carotid body is a nightmare to work on in mice. It is dispersed up and down the bifurcation and it can't be found easily.

Chandel: Does anyone do mouse work on the carotid body?

Duchen: There is published work on the P2X knockout mouse.

Gonzalez: Yes, we have performed several experiments in the mice carotid bodies and dissociated their chemoreceptor cells. However, it is really difficult to work in the mouse because we are dealing with 5 μg of tissue, and it needs to be dissected out carefully.

Duchen: There are about 100 cells! There is also a problem with the genetic manipulation of mitochondrial function. You can try knocking out various subunits but it doesn't help. Mitochondria are involved in so many processes, as soon as you start to interfere in anything you get changes in mitochondrial membrane potential, cellular biogenetics and Ca^{2+} signalling. Trying to separate these things is a nightmare.

Nurse: Unless you can do it for a limited time, using conditional knockouts.

Schumacker: Drugs have a similar problem. Nav Chandel's point is important: perhaps the mouse model isn't appropriate for every experimental paradigm we are investigating, but the genetic tools may be a cleaner way of working. If a response is retained after a gene is knocked out, then the response must not require the product of that gene. We have heard in this meeting about the variety of responses to rotenone, for example. I don't think going back, standardizing protocols and doing more experiments is ever going to get us around the fact that different people come up with incompatible results using pharmacological tools.

Sylvester: Knockouts will solve that problem?

Schumacker: I think knockouts will have their own problems, but they will get us around some of the current problems with results from drugs.

Buckler: The problem is still one of interpretation, not just of specificity of the intervention. Suppose we knock down SDH (succinate dehydrogenase) reactivity and get some data that are perfectly compatible with the mitochondrial hypothesis. Does this provide concrete evidence that SDH is directly involved in oxygen sensing?

López-Barneo: We are trying to generate $SDH^{-/-}$ cells. If they still respond to hypoxia we have answered a very clear question: mitochondrial complex II is not involved in oxygen sensing.

Gonzalez: Not necessarily, any type of compensation in the respiratory chain generated by the deficit of complex II might obscure findings and lead to erroneous conclusions.

Chandel: There is not too much compensation in mitochondria. It is one of the few non-redundant systems.

Rich: I come from a different field: I work on mitochondrial biophysics. It seems to me there is a disconnection between the way that you do your work as physiologists and the way we work as biophysicists. There needs to be a connection made. You could answer some of these questions immediately if you make a mitochondrial preparation. Someone has to do that experiment. You say you go down to 5% oxygen in a bit of tissue and you see the response occurring. The question is, is that cytochrome oxidase or not? Since it is respiring, you don't know what the percentage oxygen is in the tissue. If you make the mitochondria, this problem goes away.

Secondly, this work needs to be linked in with modern biophysics. For example, you can use near infrared technology on whole tissues and establish the oxygenation state and the level of cytochrome oxidase, looking through the intact tissue in real time. You could then tell immediately whether the response you are looking at is modulated through the change in cytochrome oxidase. I would suggest getting links with the hardcore biophysics people.

López-Barneo: I don't think it is so easy to do a good experiment in one glomus cell with patch clamp, fluorescent measurement and simultaneous measurement of cytochrome C activity.

Rich: Not in a single cell, I agree. But such techniques seem feasible, for example, in the blood vessel experiments. There are techniques existing that can answer some of these basic questions.

Sylvester: This has been done. For example, Piantadosi and his colleagues at Duke University have used near infrared absorption to assess energy state in brain and muscle (Hampson et al 1990, Boushel & Piantadosi 2000).

Archer: Probably a more useful tack is to ask whether there is something wrong with our general approach. I have one observation: labs have phenotypes and continue to do the same type of experiment over and again. This leads to a limited range of experimental results. We should collaborate with others or use a range of different techniques. If I were to make one observation about this meeting it is that a lot of the dissension and confusion would be reduced if within one's own lab there was a more complete array of models including the intact rat, as well as the rat's isolated arteries, cells, and rat's proteins.

Kumar: The readouts as we go up from a very reduced level to the whole animal are non-linear. We get responses at whole organ levels that are not related in any functional way (I'm referring to graphical function here) to the reduced level. To try to make these correlations is quite hard. We should try to do this, but we will end up with different readouts with different sensitivities. It will be hard to say that the sensitivity to isolated mitochondria and to whole animal can be correlated to demonstrate cause and effect.

Ward: Steve is right: we have followed a reductionist approach in most of the biomedical sciences, and we need to return to the integrative approach to try to get the whole lot.

Gonzalez: Along these lines, I suggest that your preparation will be ideal to check the effects of acid on HPV. In the carotid body, hypoxia and acidosis interact positively to generate an additive response.

Ward: Acidosis suppresses HPV in our hands.

Weir: For your hypothetical graduate student reading this book, there may be one element that people studying HPV can actually agree on. This is on the executive side. There are probably three components that we agree participate to some extent.

One would be K^+ channel inhibition, membrane depolarization and Ca^{2+} entry. Another would be Ca^{2+} release from the sarcoplasmic reticulum and capacitative entry, and another would be Ca^{2+} sensitization involving Rho/Rho kinase which we haven't talked about during this meeting.

Ward: It is relative components. The proportions can be argued about, but the fact that they are there is widely agreed.

Nurse: I don't think we are compelled to think that one cell must only have one oxygen sensor. We need to keep open minds about other possibilities.

Sylvester: Life is complicated. Graduate students should realise that complicated questions can enhance opportunities for funding!

Chandel: It is great that we are dazed and confused because it keeps us busy!

Duchen: Conceptually you could argue that the system is so fundamental to life that you can't afford to have just one oxygen sensor.

Chandel: Along these lines it's amazing that there is no redundancy in HIF-1α.

Buckler: We know already that there are a number of other signalling pathways that require molecular oxygen (e.g. products of lipoxygenase, cylcooxygenase and cytochrome P450 pathways and nitric oxide synthase). On top of this there are various other metabolic pathways that need oxygen. So with the right measurement any of these could be made to appear as oxygen sensors. Haem synthesis, for example, is dependent on oxygen and begins and ends in the mitochondria. There is therefore even a potential link between the mitochondrial hypothesis and the carbon monoxide synthesis hypothesis.

Schumacker: I think the question of multiple oxygen sensors in the cell is a fundamentally important issue that this group should be focused on. It is likely that there are multiple ways that a cell could sense oxygen, but it is important to distinguish between an oxygen sensor and the hypoxia-induced signalling cascade. For example, an oxygen-sensitive ion channel could be a sensor or a downstream element in a hypoxic signal transduction pathway. It may be responsive to hypoxia but not because it is an oxygen sensor, but rather because it constitutes an arm of a signal transduction pathway. I believe there are multiple ways that a cell can sense oxygen, but from a biophysical standpoint, designing a molecule or group of molecules that can interact allosterically with oxygen in a way that couples that interaction with the function of the enzyme complex is a difficult engineering project. It is easier for me to imagine that some of the systems that we think are oxygen sensors are in fact being modulated by some upstream sensor that is signalling through a more conventional type of pathway. If there are multiple sensors then they would have to be somewhat coordinated in the cell so that one isn't turned on accidentally.

Duchen: I'd like to finish by saying what a privilege it has been to be part of this meeting. My interests have moved towards mitochondrial biology, and to put all of this in perspective we must remember that the only reason we need to breathe oxygen and deliver it to tissues is to deliver oxygen to your mitochondria. All that matters, at the end of the day, is mitochondrial biology! Physiology has evolved to supply mitochondria with oxygen: as far as our mitochondria are concerned, we are just incubation chambers. Thank you all for your enthusiastic participation.

References

Boushel R, Piantadosi CA 2000 Near-infrared spectroscopy for monitoring muscle oxygenation. Acta Physiologica Scandinavica 168:615–622

Hampson NB, Camporesi EM, Stolp BW et al 1990 Cerebral oxygen availability by NIR spectroscopy during transient hypoxia in humans. J Appl Physiol 69:907–913

Index of contributors

Non-participating co-authors are indicated by asterisks. Entries in bold indicate papers; other entries refer to discussion contributions.

Subject index

E

F

G

H